STUDENT SOLUTIONS MANUAL

Volume 2

CHAPTERS 20–39

Edw. S. Ginsberg

UNIVERSITY OF MASSACHUSETTS BOSTON

CONTRIBUTING AUTHOR

Sen-Ben Liao

UNIVERSITY OF CALIFORNIA

Essential UNIVERSITY PHYSICS

Richard Wolfson

PEARSON

Addison
Wesley

San Francisco Boston New York
Cape Town Hong Kong London Madrid Mexico City
Montreal Munich Paris Singapore Sydney Tokyo Toronto

Vice President and Editorial Director: Adam Black
Project Editor: Martha N. Steele
Development Editor: Brad Patterson
Editorial Assistant: Kristin Rose
Managing Editor: Corinne Benson
Production Supervisor: Nancy Tabor
Production Management: Elm Street Production Services
Compositor/Illustrators: ITC
Manufacturing Buyer: Pam Augspurger
Director of Marketing: Christy Lawrence
Text and Cover Printer: Bigelow & Brown
Cover Photo Credit: Tyler Boley, Workbook

Many of the designations used by manufacturers and sellers to distinguish their products are claimed as trademarks. Where those designations appear in this book, and the publisher was aware of a trademark claim, the designations have been printed in initial caps or all caps.

Mastering Physics is a trademark, in the U.S. and/or other countries, of Pearson Education, Inc. or its afffiliates.

ISBN 0-8053-0405-3

1 2 3 4 5 6 7 8 9 10—B&B—10 09 08 07 06

www.aw-bc.com

CONTENTS

PREFACE

This *Student Solutions Manual* to *Essential University Physics,* by Richard Wolfson, is designed to increase your skill and confidence in solving physics problems—the key to success in your physics course. It coaches you through a range of helpful and effective problem-solving techniques and provides solutions to all the odd-numbered problems in your text. By working carefully through these techniques and solutions, you will master the proven four-step problem-solving approach used in the text (Interpret, Develop, Evaluate, Assess) and learn to successfully:

- interpret problems and identify the key physics concepts involved
- develop a plan and draw figures for your solution
- evaluate any mathematical expressions
- assess your solution to check that it makes sense and see how it adds to your broader understanding of physics

Do your best to solve each problem *before* reading the solution. When you need to refer to the solution, focus on the reasoning—make sure you understand how and why each step is taken. By pushing yourself, you'll develop and hone your problem-solving skills. Don't fall into the trap of just passively reading the solutions.

We have made every effort to ensure these solutions are accurate and correct. If you do find any errors, inconsistencies, or ambiguities, we would be delighted to hear from you. Please contact us at wolfson@aw.com.

ELECTRIC CHARGE, FORCE, AND FIELD

EXERCISES

Section 20.1 Electric Charge

15. **INTERPRET** This problem deals with quantity of charge in a typical lightning flash. We want to express the quantity in terms of elementary charge e.

DEVELOP Since the magnitude of elementary charge e is $e = 1.6 \times 10^{-19}$ C, the number of electrons involved is given by Q/e.

EVALUATE Substituting the values given in the problem statement, we find the number to be

$$N = Q/e = \frac{25 \text{ C}}{1.6 \times 10^{-19} \text{ C}} = 1.56 \times 10^{20}$$

ASSESS Since 1 coulomb is about 6.25×10^{18} elementary charges, our result has the right order of magnitude.

Section 20.2 Coulomb's Law

17. **INTERPRET** In this problem we are asked to compare the gravitational and electrical forces between a proton and an electron.

DEVELOP The gravitational force and the electrostatic force between a proton and an electron separated by a distance r are, respectively,

$$F_{grav} = \frac{Gm_p m_e}{r^2}$$

and

$$F_{elec} = \frac{ke^2}{r^2}$$

EVALUATE The ratio of the two forces is

$$\frac{F_{elec}}{F_{grav}} = \left(\frac{ke^2}{r^2}\right)\left(\frac{r^2}{Gm_p m_e}\right) = \frac{(9 \times 10^9 \text{ N} \cdot \text{m}^2/\text{C}^2)(1.6 \times 10^{-19} \text{ C})^2}{(6.67 \times 10^{-11} \text{ N} \cdot \text{m}^2/\text{kg}^2)(1.67 \times 10^{-27} \text{ kg})(9.11 \times 10^{-31} \text{ kg})}$$
$$\approx 2.3 \times 10^{39}$$

Note that the spatial dependence of both forces is the same, and cancels out.

ASSESS At all distances (for which the particles can be regarded as classical point charges), the Coulomb force is about 10^{40} times stronger than the gravitational force.

19. **INTERPRET** This problem is about comparing the gravitational and electrical forces.

DEVELOP The electric force between a proton and an electron has magnitude $F_{elec} = ke^2/r^2$, while the weight of an electron is $F_g = m_e g$.

EVALUATE When the two forces are equal, $F_{elec} = F_g$, the distance between the proton and the electron is

$$r = \sqrt{\frac{ke^2}{m_e g}} = \sqrt{\frac{(9 \times 10^9 \text{ N} \cdot \text{m}^2/\text{C}^2)(1.6 \times 10^{-19} \text{ C})^2}{(9.11 \times 10^{-31} \text{ kg})(9.8 \text{ m/s}^2)}} = 5.08 \text{ m}$$

ASSESS The distance is almost fifty billion atomic diameters (or angstroms). This demonstrates that gravity is unimportant on the molecular scale.

21. **INTERPRET** This problem is about finding the unit vector associated with the electrical force one charge exerts on the other.

DEVELOP A unit vector from $\vec{r}_q = (1 \text{ m, } 0)$, the position of charge q, to any other point $\vec{r} = (x, y)$ is

$$\hat{n} = \frac{(\vec{r} - \vec{r}_q)}{|\vec{r} - \vec{r}_q|} = \frac{(x - 1 \text{ m}, y)}{\sqrt{(x - 1 \text{ m})^2 + y^2}}$$

EVALUATE **(a)** When the other charge is at position $\vec{r} = (1 \text{ m}, 1 \text{ m})$, the unit vector is

$$\hat{n} = \frac{(0, 1 \text{ m})}{\sqrt{0 + (1 \text{ m})^2}} = (0, 1) = \hat{j}$$

(b) When $\vec{r} = (0, 0)$, $\hat{n} = \frac{(-1 \text{ m}, 0)}{\sqrt{(-1 \text{ m})^2 + 0}} = (-1, 0) = -\hat{i}$.

(c) Finally, when $\vec{r} = (2 \text{ m}, 3 \text{ m})$, the unit vector is

$$\hat{n} = \frac{(1 \text{ m}, 3 \text{ m})}{\sqrt{(1 \text{ m})^2 + (3 \text{ m})^2}} = \frac{(1, 3)}{\sqrt{10}} = 0.316\hat{i} + 0.949\hat{j}$$

The sign of q doesn't affect this unit vector, but the signs of both charges do determine whether the force exerted by q is repulsive or attractive, i.e., in the direction of $+\hat{n}$ or $-\hat{n}$.

ASSESS The unit vector always points away from the charge q located at $(1 \text{ m}, 0)$.

Section 20.3 The Electric Field

23. **INTERPRET** This problem is about calculating the electric field strength due to a source, when the force experienced by the electron is known.

DEVELOP Equation 20.2a shows that the electric field strength (magnitude of the field) at a point is equal to the force per unit charge that would be experienced by a charge at that point:

$$E = \frac{F}{q}$$

EVALUATE With $q = |e|$, we find the field strength to be

$$E = \frac{F}{|e|} = \frac{6.1 \times 10^{-10} \text{ N}}{1.6 \times 10^{-19} \text{ C}} = 3.81 \times 10^9 \text{ N/C}$$

ASSESS Since the charge of electron is negative, the force experienced by the electron is in the opposite direction of the electric field.

25. **INTERPRET** This problem is about calculating the electric field strength due to a source, when the force experienced by a charge is known.

DEVELOP Equation 20.2a shows that the electric field strength (magnitude of the field) at a point is equal to the force per unit charge that would be experienced by a charge at that point:

$$E = \frac{F}{q}$$

The equation allows us to calculate E. For part **(b)**, the force experienced by another charge q' in the same field is $F' = q'E$.

EVALUATE **(a)** With $q = 68 \text{ nC}$, we find the field strength to be

$$E = \frac{F}{|e|} = \frac{150 \text{ mN}}{68 \text{ nC}} = 2.21 \times 10^6 \text{ N/C}$$

(b) The force experienced by another charge, $q' = 35 \text{ } \mu\text{C}$, in the same field is

$$F' = q'E = (35 \text{ } \mu\text{C})(2.21 \times 10^6 \text{ N/C}) = 77.2 \text{ N}$$

ASSESS The force a test charge particle experiences is proportional to the magnitude of the test charge. In our problem, since $q' = 35 \text{ } \mu\text{C} > q = 68 \text{ nC}$, we find $F' > F$.

27. **INTERPRET** This problem is about the electric field strength due to a point source charge—the proton.

DEVELOP The electric field strength at a distance r from a point source charge q is given by Equation 20.3:

$$\vec{E} = \frac{kq}{r^2}\hat{r}$$

The proton in a hydrogen atom behaves like a point charge.

EVALUATE At a distance of one Bohr radius ($a_0 = 0.0529$ nm) away, the electric field strength is

$$E = \frac{ke}{a_0^2} = \frac{(9 \times 10^9 \text{ N} \cdot \text{m}^2/\text{C}^2)(1.6 \times 10^{-19} \text{ C})}{(5.29 \times 10^{-11} \text{ m})^2} = 5.15 \times 10^{11} \text{ N/C}$$

ASSESS The field strength at the position of the electron is enormous because of the close proximity.

Section 20.4 Fields of Charge Distributions

29. **INTERPRET** Given the magnitude of the dipole moment, we are asked to calculate the distance between the pair of opposite charges that make up the dipole.

DEVELOP As shown in Equation 20.5, the electric dipole moment p is the product of the charge q and the separation d between the two charges making up the dipole:

$$p = qd$$

EVALUATE Using the equation above, the distance separating the charges of a dipole is

$$d = \frac{p}{q} = \frac{6.2 \times 10^{-30} \text{ C} \cdot \text{m}}{1.6 \times 10^{-19} \text{ C}} = 38.8 \text{ pm} = 0.0388 \text{ nm}$$

ASSESS The distance d has the same order of magnitude as the Bohr radius ($a_0 = 0.0529$ nm).

31. **INTERPRET** In this problem we are asked about the line charge density, given the field strength at a distance from the wire.

DEVELOP If the electric field points radially toward the long wire ($L \gg 45$ cm), the charge on the wire must be negative. The magnitude of the field is given by the result of Example 20.7, $E_r = \frac{2k\lambda}{r}$.

EVALUATE Using the equation above, we find the line charge density to be

$$\lambda = \frac{E_r r}{2k} = \frac{(-260 \text{ kN/C})(0.45 \text{ m})}{2(9 \times 10^9 \text{ N} \cdot \text{m}^2/\text{C}^2)} = -6.50 \text{ } \mu\text{C/m}$$

ASSESS The electric field strength due to a line charge density decreases as $1/r$. This can be compared to the $1/r^2$ dependence due to a point charge.

Section 20.5 Matter in Electric Fields

33. **INTERPRET** This problem is about the Millikan oil drop experiment. Two forces are involved, gravitational and electrical.

DEVELOP In equilibrium, the gravitational and electrostatic forces cancel: $\vec{F}_g = -\vec{F}_e$, or $mg = qE$. The equation can be used to compute the mass m.

EVALUATE Using the equation above, we find the mass to be

$$m = \frac{qE}{g} = \frac{(10 \times 1.6 \times 10^{-19} \text{ C})(2 \times 10^7 \text{ N/C})}{(9.8 \text{ m/s}^2)} = 3.27 \times 10^{-12} \text{ kg}$$

ASSESS Because this mass is so small, the size of such a drop may be better appreciated in terms of its radius, $R = (3m/4\pi\rho_{oil})^{1/3}$. Millikan used oil of density $\rho_{oil} = 0.9199$ g/cm^3, so $R = 9.46$ μm for this drop.

35. **INTERPRET** This problem is about the motion of a proton, a charged particle, in an electric field that points to the left.

DEVELOP Choose the x axis to the right, in the direction of the proton, so that the electric field is negative to the left. If the Coulomb force on the proton is the only important one, the acceleration is

$$a_x = \frac{e(-E)}{m}$$

The negative sign means that the proton decelerates as it enters the electric field. The motion is one-dimensional kinematics.

EVALUATE   (a) Using Equation 2.11, with $v_{ox} = 3.8 \times 10^5$ m/s and $v_x = 0$, we find the maximum penetration into the field region to be

$$x - x_0 = -\frac{v_{ox}^2}{2a_x} = \frac{mv_{ox}^2}{2eE} = \frac{(1.67 \times 10^{-27}\text{ kg})(3.8 \times 10^5\text{ m/s})^2}{2(1.6 \times 10^{-19}\text{ C})(56 \times 10^3\text{ N/C})} = 1.35\text{ cm}$$

(b) The proton subsequently moves to the left, with the same constant acceleration in the field region, until it exits with the initial velocity reversed.

ASSESS   The deceleration of the proton increases with the field strength E. In addition, when E is large, the penetration is small, and the proton reverses its path rather quickly.

PROBLEMS

37.   **INTERPRET**   The problem asks for an estimate of the fraction of electrons removed from rubbing.

DEVELOP   Suppose that half the ball's mass is protons (the other half comes from neutrons). Their number is $N_{p0} = 1\text{ g}/m_p$. This is equal to the original number of electrons, N_{e0}.

EVALUATE   The number of electrons removed is $N_e = 1\ \mu C/e$, so the fraction removed is

$$\frac{N_e}{N_{e0}} = \frac{(1\ \mu C/e)}{(1\text{ g}/m_p)} = \frac{(10^{-6}\text{ C})/(1.6 \times 10^{-19}\text{ C})}{1\text{ g}/(1.67 \times 10^{-24}\text{ g})} = 1.04 \times 10^{-11}$$

ASSESS   The fraction is about a hundred billionth. Thus, only a very small amount of electrons has been removed by rubbing.

39.   **INTERPRET**   In solving this problem we follow Problem Solving Strategy 20.1 and identify the source charges as the proton and the electron.

DEVELOP   The unit vector from the proton's position to the origin is $-\hat{i}$. Using Equation 20.1, the Coulomb force of the proton on the helium nucleus is

$$\vec{F}_{P,He} = \frac{kq_p q_{He}}{r_{p,He}^2}(-\hat{i}) = \frac{(9 \times 10^9\text{ N} \cdot \text{m}^2/\text{C}^2)(e)(2e)}{(1.6\text{ nm})^2}(-\hat{i}) = (-0.180\text{ nN})\hat{i}$$

Similarly, the unit vector from the electron's position to the origin is $-\hat{j}$, so its force on the helium nucleus is

$$\vec{F}_{e,He} = \frac{kq_e q_{He}}{r_{e,He}^2}(-\hat{i}) = \frac{k(-e)(2e)}{(0.85\text{ nm})^2}(-\hat{j}) = (0.638\text{ nN})\hat{j}$$

EVALUATE   The net Coulomb force on the helium nucleus is the sum of these:

$$\vec{F}_{net} = \vec{F}_{P,He} + \vec{F}_{e,He} = (-0.180\text{ nN})\hat{i} + (0.638\text{ nN})\hat{j}$$

ASSESS   In situations where there are more than one source charge, we apply the superposition principle and add the electric forces vectorially. In the above, since the electron is closer to the He nucleus than the proton ($r_{e,He} < r_{p,He}$), we expect $|\vec{F}_{P,He}| < |\vec{F}_{e,He}|$.

41.   **INTERPRET**   Coulomb's law applies here. Since more than one source charge is involved, we make use of the superposition principle.

DEVELOP   For the force on a third charge Q to be zero, it must be placed on the x axis to the right of the (smaller) negative charge, i.e., at $x > a$. The net Coulomb force on a third charge so placed is

$$F_x = \frac{kQ(3q)}{x^2} + \frac{kQ(-2q)}{(x-a)^2}$$

We set $F_x = 0$ to solve for x.

EVALUATE   The condition $F_x = 0$ implies that $3(x-a)^2 = 2x^2$, or $x^2 - 6xa + 3a^2 = 0$. Thus,

$$x = 3a \pm \sqrt{9a^2 - 3a^2} = (3 \pm \sqrt{6})a$$

Only the solution $x = (3 + \sqrt{6})a = 5.45a$ is to the right of $x = a$.

ASSESS   At $x = (3 + \sqrt{6})a$ the forces acting on Q from $3q$ and $-2q$ exactly cancel each other. Notice that our result is independent of the sign and magnitude of the third charge Q.

43. **INTERPRET** More than one source charge is involved in this problem. Therefore, we use Coulomb's law and apply the superposition principle to find the force on q_3.

DEVELOP We denote the positions of the charges by $\vec{r}_1 = (1 \text{ m})\hat{j}$, $\vec{r}_2 = (2 \text{ m})\hat{i}$, and $\vec{r}_3 = (2 \text{ m})\hat{i} + (2 \text{ m})\hat{j}$. The unit vector pointing from q_1 toward q_3 is

$$\hat{r}_{13} = \frac{(\vec{r}_3 - \vec{r}_1)}{|\vec{r}_3 - \vec{r}_1|}$$

Similarly, the unit vector pointing from q_2 toward q_3 is $\hat{r}_{23} = \frac{(\vec{r}_3 - \vec{r}_2)}{|\vec{r}_3 - \vec{r}_2|}$. The vector form of Coulomb's law and the superposition principle give the net electric force on q_3 as:

$$\vec{F}_3 = \vec{F}_{13} + \vec{F}_{23} = \frac{kq_1 q_3 (\vec{r}_3 - \vec{r}_1)}{|\vec{r}_3 - \vec{r}_1|^3} + \frac{kq_2 q_3 (\vec{r}_3 - \vec{r}_2)}{|\vec{r}_3 - \vec{r}_2|^3}$$

EVALUATE Substituting the values given in the problem statement, we find the force acting on q_3 to be

$$\vec{F}_3 = \vec{F}_{13} + \vec{F}_{23} = \frac{kq_1 q_3 (\vec{r}_3 - \vec{r}_1)}{|\vec{r}_3 - \vec{r}_1|^3} + \frac{kq_2 q_3 (\vec{r}_3 - \vec{r}_2)}{|\vec{r}_3 - \vec{r}_2|^3}$$

$$= (9 \times 10^9 \text{ N} \cdot \text{m}^2/\text{C}^2)(15 \times 10^{-6} \text{C})\left(\frac{(68 \times 10^{-6}\text{C})(2\hat{i} + \hat{j})}{5\sqrt{5} \text{ m}^2} = + \frac{(-34 \times 10^{-6}\text{C})2\hat{j}}{8 \text{ m}^2} \right)$$

$$= (1.64\hat{i} - 0.326\hat{j}) \text{ N}$$

or $F_3 = \sqrt{F_{3x}^2 + F_{3y}^2} = 1.67$ N at an angle of $\theta = \tan^{-1}(F_{3y}/F_{3x}) = -11.2°$ to the x axis.

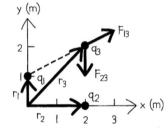

ASSESS The force between q_1 and q_3 is repulsive ($q_1 q_3 > 0$) while the force between q_2 and q_3 is attractive ($q_2 q_3 < 0$). The two forces add vectorially to give the net force on q_3.

45. Three identical charges $+q$ and a fourth charge $-q$ form a square of side a. **(a)** Find the magnitude of the electric force on a charge Q placed at the center of the square. **(b)** Describe the direction of this force.

INTERPRET Since more than one source charge is involved, we use Coulomb's law and apply the superposition principle to find the force on Q.

DEVELOP The magnitudes of the forces on Q from each of the four charges are equal to

$$F_0 = \frac{kqQ}{(\sqrt{2}a/2)^2} = \frac{2kqQ}{a^2}$$

To determine the net force and its direction, we note that the forces from the two positive charges on the same diagonal are in opposite directions, and cancel, while the forces from the positive and negative charges on the other diagonal are in the same direction (depending on the sign of Q) and add.

EVALUATE **(a)** The net force on Q has magnitude

$$F_{net} = 2F_0 = \frac{4kqQ}{a^2}$$

(b) The direction of $\vec{F}_{net}$ is toward the negative charge for $Q > 0$, and away from the negative charge for $Q < 0$.

ASSESS Even though we have four charges acting on charge Q, only two need to be considered. By the superposition principle, the force on a charge placed midway between two identical charges must add to zero.

47.   **INTERPRET**   Coulomb's law applies here. Since more than one source charge is involved, we use the superposition principle to find the point where the field strength vanishes.

DEVELOP   We first note that the field can be zero only along the line joining the charges (the x axis). To the left or right of both charges, the fields due to each are in the same direction, and cannot add to zero. Between the two, a distance $x > 0$ from the $1 \, \mu C$ charge, the electric field is

$$\vec{E} = k\left(\frac{q_1 \hat{i}}{x^2} = + \frac{q_2(-\hat{i})}{(10 \text{ cm} - x)^2} \right)$$

EVALUATE   With $q_1 = 1.0 - \mu C$ and $q_2 = 2.0 - \mu C$, the field vanishes when $1 \, \mu C/x^2 = 2 \, \mu C/(10 \text{ cm} - x)^2$, or $x = 10 \text{ cm}/(\sqrt{2} + 1) = 4.14 \text{ cm}$.

ASSESS   Since $\vec{E} = 0$ at $x = 10 \text{ cm}/(\sqrt{2} + 1)$, a charge placed at that point does not experience any force.

49.   **INTERPRET**   Coulomb's law applies here. With two source charges, we use the superposition principle to find the field strength at a point on the y axis.

DEVELOP   As in Example 20.2, we apply the symmetry argument to show that the x components of the electric field due to both charges cancel, and the net electric field points in the $+y$ direction.

EVALUATE   (a) Since electric field is the force per unit charge, from Example 20.2, we obtain

$$E_{\text{net},y} = 2\frac{kq}{r^2}\cos\theta = \frac{2kqy}{(a^2 + y^2)^{3/2}}$$

(b) The magnitude of the field, a positive function, is zero for $y = 0$ and $y = \infty$, hence it has a maximum in between. Setting the derivative equal to zero, we find

$$0 = (a^2 + y^2)^{-3/2} - \frac{3}{2}y(a^2 + y^2)^{-5/2}(2y)$$

or $a^2 + y^2 - 3y^2 = 0$. Thus, the field strength maxima are at $y = \pm\frac{a}{\sqrt{2}}$.

ASSESS   By symmetry, we expect the directions of the electric field at $y = \pm\frac{a}{\sqrt{2}}$ to be opposite.

51.   **INTERPRET**   We find the electric field on the axis of a dipole, and show that Equation 20.6b is correct. To do this we will use the equation for electric field.

DEVELOP   The spacing between the + and – charges is $2a$. We will use $E = k\frac{q}{r^2}$ for each charge to find the total field at a point $x \gg a$.

EVALUATE

$$\vec{E} = k\frac{+q}{(x - a)^2}\hat{i} + k\frac{-q}{(x + a)^2}\hat{i} = kq[(x - a)^{-2} - (x + a)^{-2}]\hat{i}$$

$$\rightarrow \vec{E} = \frac{kq}{x^2}\hat{i}\left[\left(1 - \frac{a}{x}\right)^{-2} - \left(1 + \frac{a}{x}\right)^{-2}\right]$$

For $x \gg a$. $(1 \pm \frac{a}{x})^{-2} \approx 1 \mp 2\frac{a}{x}$, so $\vec{E} \approx \frac{kq}{x^2}\hat{i}[(1 + 2\frac{a}{x}) - (1 - 2\frac{a}{x})] = \frac{kq}{x^2}[4\frac{a}{x}] = 2\frac{k(2qa)}{x^3}\hat{i}$. But $p = qd = 2qa$, so $\vec{E} = \frac{2kp}{x^3}\hat{i}$.

ASSESS   We have shown what was required.

53.   **INTERPRET**   Coulomb's law applies here. With three source charges, we use the superposition principle to find the field strength at a point on the y axis.

DEVELOP   The electric field on the y axis ($y > \sqrt{3}a/2$) due to the two charges on the x axis follows from Example 20.2:

$$\vec{E}_1 = \frac{2kqy}{(y^2 + a^2/4)^{3/2}}\hat{j}$$

On the other hand, using Equation 20.3, we find the electric field due to the charge on the y axis to be

$$\vec{E}_2 = \frac{kq}{(y - \sqrt{3}a/2)^2}\hat{j}$$

EVALUATE **(a)** For $y > \sqrt{3}\,a/2$, the total field is simply the sum of both terms:

$$\vec{E} = \vec{E}_1 + \vec{E}_2 = kq\left[\frac{2y}{(y^2 + a^2/4)^{3/2}} + \frac{1}{(y - \sqrt{3}a/2)^2}\right]\hat{j}$$

(b) For $y \gg a$, the electric field may be approximated as

$$E \approx kq\left[\frac{2y}{(y^2)^{3/2}} + \frac{1}{y^2}\right]\hat{j} = \frac{3kq}{y^2}\hat{j}$$

The field is like that due to a point charge of magnitude $3q$.

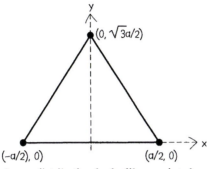

ASSESS At large distance, the charge distribution looks like a point charge located at the origin.

55. **INTERPRET** Two forces are involved in this problem: Coulomb force and spring force. The spring is stretched due to Coulomb repulsion.

DEVELOP Suppose that the Coulomb repulsion is the only force stretching the spring. When balanced with the spring force, $F_e = F_s$, or

$$\frac{kq^2}{(L_0 + x)^2} = k_s x$$

where L_0 is the equilibrium length. This cubic equation can be solved by iteration or by Newton's method.

EVALUATE Substituting the values given in the problem statement gives

$$x(L_0 + x)^2 = \frac{kq^2}{k_s} \quad \rightarrow \quad x(0.5\text{ m} + x)^2 = \frac{(9 \times 10^9\text{ N}\cdot\text{m}^2/\text{C}^2)(34\,\mu\text{C})^2}{150\text{ N/m}} = 6.94 \times 10^{-2}\text{ m}^3$$

Newton's method yields $x = 15.95$ cm.

ASSESS Our result makes sense since the amount stretched is seen to decrease with increasing spring constant k_s, and increase with the magnitude of the charge q.

57. **INTERPRET** The electron undergoes circular motion, and the centripetal force is provided by the Coulomb force.

DEVELOP The electric field of the wire is radial and falls off like $1/r$ ($E = 2k\lambda/r$, see Example 20.7). For an attractive force (negative electron encircling a positively charged wire), this is the same dependence as the centripetal acceleration. For circular motion around the wire, the Coulomb force provides the electron's centripetal acceleration:

$$a_r = -\frac{eE}{m} = -\frac{2ke\lambda}{mr} = \frac{v^2}{r}$$

The equation can be used to deduce the speed of the electron.

EVALUATE Substituting the values given, we find the speed to be

$$v = \sqrt{\frac{2ke\lambda}{m}} = \sqrt{\frac{2(9 \times 10^9\text{ N}\cdot\text{m}^2/\text{C}^2)(1.6 \times 10^{-19}\text{ C})(2.5 \times 10^{-9}\text{ C/m})}{9.11 \times 10^{-31}\text{ kg}}} = 2.81 \times 10^6\text{ m/s}$$

ASSESS We find the speed of the electron to be independent of r, the radial distance from the long wire. This is because both the electric field and the centripetal acceleration fall off as $1/r$; hence, the r dependence cancels out.

59.   **INTERPRET**   The charge undergoes circular motion, and the centripetal force is provided by the Coulomb force. We are asked to find the line charge density, given the particle's speed.

DEVELOP   The electric field of the wire is radial and falls off like l/r ($E = 2k\lambda/r$, see Example 20.7). For an attractive force (positive charge encircling a negatively charged wire), this is the same dependence as the centripetal acceleration. For circular motion around the wire, the Coulomb force provides the centripetal acceleration:

$$a_r = \frac{qE}{m} = \frac{2kq\lambda}{mr} = \frac{v^2}{r}$$

The equation can be used to deduce the line charge density, once the speed is known.

EVALUATE   The above equation gives

$$\lambda = -\frac{mv^2}{2kq} = -\frac{(6.8 \times 10^{-9} \text{ kg})(280 \text{ m/s})^2}{2(9 \times 10^9 \text{ N} \cdot \text{m}^2/\text{C}^2)(2.1 \times 10^{-9} \text{ C})} = -14.1 \text{ } \mu\text{C/m}$$

ASSESS   For the force to be attractive, we require the line charge density to be negative.

61.   **INTERPRET**   This problem is about an electric dipole placed in an external electric field.

DEVELOP   Using Equation 20.10, the energy required to reverse the orientation of such a dipole is $\Delta U = 2pE$.

EVALUATE   Using the equation above, the electric dipole moment is

$$p = \frac{\Delta U}{2E} = \frac{3.1 \times 10^{-27} \text{ J}}{2(1.2 \times 10^3 \text{ N/C})} = 1.29 \times 10^{-30} \text{ C} \cdot \text{m}$$

ASSESS   An electric dipole tends to align itself in the direction of the external electric field. Thus, energy is required to change its orientation.

63.   **INTERPRET**   This problem is about the interaction between a dipole and the electric field due to a source charge.

DEVELOP   With the x axis in the direction from Q to $\vec{p}$ and the y axis parallel to the dipole in Figure 20.30, we have $\vec{p} = (2qa)\hat{j}$ and $E = (kQ/x^2)\hat{i}$. In the limit $x \gg a$, the torque on the dipole is given by Equation 20.9, $\vec{\tau} = \vec{p} \times \vec{E}$, where $\vec{E}$ is the field from the point charge Q, at the position of the dipole.

EVALUATE   **(a)** Using Equation 20.9, we find the torque to be

$$\vec{\tau} = \vec{p} \times \vec{E} = (2qa\hat{j}) \times \left(\frac{kQ}{x^2}\hat{i}\right) = -\frac{2kQqa}{x^2}\hat{k}$$

The direction is into the page, or clockwise, to align $\vec{p}$ with $\vec{E}$.

(b) The Coulomb force obeys Newton's third law. The field of the dipole at the position of Q is (Example 20.5 adapted to new axes)

$$\vec{E}_{\text{dipole}} = -\frac{2kqa}{x^3}\hat{j}$$

Thus, the force on Q due to the dipole is

$$\vec{F}_{\text{on } Q} = Q\vec{E}_{\text{dipole}} = -\frac{2kQqa}{x^3}\hat{j}$$

The force on the dipole due to Q is the opposite of this:

$$\vec{F}_{\text{on dipole}} = -\vec{F}_{\text{on } Q} = \frac{2kQqa}{x^3}\hat{j}$$

The magnitude of $\vec{F}_{\text{on dipole}}$ is $2kQqa/x^3$.

(c) The direction of $\vec{F}_{\text{on dipole}}$ is in $+\hat{j}$, or parallel to the dipole moment.

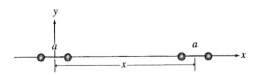

ASSESS   The net force $\vec{F}_{\text{on dipole}}$ will cause the dipole to move in the $+\hat{j}$ direction. In addition, there is a torque that tends to align p with E. So, the motion of the dipole involves both translation and rotation.

65. **INTERPRET**   This problem is about the electric field due to a charged ring.

DEVELOP   The electric field on the axis of a uniformly charged ring of radius a is calculated in Example 20.6:

$$E = \frac{kQx}{(x^2 + a^2)^{3/2}}$$

Knowing the field strengths at two different values of x allows us to deduce a and Q.

EVALUATE   The data given in the question imply

$$E_1 = 380 \text{ kN/C} = \frac{kQ(5 \text{ cm})}{[(5 \text{ cm})^2 + a^2]^{3/2}}$$

$$E_2 = 160 \text{ kN/C} = \frac{kQ(15 \text{ cm})}{[(15 \text{ cm})^2 + a^2]^{3/2}}$$

Dividing these two equations and taking the $\frac{2}{3}$ root, we get

$$\left(\frac{380 \times 15}{160 \times 5}\right)^{2/3} = 3.70 = \frac{(15 \text{ cm})^2 + a^2}{(5 \text{ cm})^2 + a^2}$$

which when solved for the radius, gives

$$a = \sqrt{\frac{(15 \text{ cm})^2 - (3.70)(5 \text{ cm})^2}{2.70}} = 7.00 \text{ cm}$$

To calculate Q, we substitute for a in either one of the field equations. This leads to

$$Q = \frac{(380 \text{ kN/C})[(5 \text{ cm})^2 + (7 \text{ cm})^2]^{3/2}}{(9 \times 10^9 \text{ N} \cdot \text{m}^2/\text{C}^3)(5 \text{ cm})} = 538 \text{ nC}$$

ASSESS   To check that our results are correct, we may substitute the values obtained for a and Q into the field equation to calculate E_1 and E_2 at $r_1 = 5$ cm and $r_2 = 15$ cm. Note that the field strength decreases as x is increased.

67. **INTERPRET**   This problem is about the electric field due to a 10-m long straight wire, which is our source charge.

DEVELOP   For a uniformly charged wire of length L and charge Q, the line density is $\lambda = Q/L$. Under the assumption that the wire is approximately infinitely long, the electric field due to the line charge can be written as (see Example 20.7):

$$E = \frac{2k\lambda}{r}$$

EVALUATE   (a) The charge density is

$$\lambda = \frac{Q}{L} = \frac{25 \text{ } \mu\text{C}}{10 \text{ m}} = 2.5 \text{ } \mu\text{C/m}$$

(b) Since $r = 15$ cm $\ll 10$ m $= L$ and the field point is far from either end, we may regard the wire as approximately infinite. Then Example 20.7 gives

$$E = \frac{2k\lambda}{r} = \frac{(2 \times 9 \times 10^9 \text{ N} \cdot \text{m}^2/\text{C}^2)(2.5 \text{ } \mu\text{C/m})}{0.15 \text{ m}} = 300 \text{ kN/C}$$

(c) At $r = 350$ m   10 m $= L$, the wire behaves approximately like a point charge, so the field strength is

$$E = \frac{kQ}{r^2} = \frac{(9 \times 10^9 \times 25 \times 10^{-6} \text{ N} \cdot \text{m}^2/\text{C})}{(350 \text{ m})^2} = 1.84 \text{ N/C}$$

ASSESS   The finite-size, line charge distribution looks like a point charge at large distances.

69. **INTERPRET**   In this problem we want to find the electric field due to a uniformly charged disk of radius R.

DEVELOP   We take the disk to be consisted of a large number of annuli. With uniform surface charge density σ, the amount of charge on an area element dA is $dq = \sigma dA$. Our strategy is to first calculate the electric field dE due to dq at a field point on the axis, simplify with symmetry argument, and then integrate over the entire disk to get E.

EVALUATE   **(a)** The area of an annulus of radii $R_1 < R_2$ is just $\pi(R_2^2 - R_1^2)$. For a thin ring, $R_1 = r$ and $R_2 = r + dr$, so the area is $\pi[(r + dr)^2 - r^2] = \pi(2rdr + dr^2)$. When dr is very small, the square term is negligible, and $dA = 2\pi r dr$. (This is equal to the circumference of the ring times its thickness.) **(b)** For surface charge density σ, $dq = \sigma dA = 2\pi\sigma r dr$. **(c)** From Example 20.6, the electric field due to a ring of radius r and charge dq is

$$dE_x = k\frac{xdq}{(x^2 + r^2)^{3/2}} = \frac{2\pi k\sigma xr}{(x^2 + r^2)^{3/2}} dr$$

which holds for x positive away from the ring's center. **(d)** Integrating from $r = 0$ to R, one finds

$$E_x = \int_0^R dE_x = 2\pi k\sigma x \int_0^R \frac{rdr}{(x^2 + r^2)^{3/2}} = 2\pi k\sigma x \frac{-1}{\sqrt{x^2 + r^2}}\bigg|_0^R = 2\pi k\sigma\left[\frac{x}{|x|} - \frac{x}{(x^2 + R^2)^{1/2}}\right]$$

For $x > 0$, $|x| = x$ and the field is

$$E_x = 2\pi k\sigma\left(1 - \frac{x}{\sqrt{x^2 + R^2}}\right)$$

On the other hand, for $x < 0$, $|x| = -x$, the electric field is

$$E_x = 2\pi k\sigma\left(-1 + \frac{|x|}{\sqrt{x^2 + R^2}}\right)$$

This is consistent with symmetry on the axis, since $E_x(x) = -E_x(-x)$.

ASSESS   One may readily verify that (see Problem 71), for $x \gg R$, $E_x \approx \frac{kQ}{x^2}$. In other words, the finite-size charge distribution looks like a point charge at large distances.

71.   **INTERPRET**   In this problem we want to show that at large distances, the electric field due to a uniformly charged disk of radius R reduces to that of a point charge.

DEVELOP   The result of Problem 69 for the field on the axis of a uniformly charged disk, of radius R, at a distance $x > 0$ along the axis (away from the disk's center) is

$$E_x = 2\pi k\sigma\left(1 - \frac{x}{\sqrt{x^2 + R^2}}\right)$$

For $R^2/x^2 \ll 1$, we use the binomial expansion in Appendix A and write

$$\left(1 + \frac{R^2}{x^2}\right)^{-1/2} \approx 1 - \frac{1}{2}\frac{R^2}{x^2} +$$

EVALUATE   Substituting the above expression into the first equation, we obtain

$$E_x = 2\pi k\sigma\left[1 - \left(1 + \frac{R^2}{x^2}\right)^{-1/2}\right] \approx 2\pi k\sigma\left[1 - \left(1 - \frac{1}{2}\frac{R^2}{x^2} + \cdots\right)\right] \approx \frac{2\pi k\sigma R^2}{2x^2} = \frac{kQ}{x^2}$$

which is the field from a point charge $Q = \pi R^2\sigma$ at a distance x.

ASSESS   The result once again demonstrates that any finite-size charge distribution looks like a point charge at large distances.

73.   **INTERPRET**   We find the position of the charge Q in Example 20.2 for which the force is a maximum. At large distances, the force will be small because of the inverse-square nature of the force. At close distances, the net force will be small because the forces from the two charges tend to cancel. Somewhere between near and far will be a maximum. The equation for force is found in the example, so we will differentiate to find the value of y where force is a maximum.

DEVELOP   We are given the equation for force: $F = \frac{2kqQy}{(a^2 + y^2)^{3/2}}\hat{j}$. We find the value of y at which this force is a maximum by setting $\frac{dF}{dy} = 0$.

EVALUATE

$$\frac{dF}{dy} = \frac{d}{dy}\left[\frac{2kqQy}{(a^2+y^2)^{3/2}}\right] = 2kqQ\frac{d}{dy}\left[\frac{y}{(a^2+y^2)^{3/2}}\right] = 2kqQ\left[\frac{1}{(a^2+y^2)^{3/2}} - \frac{3}{2}\frac{2y^2}{(a^2+y^2)^{5/2}}\right] = 0$$

$$\rightarrow \frac{(a^2+y^2)-3y^2}{(a^2+y^2)^{5/2}} = 0 \rightarrow (a^2+y^2)-3y^2 = 0 \rightarrow a^2 = 2y^2 \rightarrow y = \frac{a}{\sqrt{2}}$$

ASSESS   This is a bit less than one and a half times the distance from the center to one charge.

75.   **INTERPRET**   We find the electric field near a line of *non-uniform* charge density. This is an electric field calculation, and we will integrate to find the field.

DEVELOP   The rod has charge $\lambda = \lambda_0\left(\frac{x}{L}\right)^2$, and extends from $x = 0$ to $x = L$. We want to find the electric field at $x = -L$. We will use $d\vec{E} = \frac{k\,dq}{r^2}\hat{r}$, with $dq = \lambda\,dx$ and $r = (x+L)$.

EVALUATE

$$dE = \frac{k\,dq}{r^2}\hat{r} \rightarrow \vec{E} = k\hat{i}\int_0^L \frac{\lambda_0\left(\frac{x}{L}\right)^2}{(x+L)^2}dx = \frac{k\lambda_0}{L^2}\hat{i}\left[x - \frac{L^2}{x+L} - 2L\ln(x+L)\right]_0^L$$

$$\rightarrow \vec{E} = \frac{k\lambda_0\hat{i}}{L^2}\left[L - \frac{L^2}{2L} - 2L\ln\left(\frac{2L}{L}\right) - L\right] = \frac{k\lambda_0\hat{i}}{L^2}\left[-\frac{L}{2} - 2L\ln(2)\right] = -\frac{k\lambda_0\hat{i}}{L}\left[\frac{1}{2} + 2\ln 2\right]$$

ASSESS   Since λ_0 is charge per length, the units are correct.

77.   **INTERPRET**   We estimate the electrostatic force between two people if the charge on the electron differed from the charge on the proton by one part in a billion. We use Coulomb's law.

DEVELOP   We will assume that about half a person's mass consists of protons and the rest consists of neutrons. We will also assume that the number of electrons is equal to the number of protons, so the person's net charge is just one billionth of his total proton charge. One proton has a mass $m_p = 1.673\times10^{-27}$ kg, and the charge on a proton is $q = 1.602\times10^{-19}$ C. The magnitude of the force between two charges is $F = k\frac{q_1q_2}{r^2}$.

EVALUATE   The number of protons in a 65-kg person is about

$$N = \frac{1}{2}\frac{65\text{ kg}}{1.673\times10^{-27}\text{ kg}} = 1.9\times10^{28}$$

We multiply this by the charge of a proton, and then divide by a billion to find the charge on a person: $q = \frac{Nq}{10^9} = 3.11$ C. The force becomes $F = k\frac{q^2}{r^2} = 870$ MN, which does not qualify as "negligible."

ASSESS   This is a repulsive force nearly equal to the weight of a Nimitz-class aircraft carrier.

79.   **INTERPRET**   We are asked to find the electric field necessary to levitate small charged objects. We will assume that the only forces acting on the beads are the force of gravity and the electric force.

DEVELOP   To levitate the beads, we must arrange so that the force on the beads due to the field is equal to the force due to gravity $qE = mg$. The mass is $m = 0.010$ kg, and the charge is $q = 5.0\times10^{-6}$ C.

EVALUATE $E = \frac{mg}{q} = 19.6\times10^3$ N/C. The field should be upwards, so that the positively charged beads feel a force opposite the direction of gravity.

ASSESS   One complication to this plan is trying to keep the beads from accelerating sideways! It is very difficult to actually make a large-scale uniform field as described here.

21

GAUSS'S LAW

EXERCISES

Section 21.1 Electric Field Lines

19. **INTERPRET** This problem is about drawing field lines to represent the field strength of a charge configuration.

DEVELOP We follow the methodology illustrated in Figure 21.3. There are 16 lines emanating from charge $+2q$ (eight for each unit of $+q$). Similarly, we have 8 lines ending on $-q$.

EVALUATE The field lines of the charge configuration are shown below.

ASSESS Our sketch is similar to Fig. 21.3 (f) with twice the number of lines of force.

21. **INTERPRET** In this problem we are asked to identify the charges based on the pattern of the field lines.

DEVELOP From the direction of the lines of force (away from positive and toward negative charge) one sees that A and C are positive and B is a negative charge. Eight lines of force terminate on B, eight originate on C, but only four originate on A, so the magnitudes of B and C are equal, while the magnitude of A is half that value.

EVALUATE Based on the reasoning above, we may write $Q_C = -Q_B = 2Q_A$. The total charge is $Q = Q_A + Q_B + Q_C = Q_A$, so $Q_C = 2Q = -Q_B$.

ASSESS The magnitude of the charge is proportional to the number of field lines emerging from or terminating at the charge.

Section 21.2 Electric Flux and Field

23. **INTERPRET** This problem is about finding the electric field strength, given the flux through a surface.

DEVELOP The magnitude of the flux through a flat surface perpendicular to a uniform field is

$$|\Phi| = EA$$

EVALUATE From the equation above, we find the field strength to be

$$E = \frac{|\Phi|}{A} = \frac{65 \text{ N} \cdot \text{m}^2/\text{C}}{10^{-4} \text{ m}^2} = 650 \text{ kN/C}$$

ASSESS The general expression for the electric flux Φ is given by Equation 21.1: $\Phi = \vec{E} \cdot \vec{A} = EA\cos\theta$, where θ is the angle between the normal vector $\vec{A}$ and the electric field $\vec{E}$. We see that when $\theta = 0$, $\Phi = EA$ is a maximum, and when $\theta = 180°$, $\Phi = -EA$ is a minimum.

25. **INTERPRET** This problem is about the electric flux through the surface of a sphere.

DEVELOP The general expression for the electric flux Φ is given by Equation 21.1: $\Phi = \vec{E} \cdot \vec{A} = EA\cos\theta$, where θ is the angle between the normal vector $\vec{A}$ and the electric field $\vec{E}$. The magnitude of the normal vector $\vec{A}$ is $A = 4\pi r^2$, the surface area of the sphere.

EVALUATE For a sphere, with $\vec{E}$ parallel or anti-parallel to $\vec{A}$, Equation 21.1 gives

$$\Phi = \pm 4\pi r^2 E = \pm 4\pi (0.1 \text{ m}/2)^2 (47 \text{ kN/C}) = \pm 1.48 \text{ kN} \cdot \text{m}^2/\text{C}$$

ASSESS The flux Φ is positive if $\vec{A}$ points outward, and is negative if $\vec{A}$ points inward.

Section 21.3 Gauss's Law

27. **INTERPRET** This problem is about applying Gauss's law to find the electric flux through a closed surface.

DEVELOP Gauss's law given in Equation 21.3 states that the flux through any closed surface is proportional to the charge enclosed:

$$\Phi = \oint \vec{E} \cdot d\vec{A} = \frac{q_{enclosed}}{\varepsilon_0}$$

EVALUATE For the surfaces shown, the results are as follows:

(a) $q_{enclosed} = q + (-2q) = -q$, $\Phi = -q/\varepsilon_0$.

(b) $q_{enclosed} = q + (-2q) + (-q) + 3q + (-3q) = -2q$ and $\Phi = -2q/\varepsilon_0$.

(c) $q_{enclosed} = 0$ and $\Phi = 0$.

(d) $q_{enclosed} = 3q + (-3q) = 0$ and $\Phi = 0$.

ASSESS The flux through the closed surface depends only on the charge enclosed, and is independent of the shape of the surface.

29. **INTERPRET** This problem is about applying Gauss's law to find the electric flux through the surface of a cube which encloses a charge.

DEVELOP Gauss's law given in Equation 21.3 states that the flux through any closed surface is proportional to the charge enclosed:

$$\Phi = \oint \vec{E} \cdot d\vec{A} = \frac{q_{enclosed}}{\varepsilon_0}$$

The symmetry of the situation guarantees that the flux through one face is 1/6 the flux through the whole cubical surface.

EVALUATE The flux through one face of a cube is

$$\Phi_{face} = \frac{1}{6} \oint_{cube} \vec{E} \cdot d\vec{A} = \frac{1}{6} \frac{q_{enclosed}}{\varepsilon_0} = \frac{2.6 \ \mu C}{6(8.85 \times 10^{-12} \ C^2/N \cdot m^2)} = 49.0 \ kN \cdot m^2/C$$

ASSESS Since the flux through each surface is the same, the total flux through the cube is simply equal to $\Phi = 6\Phi_{face}$, and is proportional to $q_{enclosed}$.

Section 21.4 Using Gauss's Law

31. **INTERPRET** This problem is about applying Gauss's law to calculate electric field. Our charge distribution has spherical symmetry.

DEVELOP In computing the electric field strength, we make use of the result obtained in Example 21.1, or Equations 21.4 and 21.5:

$$E = \begin{cases} \dfrac{1}{4\pi\varepsilon_0} \dfrac{Qr}{R^3}, & r < R \\[2ex] \dfrac{1}{4\pi\varepsilon_0} \dfrac{Q}{r^2}, & r > R \end{cases}$$

EVALUATE (a) At $r = 15$ cm ($r < R = 25$ cm), the electric field is

$$E = \frac{1}{4\pi\varepsilon_0} \frac{Qr}{R^3} = \frac{(9 \times 10^9 \ N \cdot m^2/C^2)(14 \ \mu C)(15 \ cm)}{(25 \ cm)^3} = 1.21 \ MN/C$$

(b) At $r = R$,

$$E = \frac{1}{4\pi\varepsilon_0} \frac{Q}{R^2} = \frac{(9 \times 10^9 \ N \cdot m^2/C^2)(14 \ \mu C)}{(25 \ cm)^2} = 2.02 \ MN/C$$

(c) Similarly, at $r = 2R > R$, we have

$$E = \frac{1}{4\pi\varepsilon_0} \frac{Q}{(2R)^2} = \frac{(9 \times 10^9 \ N \cdot m^2/C^2)(14 \ \mu C)}{(50 \ cm)^2} = 504 \ kN/C$$

ASSESS   Inside the solid sphere where $r < R$, the electric field increases linearly with r. On the other hand, outside the sphere where $r > R$, the field strength decreases as $1/r^2$. Gauss's law can be applied in this problem because the charge configuration possesses spherical symmetry.

33. **INTERPRET**   In this problem we are given the field strength at two different points outside the charge distribution and asked to determine the symmetry possessed by the configuration.

DEVELOP   The symmetry of the charge distribution can be determined by noting that the electric field strength decreases as $1/r^2$ for a spherically symmetric charge distribution, and as $1/r$ for a line charge (see Examples 21.4 and 21.1).

EVALUATE   Let us write $E = Cr^{-n}$, for some constant C. This gives

$$\frac{E_2}{E_1} = \left(\frac{r_1}{r_2}\right)^n \quad \rightarrow \quad \ln\left(\frac{E_2}{E_1}\right) = n\ln\left(\frac{r_1}{r_2}\right)$$

Substituting the values given, we obtain

$$n = \frac{\ln(E_2/E_1)}{\ln(r_1/r_2)} = \frac{\ln(55/43)}{\ln(23/18)} = 1.00$$

Thus, we conclude that the charge distribution possesses line symmetry.

ASSESS   The $1/r$ dependence characteristic of line symmetry can be readily verified by taking the field strength to be of the form $E = 2k\lambda/r$.

35. **INTERPRET**   We find the electric field produced by a uniform sheet of charge. This problem has planar symmetry, and we will use the result for the electric field near an infinite plane of charge.

DEVELOP   The electric field near a uniform sheet of charge with charge density σ is $E = \frac{\sigma}{2\varepsilon_0}$. The charge density given in the problem is $\sigma = 87$ pC/cm^2 $= 87 \times 10^{-12}$ C/(0.01 m)^2.

EVALUATE

$$E = \frac{\sigma}{2\varepsilon_0} = 49 \times 10^3 \text{ N/C}$$

ASSESS   Although this seems at first to be a small charge per area, the resulting field is quite large. Remember that a Coulomb is a very large charge.

Section 21.5 Field of Arbitrary Charge Distribution

37. **INTERPRET**   The charge distribution has approximate line symmetry, and we apply Gauss's law to compute the electric field.

DEVELOP   Close to the rod, but far from either end, the rod appears infinite, so the electric field strength is (see Example 21.4): $E = 2k\lambda/r$.

EVALUATE   (a) Substituting the values given in the problem statement, we obtain

$$E = \frac{2k\lambda}{r} = \frac{2k(Q/l)}{r} = \frac{2(9 \times 10^9 \text{ N} \cdot \text{m}^2/\text{C}^2)(2 \text{ } \mu\text{C}/0.5 \text{ m})}{0.014 \text{ m}} = 5.14 \times 10^6 \text{ N/C}$$

(b) Far away ($r \gg L$), the rod appears like a point charge, so

$$E \approx \frac{kq}{r^2} = \frac{(9 \times 10^9 \text{ N} \cdot \text{m}^2/\text{C}^2)(2 \text{ } \mu\text{C})}{(23 \text{ m})^2} = 34.0 \text{ N/C}$$

ASSESS   Far from the finite charge distribution (line charge in this case), the field always resembles that of a point charge.

39. **INTERPRET**   We approximate the electric field strength near a charged disk and far from the charged disk. In both cases, we will choose appropriate approximations for how the disk looks.

DEVELOP   The area of the disk is given as $A = 0.14$ m^2, so the radius of the disk must be

$\pi r^2 = A \rightarrow r = \sqrt{\frac{A}{\pi}} = 21$ cm. At a point $r = 0.001$ m from the center of the disk, the disk will appear to be an infinite plane, so we can use $E = \frac{\sigma}{2\varepsilon_0} = \frac{Q}{2\varepsilon_0 A}$. At a point $r = 2.5$ m from the disk, the disk will look more like a point charge, so we can use $E = \frac{kQ}{r^2}$. The charge on the disk is $Q = 5.0 \mu$C.

EVALUATE

(a)
$$E = \frac{Q}{2\varepsilon_0 A} = 2.0 \times 10^6 \, \text{N/C}$$

(b)
$$E = \frac{kQ}{r^2} = 7.2 \times 10^3 \, \text{N/C}$$

ASSESS   If you are far enough from anything, it looks like a point charge; and if you are close enough, it looks like an infinite plane.

Section 21.6 Gauss's Law and Conductors

41.   **INTERPRET**   This problem involves finding the charge distribution of s a conductor using Gauss's law.

DEVELOP   To answer the questions, we note that the electric field within a conducting medium, in electrostatic equilibrium, is zero. In addition, the net charge must reside on the conductor surface.

EVALUATE   (a) Gauss's law implies that the net charge contained in any closed surface, lying within the metal, is zero.

(b) If the volume charge density is zero within the metal, all of the net charge must reside on the surface of the sphere. If the sphere is electrically isolated, the charge will be uniformly distributed (i.e., spherically symmetric), so

$$\sigma = \frac{Q}{4\pi R^2} = \frac{5 \, \mu\text{C}}{4\pi (1 \, \text{cm})^2} = 3.98 \times 10^{-3} \, \text{C/m}^2$$

(c) Spherical symmetry for σ depends on the proximity of other charges and conductors.

ASSESS   Since charges are mobile, the presence of other charge near the conductor will cause the charges on the surface to move and the new equilibrium configuration will not be spherically symmetric.

43.   **INTERPRET**   This problem is about finding the electric field near the surface of a conducting plate. The approximate plane symmetry of the system allows us to make use of Gauss's law.

DEVELOP   The net charge of $Q = 18 \, \mu\text{C}$ must distribute itself over the outer surface of the plate, in accordance with Gauss's law for conductors. The outer surface consists of two plane square surfaces on each face, plus the edges and corners. Symmetry arguments imply that for an isolated plate, the charge density on the faces is the same, but not necessarily uniform because the edges and corners also have charge. If the plate is thin, we could assume that the edges and corners have negligible charge and that the density on the faces is approximately uniform. Then the surface charge density is the total charge divided by the area of both faces,

$$\sigma = \frac{Q}{2A} = \frac{18 \, \mu\text{C}}{2(75 \, \text{cm})^2} = 16.0 \, \mu\text{C}$$

EVALUATE   Using Equation 21.8, we find the field strength near the plate (but not near an edge) to be

$$E = \frac{\sigma}{\varepsilon_0} = 1.81 \, \text{MN/C}$$

ASSESS   Note the distinction between a charged conducting plate and a uniformly charged plate. In the latter, charges are not free to move and the electric field is (see Example 21.6) $E = \sigma/2\varepsilon_0$.

PROBLEMS

45.   **INTERPRET**   This problem is about finding the electric flux through a given surface.

DEVELOP   The electric flux through a surface is given by Equation 21.2:

$$\Phi = \int \vec{E} \cdot d\vec{A}$$

Since the electric field depends only on y, we break up the square into strips of area $d\vec{A} = \pm a \, dy \, \hat{k}$ of length a parallel to the x axis and width dy. The normal to the surface could be $\pm \hat{k}$.

EVALUATE   The integral of Equation 21.2 gives

$$\Phi = \int \vec{E} \cdot d\vec{A} = \int_0^a (E_0 y/a)\hat{k} \cdot (\pm a \, dy \, \hat{k}) = \pm E_0 \int_0^a y \, dy = \pm \frac{1}{2} E_0 a^2$$

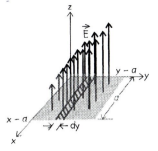

ASSESS   Our result can be compared to the case where the field strength is constant. In that case, the flux through the surface would be $\Phi = \pm E_0 a^2$.

47. **INTERPRET**   The charge distribution has spherical symmetry, so we can apply Gauss's law to find the electric field.
DEVELOP   The balloon can be regarded as a spherical shell with charge residing on the outer surface. For (**b**), we note that outside the shell, the field is like that of a point charge, with total charge at the center. Therefore,

$$E_1 r_1^2 = E_2 r_2^2$$

EVALUATE   (**a**) Since $R = 70$ cm, the point at $r = 50$ cm is inside the uniformly charged spherical shell. Therefore, $q_{enclosed} = 0$ and by Gauss's law, the electric field is zero (see Example 21.2).
(**b**) Let the electric field at $R = 70$ cm be E_0. The electric field at $r = 190$ cm is

$$E = E_0 \left(\frac{R}{r}\right)^2 = (26 \text{ kN/C})\left(\frac{70 \text{ cm}}{190 \text{ cm}}\right)^2 = 3.53 \text{ kN/C}$$

(**c**) Using the given field strength at the surface, we find a net charge

$$Q = \frac{E_0 R^2}{k} = \frac{(26 \text{ kN/C})(0.70 \text{ m})^2}{9 \times 10^9 \text{ N} \cdot \text{m}^2/\text{C}^2} = 1.42 \ \mu\text{C}$$

ASSESS   The electric field inside a spherical shell is identically zero. Outside the shell, the field decreases as $1/r^2$.

49. **INTERPRET**   The charge distribution has spherical symmetry, so we can apply Gauss's law to find the electric field.
DEVELOP   The total electric field is the superposition of the fields due to the point charge and the spherical shell. The field is spherically symmetric about the center.
EVALUATE   (**a**) At $r = R/2 < R$ (inside shell), the electric field is

$$E = E_{pt} + E_{shell} = \frac{k(-2Q)}{(R/2)^2} + 0 = -\frac{8 kQ}{R^2}$$

Note that the minus sign means the direction is radially inward.
(**b**) At $r = 2R > R$ (outside shell), the field strength is

$$E = E_{pt} + E_{shell} = \frac{k(-2Q + Q)}{(2R)^2} = -\frac{kQ}{4R^2}$$

Again the direction of the field is radially inward.
(**c**) If $Q_{shell} = 2Q$, the field inside would be unchanged, but the field outside would be zero since $q_{enclosed} = q_{shell} + q_{pt} = 2Q - 2Q = 0$.
ASSESS   By Gauss's law, the shell produces no electric field in its interior. The field outside a spherically symmetric distribution is the same as if all the charges were concentrated at the center of the sphere.

51. **INTERPRET**   The charge distribution has spherical symmetry, so we can apply Gauss's law to find the electric field.
DEVELOP   The total electric field is the superposition of the fields due to the point charge and the spherical shell. The field is spherically symmetric about the center.
EVALUATE   (**a**) The field due to the shell is zero inside, so at $r = 5$ cm, the field is due to the point charge only. Thus,

$$E = \frac{kq}{r^2} = \frac{(9 \times 10^9 \text{ N} \cdot \text{m}^2/\text{C}^2)(1 \ \mu\text{C})}{(0.05 \text{ m})^2} = 3.60 \times 10^6 \text{ N/C}$$

The field points radially outward.

(b) Outside the shell, its field is like that of a point charge, so at $r = 45$ cm, the field strength is

$$E = \frac{k(q+Q)}{r^2} = \frac{(9 \times 10^9 \text{ N} \cdot \text{m}^2/\text{C}^2)(86 \ \mu\text{C})}{(0.45 \text{ m})^2} = (3.82 \times 10^6 \text{ N/C})$$

The direction of the field is radially outward.

(c) If the charge on the shell were doubled, the field inside would be unaffected, while the field outside would be

$$E = \frac{k(1.0 \ \mu\text{C} + 2 \times 85 \ \mu\text{C})}{(45 \text{ cm})^2} = 7.60 \text{ MN/C}$$

which is approximately doubled,

ASSESS By Gauss's law, the shell produces no electric field in its interior. The field outside a spherically symmetric distribution is the same as if all the charges were concentrated at the center of the sphere.

53. **INTERPRET** The charge distribution has approximate line symmetry, and we apply Gauss's law to compute the electric field.

DEVELOP With the assumption that the electric field is approximately that from an infinitely long, line symmetric charge distribution, we have

$$E = \frac{2k\lambda_{\text{enclosed}}}{r}$$

where $k = \frac{1}{4}\pi\varepsilon_0$ and $\lambda_{\text{enclosed}}$ is the charge inside a unit length of cylindrical Gaussian surface of radius r about the symmetry axis.

EVALUATE   **(a)** For $r = 0.5$ cm $= 0.005$ m, between the wire and the pipe, the enclosed charge per unit length is $\lambda_{\text{enclosed}} = \lambda_{\text{wire}}$, and

$$E = \frac{2k\lambda_{\text{enclosed}}}{r} = \frac{2(9 \times 10^9 \text{ N} \cdot \text{m}^2/\text{C}^2)(5.6 \text{ nC/m})}{(0.005 \text{ m})} = 20.2 \text{ kN/C}$$

The field is (positive) radially away from the axis of symmetry, i.e., the wire.

(b) For $r = 1.5$ cm $= 0.015$ m, the enclosed charge per unit length is $\lambda'_{\text{enclosed}} = \lambda_{\text{wire}} + \lambda_{\text{pipe}}$, and

$$E' = \frac{2k\lambda'_{\text{enclosed}}}{r} = \frac{2(9 \times 10^9 \text{ N} \cdot \text{m}^2/\text{C}^2)(5.6 \text{ nC/m} - 4.2 \text{ nC/m})}{(0.015 \text{ m})} = 1.68 \text{ kN/C}$$

The field is in the same direction as the field in part **(a)**.

ASSESS Between the wire and the pipe, the enclosed charge per unit length is $\lambda_{\text{enclosed}} = \lambda_{\text{wire}}$, while outside the pipe, the enclosed charge is $\lambda'_{\text{enclosed}} = \lambda_{\text{wire}} + \lambda_{\text{pipe}}$. Since the pipe and the wire carry opposite charges, $\lambda'_{\text{enclosed}} < \lambda_{\text{enclosed}}$, and $E' < E$, as we have shown.

55. **INTERPRET** The charge distribution has approximate line symmetry, and we apply Gauss's law to compute the electric field.

DEVELOP We assume that the rod is long enough and approximate its field using line symmetry:

$$E = \frac{2k\lambda_{\text{enclosed}}}{r}$$

where $k = \frac{1}{4}\pi\varepsilon_0$ and $\lambda_{\text{enclosed}}$ is the charge inside a unit length of cylindrical Gaussian surface of radius r about the symmetry axis. Since the charge is uniformly distributed throughout the solid rod, the line charge density is simply equal to the volume charge density times the cross-sectional area of the rod: $\lambda_{\text{enclosed}} = \rho A = \rho\pi r^2$. Combining the two equations yields

$$E = \frac{2k\lambda_{\text{enclosed}}}{r} = \frac{2k(\rho\pi r^2)}{r} = 2\pi k\rho r$$

This equation (valid for $r \leq R$, the radius of the rod) allows us to solve for ρ.

EVALUATE With the electric field at $r = R$ given, we find the volume charge density to be

$$\rho = \frac{E}{2\pi kR} = \frac{2\varepsilon_0 E}{R} = \frac{2(8.85 \times 10^{-12} \text{ C}^2/\text{N} \cdot \text{m}^2)(16 \text{ kN/C})}{0.045 \text{ m}} = 6.29 \ \mu\text{C/m}^3$$

This is the magnitude of ρ, since the direction of the field at the surface, radially inward or outward, was not specified.

ASSESS   In case we are given the field strength at a point $r > R$, then

$$E = \frac{2k\lambda_{\text{enclosed}}}{r} = \frac{2k(\rho\pi R^2)}{r}$$

and the volume charge density would be $\rho = \frac{Er}{2\pi k R^2} = \frac{2\varepsilon_0 Er}{R^2}$.

57.   **INTERPRET**   The infinitely large slab has plane symmetry, and we can apply Gauss's law to compute the electric field.

DEVELOP   When we take the slab to be infinitely large, the electric field is everywhere normal to it (the x direction, as shown in the figure) and symmetrical about the center plane. We follow the approach outlined in Example 21.6 to compute the electric field everywhere.

EVALUATE   **(a)** For points inside the slab $|x| \le d/2$, the charge enclosed by our Gaussian cylinder is

$$q_{\text{enclosed}} = \rho V_{\text{enclosed}} = \rho A(2x)$$

Thus, Gauss's law gives

$$\Phi = \int \vec{E} \cdot d\vec{A} = \frac{q_{\text{enclosed}}}{\varepsilon_0} \quad \rightarrow \quad E(2A) = \frac{\rho A(2x)}{\varepsilon_0}$$

or $E = \frac{\rho x}{\varepsilon_0}$. The direction of $\vec{E}$ is away from (toward) the central plane for positive (negative) charge density.

(b) For points outside the slab $|x| > d/2$, the enclosed charge is

$$q_{\text{enclosed}} = \rho V_{\text{enclosed}} = \rho A d$$

Thus, applying Gauss's law leads to

$$E(2A) = \frac{\rho A d}{\varepsilon_0}$$

or $E = \frac{\rho d}{2\varepsilon_0}$. Again, the direction of $\vec{E}$ is away from (toward) the central plane for positive (negative) charge density.

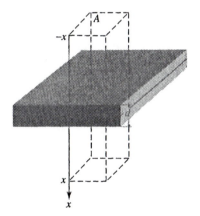

ASSESS   Inside the slab, the charge distribution is equivalent to a sheet with $\sigma = \rho x$. On the other hand, outside the slab, it is equivalent to a sheet with $\sigma = \rho d$.

59.   **INTERPRET**   The square plate has approximate plane symmetry, and we can apply Gauss's law to compute the electric field.

DEVELOP   The electric field strength close to the plate ($x = 1$ cm $\ll 75$ cm $= a$) has approximate plane symmetry, and is given by (see Equation 21.7)

$$E = \frac{\sigma}{2\varepsilon_0}$$

Therefore, the charge on the plate is $q = \sigma A = 2\varepsilon_0 EA = 2\varepsilon_0 Ea^2$, where a is the length of the square plate. At a point sufficiently far from the plate ($r \gg a$) the field strength will resemble that from a point charge, $E = kq/r^2$.

EVALUATE   Substituting the values given, we find the charge on the plate to be

$$q = 2\varepsilon_0 Ea^2 = 2(8.85 \times 10^{-12} \text{ C}^2/\text{N} \cdot \text{m}^2)(45 \text{ kN/C})(0.75 \text{ m})^2 = 448 \text{ } n\text{C}$$

The field strength at $r = 15$ m $\gg 0.75$ m is like that from a point charge:

$$E = \frac{kq}{r^2} = \frac{(9 \times 10^9 \text{ N} \cdot \text{m}^2/\text{C}^2)(448 \text{ nC})}{(15 \text{ m})^2} = 17.9 \text{ N/C}$$

ASSESS   Far from the finite charge distribution (plane charge in this case), the field always resembles that of a point charge.

61.   **INTERPRET**   This problem involves a conductor in electrostatic equilibrium.

DEVELOP   To answer the questions, we note that the electric field within a conducting medium, in electrostatic equilibrium, is zero. In addition, the net charge must reside on the conductor surface.

EVALUATE   **(a)** When there is no charge inside the cavity, the flux through any closed surface within the cavity (S_1) is zero, hence the electric field is also zero. Note that the argument depends on the conservative nature of the electrostatic field, for then positive flux on one part of S_1 canceling negative flux on another part is ruled out.

(b) If the surface charge density on the outer surface (and also the electric field there) is to vanish, then the net charge inside a Gaussian surface containing the conductor (S_2) is zero. Thus, the point charge in the cavity must equal $-Q$, so that $q_{\text{enclosed}} = Q + (-Q) = 0$.

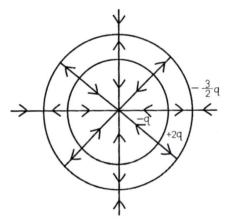

ASSESS   An alternative approach is to say that the total charge on the conductor is $q_c = q_{\text{inner}} + q_{\text{outer}} = Q$. Requiring $q_{\text{outer}} = 0$ means that $q_{\text{inner}} = Q$. Since electric field inside the cavity vanishes, the point charge placed inside must be $q = -q_{\text{inner}} = -Q$.

63.   **INTERPRET**   This is a spherically symmetric charge distribution, so Gauss's law can be applied in this problem.

DEVELOP   The field from the given charges is spherically symmetric, so (from Gauss's law) is like that of a point charge, located at the center, with magnitude equal to the net charge enclosed by a sphere of radius equal to the distance to the field point.

EVALUATE   Thus, the electric field is $E = -\frac{kq}{r^2}$ inside the first shell (8 lines radially inward), $E = +\frac{kq}{r^2}$ between the first and second shells (8 lines radially outward), and $E = -\frac{kq}{2r^2}$ outside the second shell (4 lines radially inward).

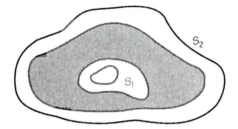

ASSESS   The direction of the electric field depends on the net charge enclosed. When $q_{\text{enclosed}} > 0$, $\vec{E}$ is radially outward. On the other hand, $\vec{E}$ is radially inward when $q_{\text{enclosed}} < 0$. The discontinuity in electric field across the shell is due to a net surface charge density on the shell.

65. **INTERPRET** The charge distribution has spherical symmetry, so we can apply Gauss's law. Also, since the charge distribution is non-uniform, integration is needed to find the electric field.

DEVELOP The charge inside a sphere of radius $r \leq a$ is $q(r) = \int_0^r \rho \, dV$. For volume elements, take concentric shells of radius r and thickness dr, so $dV = 4\pi r^2 dr$.

EVALUATE (a) Integrating over r, the amount of charge enclosed by a Gaussian sphere of radius r is

$$q(r) = 4\pi \int_0^r \rho r^2 dr = 4\pi (\rho_0/a) \int_0^r r^3 \, dr = \frac{\pi \rho_0 r^4}{a}$$

For $r = a$, the total charge is $Q = \pi \rho_0 a^3$.

(b) For spherical symmetry, Gauss's law and Equation 21.5 give

$$4\pi r^2 E(r) = \frac{q(r)}{\varepsilon_0} = \frac{\pi \rho_0 r^4}{\varepsilon_0 a} \quad \rightarrow \quad E(r) = \frac{1}{4\pi\varepsilon_0} \frac{\pi \rho_0 r^2}{a}$$

ASSESS The r^2 dependence of E inside the sphere can be contrasted with the r dependence in the case (see Example 21.1) where the charge distribution is uniform.

67. **INTERPRET** The charge distribution has spherical symmetry, so we can apply Gauss's law. Also, since the charge distribution is non-uniform, integration is needed to find the condition which gives zero electric field outside the sphere.

DEVELOP Using Gauss's law, we see that the field outside the sphere will be zero if the total charge within the volume is zero.

EVALUATE Thus, using thin concentric shells of thickness dr and volume $dV = 4\pi r^2 dr$ as our charge elements, we require that

$$0 = q_{enclosed} = \int_{sphere} \rho \, dV = \int_0^R (\rho_0 - ar^2) 4\pi r^2 dr = 4\pi \left(\frac{1}{3} \rho_0 R^3 - \frac{1}{5} a R^5 \right)$$

or $a = \frac{5\rho_0}{3R^2}$.

ASSESS The charge density starts out from the center of the sphere as ρ_0 and decreases as r^2. At $r = R$, the density is

$$\rho(R) = \rho_0 \left(1 - \frac{5}{3} \right) = -2\rho_0/3$$

Note that $\rho(r)$ must change sign (from positive to negative) in order for $q_{enclosed}$ to be zero at $r = R$.

69. **INTERPRET** Given a charge density in a sphere, we find the electric field at the surface of the sphere. We use Gauss's law, and integrate the charge density to find the charge enclosed.

DEVELOP The charge density is given by $\rho = \rho_0 e^{r/R}$, where r is the distance from the center and R is the radius of the sphere. Gauss's law tells us that $\oint \vec{E} \cdot d\vec{A} = \frac{q_{enclosed}}{\varepsilon_0}$.

EVALUATE

$$\oint \vec{E} \cdot d\vec{A} = \frac{q_{enclosed}}{\varepsilon_0} \rightarrow E \oint dA = \frac{1}{\varepsilon_0} \int_V dq$$

$$E(4\pi R^2) = \frac{1}{\varepsilon_0} \int_0^{2\pi} \int_0^{\pi} \int_0^R \rho_0 e^{r/R} r^2 \sin\theta \, dr \, d\theta \, d\phi$$

$$E = \frac{1}{4\pi\varepsilon_0} 4\pi \rho_0 \int_0^R r^2 e^{r/R} dr = \frac{\rho_0}{\varepsilon_0} (e^{r/R} R(r^2 - 2rR + 2R^2))_0^R$$

$$E = \frac{\rho_0}{\varepsilon_0} [eR(R^2) - R(2R^2)] = R^3 \frac{\rho_0}{\varepsilon_0} (e - 2)$$

ASSESS This answer is positive: the field points away from the sphere. It is also linear with ρ_0, as we would expect.

71. **INTERPRET** We use the gravitational equivalent of Gauss's law to find the gravitational field inside the Earth.

DEVELOP We are told that $\oint \vec{g} \cdot d\vec{A} = -4\pi GM_{enclosed}$. We can use the same approach to solve for $\vec{g}$ that we use to solve for $\vec{E}$ in $\oint \vec{E} \cdot d\vec{A} = \frac{q_{enclosed}}{\varepsilon_0}$.

EVALUATE   Assume that the Earth is spherically symmetric, so $\vec{g}$ is parallel to $d\vec{A}$. We will also assume that the density is approximately constant, so $M_{enclosed} = \frac{4}{3}\pi r^3 \rho$, where $\rho = \frac{m_E}{\frac{4}{3}\pi R_E^3}$. Then at some internal radius r,

$$g \oint dA = -4\pi G M_{enclosed} \rightarrow g(4\pi r^2) = -4\pi G \left[\frac{4}{\cancel{3}}\cancel{\pi} r^3 \left(\frac{m_E}{\frac{4}{\cancel{3}}\cancel{\pi} R_E^3} \right) \right]$$

$$\rightarrow g = G\frac{m_E r}{R_E^3}$$

ASSESS   Coulomb's law and Newton's law of universal gravitation have the same form, so it's not surprising that we can use some of the same mathematics on both. Compare this result with the result for the electric field inside a uniformly charged sphere, $E = k\frac{Qr}{R^3}$.

ELECTRIC POTENTIAL

EXERCISES

Section 22.1 Electric Potential Difference

17. **INTERPRET** This problem is about the energy gained by an electron as it moves through a potential difference ΔV.
DEVELOP We assume that the electron is initially at rest. When released from the negative plate, it moves toward the positive plate, and the kinetic energy gained is $\Delta U = |q\Delta V|$.
EVALUATE As the electron moves from the negative side to the positive side (i.e., against the direction of the electric field), the *kinetic* energy it gains is

$$\Delta K = |(-e)\,\Delta V| = 120 \text{ eV} = (1.6 \times 10^{-19}\,\text{C})(120 \text{ V}) = 1.92 \times 10^{-17}\,\text{J}$$

ASSESS Moving a negative charge through a positive potential difference is like going downhill; potential energy decreases. However, the kinetic energy of the electron is increased.

19. **INTERPRET** This problem is about conversion of units.
DEVELOP By definition, 1 volt = 1 joule/coulomb (1 V = 1 J/C). On the other hand, 1 joule = 1 newton-meter (1 J = 1 N · m).
EVALUATE Combining the two expressions gives 1 V = 1 J/C = 1 N · m/C. It follows that 1 V/m = 1 N/C.
ASSESS These are the units for the electric field strength.

21. **INTERPRET** This problem is about the work done by the battery to move the charge from the positive terminal to the negative terminal.
DEVELOP The work done by the battery is equal to the kinetic energy gained by the charge, and is equal to $W = |q\Delta V|$.
EVALUATE Substituting the values given, we have

$$W = q\Delta V = (3.1 \text{ C})(9.0 \text{ V}) = 27.9 \text{ J}$$

ASSESS The charged particle gains kinetic energy as it moves toward the negative plate (in the direction of the electric field). The battery is needed to maintain the potential difference between the plates.

Section 22.2 Calculating Potential Difference

23. **INTERPRET** In this problem we are given a uniform electric field and asked to calculate the potential difference between two points.
DEVELOP For a uniform field, the potential difference between two points a and b is given by Equation 22.1b:

$$\Delta V_{ab} = V_b - V_a = -\vec{E} \cdot \Delta\vec{r}$$

where $\Delta\vec{r}$ is a vector from a to b.
EVALUATE With $\Delta\vec{r} = \vec{r}_b - \vec{r}_a = y\hat{j}$, we obtain

$$V(y) - V(0) = V(y) = -\vec{E} \cdot \Delta\vec{r} = -(E_0\hat{j}) \cdot (y\hat{j}) = -E_0 y$$

ASSESS The electric potential decreases in the direction of the electric field. In other words, electric field lines always point in the direction of decreasing potential.

25. **INTERPRET** We are asked to find the charge on a sphere, given the potential at the surface. We will assume that the charge is spherically symmetric, and use the equation for potential of a point charge.

DEVELOP The equation for a point charge is $V(r) = \frac{kq}{r}$. Since any spherically symmetric charge distribution looks like a point charge from outside, we will solve this for q. The potential at the surface of the sphere is $V = 4.8$ kV, and the radius is $r = 0.10$ m.

EVALUATE

$$V(r) = \frac{kq}{r} \rightarrow q = \frac{Vr}{k} = 53.3 \text{ nC}$$

ASSESS The key is the recognition that spherically symmetric charge distributions look like point charges from the outside. This is the same as the case for gravitational potential.

27. **INTERPRET** This problem is about the electric potential of a spherically symmetric charge distribution.

DEVELOP Since the electric field outside the spherical charge distribution is the same as that of a point charge, the electric potential outside the metal sphere ($r \geq R$) is given by Equation 22.3:

$$V(r) = \frac{kQ}{r}$$

Note that we have taken the zero of the potential to be at infinity.

EVALUATE **(a)** An isolated metal sphere has a uniform surface charge density, so the potential at its surface is

$$V(R) = \frac{kQ}{R} = \frac{(9 \times 10^9 \text{ N} \cdot \text{m}^2/\text{C}^2)(0.86 \text{ } \mu\text{C})}{(0.035 \text{ m})/2} = 442 \text{ kV}$$

(b) The work done by the repulsive electrostatic field (the negative of the change in the proton's potential energy) equals the proton's kinetic energy at infinity:

$$W = -e(V_\infty - V(R)) = eV(R) = \frac{1}{2}mv^2$$

Thus, the speed of the proton far from the sphere is

$$v = \sqrt{\frac{2eV(R)}{m}} = \sqrt{\frac{2(1.6 \times 10^{-19} \text{ C})(442 \text{ kV})}{1.67 \times 10^{-27} \text{ kg}}} = 9.21 \times 10^6 \text{ m/s}$$

ASSESS As the proton moves away from the metal sphere, its potential energy decreases. However, by energy conservation, its kinetic energy increases.

Section 22.3 Potential Difference and the Electric Field

29. **INTERPRET** This problem is about calculating electric field from electric potential.

DEVELOP Given electric potential $V(x)$, the x component of the electric field may be obtained as $E_x = -dV/dx$. This equation is what we shall use to estimate E_x for the seven straight-line segments shown in Fig. 22.21.

EVALUATE Using the equation above, we find $E_x = 0$ for $x = 0$ to 2 m. Similarly, for $x = 2$ m to 4 m, $E_x = -(-2 \text{ V} - 2 \text{ V})/(4 \text{ m} - 2 \text{ m}) = 2$ V/m. The field strength in other regions can be calculated in a similar manner. The result is sketched below.

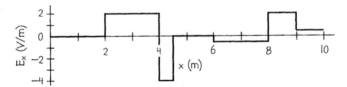

ASSESS The field component $E_x = -dV/dx$ is the negative of the rate of change of V with respect to x. The negative sign means that if we move in the direction of increasing potential, then we're moving against the electric field.

31. **INTERPRET** This problem is about calculating electric field from electric potential.

DEVELOP Given the electric potential V, the corresponding electric field is (see Equation 22.9)

$$\vec{E} = E_x\hat{i} + E_y\hat{j} + E_z\hat{k} = -\left(\frac{\partial V}{\partial x}\hat{i} + \frac{\partial V}{\partial y}\hat{j} + \frac{\partial V}{\partial z}\hat{k}\right)$$

Thus, taking the partial derivatives of V allows us to get the field components.

EVALUATE   (a) Direct substitution gives

$$V(P) = 2xy - 3zx + 5y^2 = 2(1)(1) - 3(1)(1) + 5(1)^2 = 4 \text{ V}$$

(**b**) Use of Equation 22.9 gives

$$E_x = -\frac{\partial V}{\partial x} = -2y + 3z$$

$$E_y = -\frac{\partial V}{\partial y} = -2x - 10y$$

$$E_z = -\frac{\partial V}{\partial z} = 3x$$

At $P(x = y = z = 1)$, we obtain $E_x = 1$ V/m, $E_y = -12$ V/m and $E_z = 3$ V/m.

ASSESS   Electric field is strong in the region where the potential changes rapidly. At $P(x = y = z = 1)$, the potential changes most rapidly in the direction of the electric field

$$\vec{E} = (\hat{i} - 12\hat{j} + 3\hat{k})\text{V/m}$$

Section 22.4 Charged Conductors

33.   **INTERPRET**   This problem is about finding the minimum potential which leads to a dielectric breakdown in air.

DEVELOP   We shall treat the field from the central electrode as that from an isolated sphere. Then $E = kq/R^2$ and $V = kq/R$, so that $V = RE$.

EVALUATE   Breakdown of air occurs at a field strength of $E = 3 \times 10^6$ V/m. Therefore, dielectric breakdown in air would occur for potentials exceeding

$$V = RE = (1 \times 10^{-3} \text{ m})(3 \times 10^6 \text{ V/m}) = 3 \text{ kV}$$

ASSESS   The result means that if we attempt to raise the potential of the electrode in air above 3 kV, then the surrounding air would become ionized and conductive; the extra added charge would leak into the air, resulting in plug sparks.

PROBLEMS

35.   **INTERPRET**   This problem is about finding the electric field strength, given the potential difference between two points.

DEVELOP   Since the field $\vec{E}$ is uniform, Equation 22.1b, $\Delta V_{AB} = -\vec{E} \cdot \Delta \vec{r}$, can be used to relate $\vec{E}$ to the potential difference ΔV_{AB}. Since the path AB is parallel to $\vec{E}$, we write

$$|\Delta V_{AB}| = E\Delta r$$

where E is the field strength, and Δr is the separation between points A and B.

EVALUATE   The field strength is

$$E = \frac{|\Delta V_{AB}|}{\Delta r} = \frac{840 \text{ V}}{0.15 \text{ m}} = 5.60 \text{ kV/m}$$

ASSESS   Since $dV = -\vec{E} \cdot d\vec{r}$ the potential always decreases in the direction of the electric field.

37.   **INTERPRET**   This problem is about finding the potential difference between two points, given the electric field strength.

DEVELOP   Since the field $\vec{E}$ is uniform, Equation 22.1b, $\Delta V = -\vec{E} \cdot \Delta \vec{r}$, can be used to relate $\vec{E}$ to the potential difference ΔV across the membrane. For a uniform electric field parallel to the thickness of the membrane, we write

$$|\Delta V| = E\Delta r$$

where E is the field strength, and Δr is the thickness of the membrane.

EVALUATE   Using the equation above, the potential difference is

$$|\Delta V| = E\Delta r = (8.0 \text{ MV/m})(10 \text{ nm}) = 80 \text{ mV}$$

ASSESS   Since $dV = -\vec{E} \cdot d\vec{r}$ the potential always decreases in the direction of the electric field.

39. **INTERPRET** The problem asks for the speed of the electrons which have been accelerated through a potential difference.

DEVELOP According to the work-energy theorem, the work done on an electron equals the change in its kinetic energy:

$$W = |e\Delta V| = \frac{1}{2}mv^2$$

The electron starts from rest. Knowing ΔV allows us to determine the speed of the electron.

EVALUATE Using the equation above, we find the electron speed to be

$$v = \sqrt{\frac{2\,|e\Delta V|}{m}} = \sqrt{\frac{2(1.6\times10^{-19}\text{ C})(25\times10^3\text{ V})}{9.11\times10^{-31}\text{ kg}}} = 9.37\times10^7 \text{ m/s}$$

ASSESS The electron speed is about $\frac{1}{3}$ of the speed of light. This value is typical of electrons in the cathode ray tube (CRT) TV.

41. **INTERPRET** This problem is about the potential difference between two conducting plates separated by a distance d and having opposite charge densities.

DEVELOP We first calculate the electric field between the plates. Using the result obtained in Example 21.6 for one sheet of charge, and applying the superposition principle, the electric field strength between the plates is $E = \sigma/\varepsilon_0$, and is directed from the positive to the negative plates. Once E is known, we can use Equation 22.1b to calculate V.

EVALUATE Equation 22.1b gives

$$V = V_+ - V_- = -\vec{E}\cdot\Delta\vec{r} = -\left(\frac{\sigma}{\varepsilon_0}\right)(-d) = \frac{\sigma d}{\varepsilon_0}$$

ASSESS The displacement from the negative to the positive plate is opposite to the field direction. In other words, the potential always decreases in the direction of the electric field.

43. **INTERPRET** The problem asks for the charge of the particle which has been accelerated through a potential difference.

DEVELOP The speed acquired by a charge q, starting from rest at point A and moving through a potential difference of V, is given by

$$\frac{1}{2}mv^2 = q(V_A - V_B) = qV \rightarrow v = \sqrt{\frac{2qV}{m}}$$

This is the work-energy theorem for the electric force. A positive charge is accelerated in the direction of decreasing potential. If we have two masses moving through the same potential difference, the ratio of their speed would be

$$\frac{v_2}{v_1} = \sqrt{\frac{2q_2V/m_2}{2q_1V/m_1}} = \sqrt{\frac{q_2}{q_1}\frac{m_1}{m_2}}$$

EVALUATE If the second object acquires twice the speed of the first object ($v_2/v_1 = 2$), moving through the same potential difference, from the equation above, we find its charge to be

$$q_2 = \left(\frac{m_2}{m_1}\right)\left(\frac{v_2}{v_1}\right)^2 q_1 = \left(\frac{2g}{5g}\right)(2)^2(3.8\ \mu\text{C}) = 6.08\ \mu\text{C}$$

ASSESS The speed of the particle moving through a potential difference is proportional to the square root of its charge, and inversely proportional to the square root of its mass.

45. **INTERPRET** The positively charged proton is attracted to the negatively charged sphere via Coulomb interaction. Work must be done to pull the proton away from the sphere.

DEVELOP The work done by the electric field, when a proton escapes from the surface to an infinite distance, equals the change in kinetic energy, or

$$-e(V_\infty - V_{surf}) = eV_{surf} = K_\infty - K_{surf} = -\frac{1}{2}mv^2$$

where we have assumed zero kinetic energy for the proton at infinity, and that the sphere is stationary.

EVALUATE   For a uniformly negatively charged sphere, $V_{surf} = -kQ/R$ (see Example 22.3), Thus, the escape speed is

$$v = \sqrt{\frac{-2eV_{surf}}{m}} = \sqrt{\frac{2keQ}{mR}}$$

ASSESS   The escape speed of a proton from the electric field of the charged sphere in this problem is analogous to the escape speed of a rocket from the Earth's gravitational field.

47.   **INTERPRET**   The problem is about the potential difference between the center of a charged sphere and a point on its surface.

DEVELOP   From Equation 22.1a, we see that the potential difference from point A to point B is given by

$$\Delta V_{AB} = V_B - V_A = -\int_A^B \vec{E} \cdot d\vec{r}$$

As shown in Example 21.1, the electric field inside a uniformly charged sphere is $E = kQr/R^3$ and is radially symmetric.

EVALUATE   The integration above gives

$$V(R) - V(0) = -\int_0^R \left(\frac{kQr}{R^3}\right)dr = -\left.\frac{kQr^2}{2R^3}\right|_0^R = -\frac{kQ}{2R}$$

ASSESS   The potential is higher at the center if Q is positive.

49.   **INTERPRET**   The problem is about the potential difference between two conductors that make up the coaxial cable.

DEVELOP   If the length of the cable is long compared to its outer diameter, the electric field inside, away from the ends, will be due to the inner conductor only (apply cylindrical symmetry and Gauss's law). The potential difference is given by Equation 22.4:

$$\Delta V_{AB} = V_B - V_A = \frac{\lambda}{2\pi\varepsilon_0}\ln\left(\frac{r_A}{r_B}\right) = 2k\lambda\ln\left(\frac{r_A}{r_B}\right)$$

EVALUATE   Substituting the values given in the problem statement, we find the potential difference to be

$$\Delta V_{AB} = 2k\lambda\ln\left(\frac{r_A}{r_B}\right) = 2(9\times10^9 \text{ N}\cdot\text{m}^2/\text{C}^2)(0.56\times10^{-9} \text{ C/m})\ln\left(\frac{1.6 \text{ cm}}{0.2 \text{ cm}}\right) = 21.0 \text{ V}$$

ASSESS   In this problem, we have chosen B to be a point on the inner conductor, and A to be a point on the outer conductor. The inner conductor has a positive charge density while the outer one has a negative charge density. We expect $V_B > V_A$ since the electric field always points in the direction of decreasing potential.

51.   **INTERPRET**   This problem is about the electric potential at a point due to a system of charges.

DEVELOP   The electric potential at a point P due to a collection of charges is given by Equation 22.5:

$$V_P = \sum_i \frac{kq_i}{r_i}$$

This is the equation we shall use to compute the electric potential at the center of a triangle.

EVALUATE   The center is equidistant from each vertex, with

$$r = \frac{a}{2\cos30°} = \frac{a}{\sqrt{3}}$$

Since each charge contributes equally to the potential, the potential at the center is

$$V = \frac{3kq}{r} = \frac{3\sqrt{3}kq}{a}$$

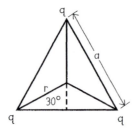

ASSESS   Electric potential is a scalar, so there is no need to consider angles, vector components, or unit vectors.

53.   **INTERPRET**   This problem is about the electric potential at a point due to a system of charges.

DEVELOP   The electric potential at a point P due to a collection of charges is given by Equation 22.5:

$$V_P = \sum_i \frac{kq_i}{r_i}$$

Consider a point $P(x,y)$. The distance from P to $(0, a)$ is $r_+ = \sqrt{(x-a)^2 + y^2}$. Similarly, the distance from P to $(0, a)$ is $r_- = \sqrt{(x+a)^2 + y^2}$.

EVALUATE   (a) The equations above give

$$V_P = \frac{kq}{r_+} + \frac{kq}{r_-} = kq\left(\frac{1}{\sqrt{(x-a)^2 + y^2}} + \frac{1}{\sqrt{(x+a)^2 + y^2}} \right)$$

(b) If $r = \sqrt{x^2 + y^2} \gg a$, then a can be neglected relative to x or y, so

$$V_P = kq\left(\frac{1}{\sqrt{(x-a)^2 + y^2}} + \frac{1}{\sqrt{(x+a)^2 + y^2}} \right) \approx kq\left(\frac{1}{\sqrt{x^2 + y^2}} + \frac{1}{\sqrt{x^2 + y^2}} \right) = \frac{2kq}{r}$$

which is the potential of a point charge of magnitude $2q$.

ASSESS   At a distance much greater than the separation of two charges q_1 and q_2, the electric potential is like that due to one single point charge $q_1 + q_2$. Note that electric potential is a scalar, so there is no need to consider angles, vector components, or unit vectors.

55.   **INTERPRET**   The problem is about finding the electric potential due to a continuous charge distribution.

DEVELOP   From Example 22.6, we see that the electric potential at the center of a charged ring of radius a is

$$V = \frac{kQ}{a}$$

EVALUATE   (a) The radius of the circle is $a = L/2\pi$, where L is the length of the rod. Therefore, the potential at the center of a uniformly charged ring is

$$V = \frac{kQ}{a} = \frac{(9 \times 10^9 \text{ N} \cdot \text{m}^2/\text{C}^2)(3.2 \text{ nC})}{0.20 \text{ m}/2\pi} = 905 \text{ V}$$

(b) Since each charge element, dq, of a circular arc, in Fig. 22.12, is the same distance from the center of the arc, Example 22.6 also gives the potential at the center of a uniformly charged semicircular ring. However, the radius is now $a' = 20 \text{ cm}/\pi$, or twice the value in part (a). Thus, the potential is

$$V' = \frac{1}{2}V = \frac{1}{2}(905 \text{ V}) = 452 \text{ V}$$

ASSESS   Electric potential is a scalar, so there is no need to consider angles, vector components, or unit vectors.

57.   **INTERPRET**   This problem is about the electric potential due to a charged ring, which is a continuous charge distribution.

DEVELOP   From Example 22.6, we see that the electric potential at the center of a charged ring of radius a is

$$V(0) = \frac{kQ}{a}$$

At a distance x along the ring axis from the center of the ring, the potential is

$$V(x) = \frac{kQ}{\sqrt{x^2 + a^2}}$$

These two equations allow us to determine the radius a and the total charge Q.

EVALUATE   Substituting the values given in the problem statement yields

$$V(0) = \frac{kQ}{a} = 45 \text{ kV}, \quad \text{and} \quad V(0.15 \text{ m}) = \frac{kQ}{\sqrt{(0.15 \text{ m})^2 + a^2}} = 33 \text{ kV}$$

Thus, we find

$$\frac{33 \text{ kV}}{45 \text{ kV}} = \frac{a}{\sqrt{(0.15 \text{ m})^2 + a^2}} \quad \rightarrow \quad a = (0.15 \text{ m})\left(\frac{33}{45}\right)\frac{1}{\sqrt{1 - (33/45)^2}} = 0.162 \text{ m}$$

The charge is $Q = \frac{V(0)a}{k} = \frac{(45 \text{ kV})(0.162 \text{ m})}{9 \times 10^9 \text{ N·m}^2/\text{C}^2} = 809$ nC.

ASSESS   In this problem, we are given two conditions, which allow us to solve for two unknowns—the radius and the charge of the ring. Note that the electric potential is the greatest at the center of the ring and falls off as x increases. When $x \gg a$, the potential resembles that of a point charge: $V(x) \approx kQ/x$.

59.   INTERPRET   This problem is about calculating the electric potential from the electric field.

DEVELOP   Equation 22.2b gives the potential for a uniform field. We take the zero of potential at the origin (point A in Equation 22.2b) and let $\vec{r} = x\hat{i} + y\hat{j} + z\hat{k}$ be the vector from the origin to the field point (point B in Equation 22.2b).

EVALUATE   (a) The potential difference between A and B is

$$\Delta V_{AB} = V_B - V_A = V(\vec{r}) - 0 = -E_0(\hat{i} + \hat{j}) \cdot \vec{r} = -E_0(x + y)$$

Since the potential is independent of z, it can be written as $V(\vec{r}) = V(x, y)$.

(b) Using the result obtained in (a), we see that the potential difference between (x_2, y_2) and (x_1, y_1) is

$$\Delta V_{12} = -E_0(x_2 + y_2) + E_0(x_1 + y_1) = -E_0(x_2 + y_2 - x_1 - y_1)$$

Thus, we obtain

$$V(3.5 \text{ m}, -1.5 \text{ m}) - V(2.0 \text{ m}, 1.0 \text{ m}) = -(150 \text{ V/m})(3.5 \text{ m} - 1.5 \text{ m} - 2.0 \text{ m} - 1.0 \text{ m}) = 150 \text{ V}$$

ASSESS   An alternative approach is to note that the displacement vector from $(2.0 \text{ m}, 1.0 \text{ m})$ to $(3.5 \text{ m}, -1.5 \text{ m})$ is

$$\Delta\vec{r} = (3.5 \text{ m} - 2.0 \text{ m})\hat{i} + (-1.5 \text{ m} - 1.0 \text{ m})\hat{j} = (1.5 \text{ m})\hat{i} - (2.5 \text{ m})\hat{j}$$

Using Equation 22.1b, the potential difference is then equal to

$$\Delta V = -\vec{E} \cdot \Delta\vec{r} = -(150 \text{ V/m})(\hat{i} + \hat{j}) \cdot [(1.5 \text{ m})\hat{i} - (2.5 \text{ m})\hat{j}] = -(150 \text{ V/m})(1.5 \text{ m} - 2.5 \text{ m}) = 150 \text{ V}$$

Thus, the same result is obtained.

61.   INTERPRET   In this problem we want to use the expression for the electric dipole potential to find the electric field at a point on the perpendicular bisector of the dipole.

DEVELOP   The dipole potential is given by Equation 22.6:

$$V(r, \theta) = \frac{kp\cos\theta}{r^2}$$

Using Equation 22.9, the general expressions for the r and q components of the electric fields are

$$E_r = -\frac{\partial V}{\partial r} = \frac{2kp\cos\theta}{r^3}$$

$$E_\theta = -\frac{1}{r}\frac{\partial V}{\partial \theta} = \frac{kp\sin\theta}{r^3}$$

EVALUATE   On the bisecting plane, $\theta = 90°$, which yields $E_r = 0$ and $E_\theta = kp/r^3$, or $\vec{E} = E_\theta\hat{\theta} = (kp/r^3)\hat{\theta}$. To compare with Equation 20.6a, we take the origin at the center of the dipole, the dipole moment along the x axis ($\vec{p} = p\hat{i}$), and the y axis up in Fig. 22.10, so $\hat{\theta} = -\hat{i}\sin 90° + \hat{j}\cos 90° = -\hat{i}$ and $r = \sqrt{0^2 + y^2} = y$ on the bisecting plane. This leads to $\vec{E} = -(kp/y^3)\hat{i}$.

ASSESS   Instead of using polar coordinates, one could first express V in terms of x and y (using $x = r\cos\theta$ and $y = r\sin\theta$):

$$V(x,y) = \frac{kpx}{(x^2 + y^2)^{3/2}}$$

and then differentiate, $E_x = -\partial V/\partial x$ and $E_y = -\partial V/\partial y$. The result is the same.

63.   **INTERPRET**   We are given the electric potential and asked to find the corresponding electric field.

DEVELOP   We first note that the potential $V(r) = -(V_0/R)r$ depends only on r. This implies that the electric field is spherically symmetric and points in the radial direction. The field can be calculated using Equation 22.9.

EVALUATE   Using Equation 22.9, we obtain $E_r = -\frac{dV}{dr} = \frac{V_0}{R}$. In vectorial form, the field can be written as $\vec{E} = \frac{V_0}{R}\hat{r}$, where $\hat{r}$ is a unit vector that points radially outward.

ASSESS   The electric field is uniform, but the potential is linear in r. The difference of one power in r is because the potential is an integral of the field over distance.

65.   **INTERPRET**   We are given two charge-carrying conducting spheres, and we want to find the electric potential and electric field at various points.

DEVELOP   Since the spheres are small and widely separated, they behave like isolated spheres, and their charge distributions are essentially spherical.

EVALUATE   (a) Using the result obtained in Example 22.3, at the surface of either sphere, the potential is

$$V(R) = \frac{kq}{R} = \frac{(9\times10^9\ \text{N}\cdot\text{m}^2/\text{C}^2)(1.2\times10^{-7}\ \text{C})}{0.025\ \text{m}} = 43.2\ \text{kV}$$

(b) The electric field at the surface of each sphere is

$$E(R) = \frac{kq}{R^2} = \frac{V(R)}{R} = \frac{43.2\ \text{kV}}{0.025\ \text{m}} = 1.73\times10^6\ \text{V/m}$$

(c) Midway between the spheres, the potential from each one is the same, so

$$V_{\text{mid-pt.}} = \frac{2kq}{r} = \frac{2(9\times10^9\ \text{V}\cdot\text{m/C})(1.2\times10^{-7}\ \text{C})}{4\ \text{m}} = 540\ \text{V}$$

(d) Since the spheres are at the same potential, the difference is zero.

ASSESS   In this problem, the two conducting spheres can be treated as being isolated and the superposition principle applies because they are far apart ($r \gg R$). If they were brought close to each other, then the charge distribution would no longer be spherical.

67.   **INTERPRET**   Our coaxial cable consists of an inner conductor and an outer conductor, each carrying different charges.

DEVELOP   For a long straight cable, the field between the two conductors is approximately cylindrically symmetric and depends only on the charge density of the wire. In fact, the field strength in the region between the conductors is $E = \frac{2k\lambda_{\text{wire}}}{r}$, and its direction is radially outward. The potential difference can be calculated using Equation 22.4:

$$\Delta V_{AB} = -\int_{r_A}^{r_B} \vec{E}\cdot d\vec{r} = 2k\lambda_{\text{wire}}\ln\left(\frac{r_A}{r_B}\right)$$

where r_A and r_B are the radius of the inner and outer conductors, respectively.

EVALUATE   (a) Substituting the values given in the problem, we find the potential difference to be

$$\Delta V_{AB} = 2k\lambda_{\text{wire}}\ln\left(\frac{r_A}{r_B}\right) = 2(9\times10^9\ \text{N}\cdot\text{m}^2/\text{C}^2)(75\times10^{-9}\ \text{C/m})\ln\left(\frac{2\ \text{mm}}{10\ \text{mm}}\right) = -2.17\ \text{kV}$$

(b) The potential difference in the region between the conductors does not depend on the charge density of the outer conductor. Therefore, the result does not change.

ASSESS   The negative sign in ΔV_{AB} means that the outer conductor is at a lower potential than the inner one. This is precisely what we expect because the electric field points radially outward (in the direction of decreasing potential).

69. **INTERPRET** This problem is about an electric potential function and its corresponding electric field.

DEVELOP We first note that the potential can be rewritten as

$$V(x) = x(x+3)(1-x)$$

Since the potential V is independent of y and z, the electric field has only an x component, which can be computed using Equation 22.9: $E_x = -\partial V/\partial x$.

EVALUATE (a) From the equation above, we see that $V(x) = 0$ at $x = 0, 1$ m and 3 m.

(b) Differentiating V with respect to x, we find $E_x = -\frac{dV}{dx} = 3x^2 + 4x - 3$.

(c) Solving the quadratic equation, the solutions for $E_x = 0$ are

$$x = \frac{-4 \pm \sqrt{16 - 4(3)(-3)}}{2(3)} = 0.535 \text{ m and } -1.87 \text{ m}$$

ASSESS The component of the electric field is the negative of the rate of change of the potential. So the points where $E_x = 0$ are where the slope of $V(x)$ vanishes (see the figure).

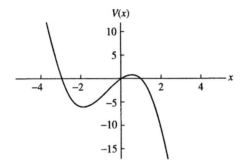

71. **INTERPRET** This problem is about the electric field of a charged disk. We want to show that the expression found in Example 22.8 has the correct asymptotic behavior.

DEVELOP From Example 22.8, the electric field on the axis ($x > 0$) of a charged disk of radius a is

$$E_x = \frac{2kQ}{a^2}\left(1 - \frac{x}{\sqrt{x^2 + a^2}}\right)$$

In the limit where $x \gg a$, the quantity $x/\sqrt{x^2 + a^2}$ may be approximated as

$$\frac{x}{\sqrt{x^2 + a^2}} = [1 + (a/x)^2]^{-1/2} \approx 1 - \frac{1}{2}\frac{a^2}{x^2} + \cdots$$

EVALUATE Substituting the above expression into the first one, we obtain

$$E_x = \frac{2kQ}{a^2}\left(1 - \frac{x}{\sqrt{x^2 + a^2}}\right) \approx \frac{2kQ}{a^2}\left(1 - 1 + \frac{1}{2}\frac{a^2}{x^2}\right) = \frac{kQ}{x^2}$$

which is like that of a point charge.

ASSESS As always, a finite-size charge distribution looks like a point charge at a large distance.

73. **INTERPRET** This problem is about the work done by the electric field in separating the thorium nucleus and the alpha particle.

DEVELOP The work done by the Coulomb repulsion as the thorium nucleus and the alpha particle separate is

$$W = \Delta U_{R\to\infty} = q_\alpha \Delta V_{\text{Th},R\to\infty} = q_\alpha\left(\frac{kq_{\text{Th}}}{R}\right) = \frac{kq_\alpha q_{\text{Th}}}{R}$$

where $R = 7.4$ fm is the initial separation. (The final separation is essentially infinite.) On the other hand, with the neglect of other interactions, the work-energy theorem requires that this equal the change in the kinetic energy of the two particles:

$$W = \frac{1}{2}m_\alpha v_\alpha^2 + \frac{1}{2}m_{\text{Th}}v_{\text{Th}}^2$$

Note that the two particles start from rest. The conservation of momentum (under the same assumptions) requires also that $0 = m_\alpha v_\alpha + m_{\text{Th}}v_{\text{Th}}$ (since the total momentum is zero initially), so that v_α and v_{Th} can be determined.

EVALUATE   Combining the two expressions, noting that $|v_{Th}| = (4/234) |v_\alpha|$ gives

$$\frac{k(2e)(90e)}{R} = \frac{1}{2}\left(234\ u\ v_{Th}^2 + 4\ u\ v_\alpha^2\right) = \frac{1}{2}\left(\frac{16}{234}\ u + 4\ u\right)v_\alpha^2$$

The speeds are $v_\alpha = 4.07 \times 10^7$ m/s and $v_{Th} = 6.96 \times 10^5$ m/s, where we have used $1\ u = 1.66 \times 10^{-27}$ kg.

ASSESS   Since $m_{Th} \gg m_\alpha$, we expect $v_\alpha \gg v_{Th}$. This indeed is what we have found.

75. **INTERPRET**   The cylinder is a continuous charge distribution, and we want to find the potential at its center on the axis.

DEVELOP   Following the hint given in the problem, we consider the cylinder to be composed of rings of radius a, width dx, and charge $dq = (q/2a)dx$. The potential at the center of the cylinder (which we take as the origin, with x axis along the cylinder axis) due to a ring at $x(-a \le x \le a)$ is (see Example 22.6)

$$dV_{cyl} = \frac{k\,dq}{\sqrt{x^2 + a^2}} = \frac{kq\,dx}{2a\sqrt{x^2 + a^2}}$$

EVALUATE   The whole potential at the center follows from integration:

$$V_{cyl} = \int_{-a}^{a} \left(\frac{kq}{2a}\right)\frac{dx}{\sqrt{x^2 + a^2}} = \frac{kq}{2a}\ln\left(\frac{a + \sqrt{2}a}{-a + \sqrt{2}a}\right) = \frac{kq}{a}\ln(1 + \sqrt{2}) = 0.881\frac{kq}{a}$$

ASSESS   The result can be compared with that at the center of a charged ring of radius a: $V_{ring} = kq/a$. In the cylinder case, the charge elements generally are farther away from the center compared to the ring. Hence, we expect $V_{cyl} < V_{ring}$.

77. **INTERPRET**   We need to find the potential due to a line of charge with a non-constant charge density. The problem is one-dimensional: the line and the point of interest are all on the x axis. We will use the integral expression for potential.

DEVELOP   We start with $V = \int\frac{k\,dq}{r}$, where $dq = \lambda\,dx' = \lambda_0\left(\frac{x'}{L}\right)^2 dx'$ and $r = x - x'$. We will integrate from $x' = -\frac{L}{2}$ to $x' = \frac{L}{2}$.

We must also check to see that our expression reduces to the expected result for $x \gg L$.

EVALUATE

$$V = \int\frac{k\,dq}{r} = \int_{-L/2}^{L/2}\frac{k\lambda_0}{x - x'}\left(\frac{x'}{L}\right)^2 dx' = \frac{k\lambda_0}{L^2}\int_{-L/2}^{L/2}\frac{x'^2}{x - x'}dx'$$

$$V = \frac{k\lambda_0}{L^2}\left[-\frac{x'^2}{2} + x'x - x^2\ln(x - x')\right]_{-L/2}^{L/2}$$

$$V = -\frac{k\lambda_0}{L^2}\left[Lx + x^2\ln\left(\frac{2x - L}{2x + L}\right)\right]$$

For $x \gg L$, we rearrange slightly and use the approximation $\ln(1 + \xi) \approx \xi$ for small ξ:

$$V = -\frac{k\lambda_0}{L^2}x^2\left[\frac{L}{x} + \ln\left(1 - \frac{L}{2x}\right) - \ln\left(1 + \frac{L}{2x}\right)\right]$$

$$V \approx -\frac{k\lambda_0}{L^2}x^2\left[\frac{L}{x} + \left(-\frac{L}{2x} - \frac{1}{2}\left(\frac{L}{2x}\right)^2 + \frac{1}{3}\left(\frac{L}{2x}\right)^3\right) - \left(\frac{L}{2x} - \frac{1}{2}\left(\frac{L}{2x}\right)^2 + \frac{1}{3}\left(\frac{L}{2x}\right)^3\right)\right]$$

$$\rightarrow V \approx -\frac{k\lambda_0}{L^2}x^2\left[\frac{L}{x} - \frac{L}{x} - \frac{L^2}{8x^2} + \frac{L^2}{8x^2} - \frac{L^3}{24x^3} - \frac{L^3}{24x^3}\right] = \frac{k\lambda_0}{L^2}x^2\frac{L^3}{12x^3} = \frac{k\lambda_0 L}{12x}$$

The total charge on the rod is

$$q = \int dq = \int_{-L/2}^{L/2}\lambda_0\left(\frac{x'}{L}\right)^2 dx' = \frac{\lambda_0}{L^2}\int_{-L/2}^{L/2}x'^2 dx' = \frac{\lambda_0}{L^2}\left[\frac{x'^3}{3}\right]_{-L/2}^{L/2} = \frac{\lambda_0}{3L^2}\left[\frac{2L^3}{8}\right]$$

$$q = \frac{\lambda_0 L}{12}$$

so in the limit of $x \gg L$, $V = \frac{kq}{x}$.

ASSESS   The answer we obtain appears to be a point charge from a large distance, as we would hope.

79.   **INTERPRET**   Behind all the details of this problem, the actual question is to determine the potential a distance r from the center of a uniform line of charge. We will use the integral form for potential, and verify whether the given equation is correct.

DEVELOP   We start with $V = \int \frac{k\, dq}{r'}$, where $dq = \frac{Q}{L} dz'$ and $r' = \sqrt{r^2 + z'^2}$. We will integrate from $z' = 0$ to $z' = \frac{L}{2}$, and multiply by 2.

Hopefully, the result of our calculation will be $V(r) = \frac{2kQ}{L} \ln\left(\frac{L}{2r} + \sqrt{1 + \frac{L^2}{4r^2}} \right)$.

EVALUATE

$$V = \int \frac{k\, dq}{r'} = 2\int_0^{L/2} \frac{k}{r'} \frac{Q}{L} dz' = \frac{2kQ}{L} \int_0^{L/2} \frac{dz'}{\sqrt{r^2 + z'^2}}$$

$$\rightarrow V = \frac{2kQ}{L} \ln(z' + \sqrt{r^2 + z'^2})\Big|_0^{L/2} = \frac{2kQ}{L}\left[\ln\left(\frac{L}{2} + \frac{1}{2}\sqrt{L^2 + 4r^2} \right) - \ln(r) \right]$$

$$\rightarrow V = \frac{2kQ}{L} \ln\left(\frac{L + \sqrt{L^2 + 4r^2}}{2r} \right) = \frac{2kQ}{L} \ln\left(\frac{L}{2r} + \sqrt{\frac{L^2}{4r^2} + 1} \right)$$

ASSESS   The equation is correct.

ELECTROSTATIC ENERGY AND CAPACITORS

23

EXERCISES

Section 23.1 Electrostatic Energy

15. **INTERPRET** We find the electrostatic energy of a collection of point charges, using the method outlined in Section 23.1.

DEVELOP The point charges are arranged on the corners of a square of side length a. Three of them are the same: $q_1 = q_2 = q_3 = +q$. The fourth has charge $q_4 = -\frac{1}{2}q$. The first charge takes no work to bring into position, since there is no field present initially.

$$W_1 = 0$$

The work to bring in charge 2 takes the potential energy the charge will have due to charge 1:

$$W_2 = k\frac{q_1 q_2}{a}$$

The work to bring in charge 3 takes the potential energy due to the presence of charges 1 and 2:

$$W_3 = k\frac{q_1 q_3}{a} + k\frac{q_2 q_3}{a\sqrt{2}}$$

We've assumed that charges 2 and 3 are diagonally opposite on the square, thus the $\sqrt{2}$ term. The work to bring in charge 4 is the potential energy due to the presence of charges 1, 2, and 3:

$$W_4 = k\frac{q_1 q_4}{a\sqrt{2}} + k\frac{q_2 q_4}{a} + k\frac{q_3 q_4}{a}$$

Again, charges 1 and 4 are diagonally opposite each other, while charges 2 and 3 are separated from charge 4 by only the side length a. The energy of the configuration is the sum of these works.

EVALUATE

$$E = W_1 + W_2 + W_3 + W_4 = \frac{k}{a}\left(q_1 q_2 + q_1 q_3 + \frac{q_2 q_3}{\sqrt{2}} + \frac{q_1 q_4}{\sqrt{2}} + q_2 q_4 + q_3 q_4 \right)$$

$$\rightarrow E = \frac{kq^2}{a}\left(1 + 1 + \frac{1}{\sqrt{2}} - \frac{1}{2\sqrt{2}} - \frac{1}{2} - \frac{1}{2} \right) = \frac{kq^2}{a}\left(1 + \frac{1}{2\sqrt{2}} \right)$$

ASSESS This is not a particularly convenient method of calculating energies. Fortunately, the charge on a single electron is small enough that we can usually approximate real charge distributions as continuous and integrate.

17. **INTERPRET** We find the speed of the charges in an "electrostatic explosion." We use conservation of energy to solve this problem.

DEVELOP The three charges are initially in a symmetric arrangement, as shown in Figure 23.1. They have the same charge q, and as shown in Section 23.1, the work done to assemble the configuration is $W = k\frac{q_1 q_2}{a} + k\frac{q_1 q_3}{a} + k\frac{q_2 q_3}{a} = \frac{3kq^2}{a}$. This is the initial energy of the configuration. From the symmetry of the initial configuration, we can see that they will each gain $\frac{1}{3}$ of that energy when they fly apart, and we can find the speed v from their kinetic energy.

EVALUATE The initial energy is $E_i = \frac{3kq^2}{a}$. The final energy is $E_f = (3)\frac{1}{2}mv^2$. These energies must be equal, so $\frac{3}{2}mv^2 = \frac{3kq^2}{a} \rightarrow v = q\sqrt{\frac{2k}{am}}$.

ASSESS   Conservation of energy applies to more than just mechanical systems. We use this technique throughout all areas of physics.

Section 23.2 Capacitors

19.   **INTERPRET**   This problem is about a parallel-plate capacitor. We are given the plate separation and the charges on the plates, and asked to find the electric field between the plates, the potential difference, and the energy stored.

   DEVELOP   The electric field between two closely spaced, oppositely charged, parallel conducting plates is approximately uniform (directed from the positive to the negative plate), with strength (See the last paragraph of Section 22.6.)

$$E = \frac{\sigma}{\varepsilon_0} = \frac{q}{\varepsilon_0 A}$$

   Since the electric field is uniform, the potential difference between the plates is given by Equation 22.1b, $V = Ed$, where d is the plate separation. Finally, the energy stored in the capacitor can be calculated using Equation 23.3: $U = CV^2/2 = qV/2$, where $q = CV$.

   EVALUATE   (a) Using the equation above, we find the electric field to be

$$E = \frac{q}{\varepsilon_0 A} = \frac{(1.1\ \mu C)}{(8.85 \times 10^{-12}\ C^2/N \cdot m^2)(0.25\ m)^2} = 1.99\ MV/m$$

   (b) The potential difference is

$$V = Ed = (1.99\ MV/m)(5\ mm) = 9.94\ kV$$

   (c) The energy stored is

$$U = \frac{1}{2}CV^2 = \frac{1}{2}qV = \frac{1}{2}(1.1\ \mu C)(9.94\ kV) = 5.47\ mJ$$

   ASSESS   For completeness, the capacitance of the capacitor is

$$C = \frac{\varepsilon_0 A}{d} = \frac{(8.85 \times 10^{-12}\ C^2/N \cdot m^2)(0.25\ m)^2}{5 \times 10^{-3}\ m} = 1.1 \times 10^{-10}\ F = 11\ nF$$

   The value is typical of a capacitor.

21.   **INTERPRET**   This problem is about the energy stored in a parallel-plate capacitor.

   DEVELOP   Using Equations 23.1–3, the energy stored in a parallel-plate capacitor is related to the charges on the plates as

$$U = \frac{1}{2}CV^2 = \frac{1}{2}C\left(\frac{Q}{C}\right)^2 = \frac{1}{2}\frac{Q^2}{C} = \frac{1}{2}\frac{Q^2}{C} = \frac{1}{2}\frac{Q^2}{\varepsilon_0 A/d} = \frac{Q^2 d}{2\varepsilon_0 A} \quad \rightarrow \quad Q = \sqrt{\frac{2\varepsilon_0 A U}{d}}$$

   Once Q is found, the potential difference V between the plates is simply given by

$$U = \frac{1}{2}CV^2 = \frac{1}{2}(CV)V = \frac{1}{2}QV \quad \rightarrow \quad V = \frac{2U}{Q}$$

   EVALUATE   (a) Using the values given in the problem statement, we find the charge to be

$$Q = \sqrt{\frac{2\varepsilon_0 A U}{d}} = \sqrt{\frac{2(8.85 \times 10^{-12}\ C^2/N \cdot m^2)(0.05\ m)^2(15 \times 10^{-3}\ J)}{1.2 \times 10^{-3}\ m}} = 0.744\ \mu C$$

   (b) The potential difference is $V = \frac{2U}{Q} = \frac{2(15 \times 10^{-3}\ J)}{0.744 \times 10^{-6}\ C} = 40.3\ kV$.

   ASSESS   Since the electric field between the plates is $E = \sigma/\varepsilon_0$, the potential can also be calculated as

$$V = \frac{E}{d} = \frac{\sigma}{\varepsilon_0 d} = \frac{Qd}{\varepsilon_0 A} = 40.3\ kV$$

23.   **INTERPRET**   This problem deals with the electrostatic energy stored in a capacitor.

   DEVELOP   Equation 23.1, $C = Q/V$, provides the connection between capacitance C, charge Q, and potential difference V. This is the equation we shall use to solve for Q.

EVALUATE   The magnitude of the charge on either plate is

$$Q = CV = (4.0\ \text{F})(3.5\ \text{V}) = 14.0\ \text{C}$$

ASSESS   This is a very large capacitor, since the capacitance of most capacitors falls in the range of 1 pF to 1 F.

25. **INTERPRET**   This problem is about the units of ε_0, the permittivity constant.

    DEVELOP   The value of ε_0 is $8.85 \times 10^{-12}\ \text{C}^2/\text{N} \cdot \text{m}^2$, with units $\text{C}^2/\text{N} \cdot \text{m}^2$. On the other hand, since $C = Q/V$, 1 F (farad) is equivalent to 1 C/V.

    EVALUATE   From the above, we see that the units of ε_0 are

    $$\text{C}^2/\text{N} \cdot \text{m}^2 = (\text{C/N} \cdot \text{m})(\text{C/m}) = (\text{C/J})(\text{C/m}) = \text{C/V} \cdot \text{m} = \text{F/m}$$

    ASSESS   To see that the result makes sense, we can use Equation 23.2 and write

    $$\varepsilon_0 = \frac{Cd}{A}$$

    The units of C are F, while the units of d/A are 1/m.

27. **INTERPRET**   The configuration is a parallel-plate capacitor, and the object of interest is the area of each plate.

    DEVELOP   Using Equations 23.1 and 23.2 for capacitance, we have

    $$C = \frac{Q}{V} = \frac{\varepsilon_0 A}{d} \quad \rightarrow \quad A = \frac{Qd}{\varepsilon_0 V}$$

    This is the equation we shall use to solve for the area A.

    EVALUATE   Substituting the values given, we find the area to be

    $$A = \frac{Qd}{\varepsilon_0 V} = \frac{(2.3 \times 10^{-6}\ \text{C})(1.1 \times 10^{-3}\ \text{m})}{(8.85 \times 10^{-12}\ \text{C}^2/\text{N} \cdot \text{m}^2)(150\ \text{V})} = 1.91\ \text{m}^2$$

    ASSESS   If we take the plate to be square, then each side has a length of 1.38 m, which indeed is much greater than the distance of 1.1 mm between the plates.

29. **INTERPRET**   The configuration is a parallel-plate capacitor, and the object of interest is the capacitance, given the energy stored in the capacitor.

    DEVELOP   Equation 23.3, $U = CV^2/2$, provides the connection between the stored energy U, the capacitance C, and the potential V. This is the equation we shall use to solve for C.

    EVALUATE   Substituting the values given, we find the capacitance to be

    $$C = \frac{2U}{V^2} = \frac{2(350\ \mu\text{J})}{(100\ \text{V})^2} = 70\ \text{nF}$$

    ASSESS   The value is within the typical range of capacitance ($10^{-12}\ \text{F}$ to $10^{-6}\ \text{F}$).

Section 23.3 Using Capacitors

31. **INTERPRET**   This problem is about connecting two capacitors in series.

    DEVELOP   For two capacitors connected in series, the total voltage is the sum of the voltages across each one, $V = V_1 + V_2$, whereas the charge on each capacitor is the same,

    $$Q_1 = Q_2 \quad \rightarrow \quad C_1 V_1 = C_2 V_2$$

    Given V, V_1, and V_2, we can use the above two equations to solve for C_1 and C_2.

    EVALUATE   The above equations can be combined to give $V = V_1 + (C_1/C_2)V_1$, or

    $$V_1 = V \frac{C_2}{C_1 + C_2}$$

    Similarly, we find $V_2 = \frac{VC_1}{C_1 + C_2}$. For $V_1 = V_2 = V/2 = 50\ \text{V}$ as in this problem, either equation implies $C_1 = C_2$.

    ASSESS   The equivalent capacitance of two capacitors connected in series is

    $$\frac{1}{C} = \frac{1}{C_1} + \frac{1}{C_2} \quad \rightarrow \quad C = \frac{C_1 C_2}{C_1 + C_2}$$

In our case, $C = C_1/2$ since $C_1 = C_2$, and the voltage across C is

$$V = \frac{Q}{C} = \frac{Q_1}{C_1/2} = \frac{2Q_1}{C_1} = 2V_1$$

This is precisely the condition given in the problem statement.

33. **INTERPRET** This problem is about all possible values of equivalent capacitance that could be obtained by connecting three capacitors.

DEVELOP The equivalent capacitance has a maximum value when all three capacitors are connected in parallel, with

$$C = C_1 + C_2 + C_3$$

On the other hand, C is a minimum when the capacitors are connected in series:

$$\frac{1}{C} = \frac{1}{C_1} + \frac{1}{C_2} + \frac{1}{C_3} \quad \rightarrow \quad C = \frac{C_1 C_2 C_3}{C_1 C_2 + C_1 C_3 + C_2 C_3}$$

Intermediate values are obtained when one is in parallel with the other two in series, or one in series with the other two in parallel.

EVALUATE (a) When all capacitors are in parallel, we have

$$C = C_1 + C_2 + C_3 = 1 + 2 + 3 = 6 \ \mu F$$

(b) When all are in series, the equivalent capacitance is

$$C = \frac{C_1 C_2 C_3}{C_1 C_2 + C_1 C_3 + C_2 C_3} = \frac{(1 \ \mu F)(2 \ \mu F)(3 \ \mu F)}{(1 \ \mu F)(2 \ \mu F) + (1 \ \mu F)(3 \ \mu F) + (2 \ \mu F)(3 \ \mu F)} = \frac{6}{11} \ \mu F = 0.545 \ \mu F$$

(c) When one is in parallel with the other two in series, the possible values are

$$C = C_1 + \frac{C_2 C_3}{C_2 + C_3} = 1 \ \mu F + \frac{(2 \ \mu F)(3 \ \mu F)}{2 \ \mu F + 3 \ \mu F} = \frac{11}{5} \ \mu F = 2.20 \ \mu F$$

$$C = C_2 + \frac{C_1 C_3}{C_1 + C_3} = 2 \ \mu F + \frac{(1 \ \mu F)(3 \ \mu F)}{1 \ \mu F + 3 \ \mu F} = \frac{11}{4} \ \mu F = 2.75 \ \mu F$$

$$C = C_3 + \frac{C_1 C_2}{C_1 + C_2} = 3 \ \mu F + \frac{(1 \ \mu F)(2 \ \mu F)}{1 \ \mu F + 2 \ \mu F} = \frac{11}{3} \ \mu F = 3.67 \ \mu F$$

Similarly, when one is in series with the other two in parallel, the equivalent capacitance is

$$\frac{1}{C} = \frac{1}{C_i} + \frac{1}{C_j + C_k} \quad \rightarrow \quad C = \frac{C_i(C_j + C_k)}{C_i + C_j + C_k}$$

Therefore, the possible values are

$$C = \frac{C_1(C_2 + C_3)}{C_1 + C_2 + C_3} = \frac{(1 \ \mu F)(2 \ \mu F + 3 \ \mu F)}{1 \ \mu F + 2 \ \mu F + 3 \ \mu F} = \frac{5}{6} \ \mu F = 0.933 \ \mu F$$

$$C = \frac{C_2(C_1 + C_3)}{C_2 + C_1 + C_3} = \frac{(2 \ \mu F)(1 \ \mu F + 3 \ \mu F)}{2 \ \mu F + 1 \ \mu F + 3 \ \mu F} = \frac{4}{3} \ \mu F = 1.33 \ \mu F$$

$$C = \frac{C_3(C_1 + C_2)}{C_3 + C_1 + C_2} = \frac{(3 \ \mu F)(1 \ \mu F + 2 \ \mu F)}{3 \ \mu F + 1 \ \mu F + 2 \ \mu F} = \frac{3}{2} \ \mu F = 1.50 \ \mu F$$

ASSESS With three capacitors, each having two options (parallel or series), there are eight possible outcomes.

Section 23.4 Energy in the Electric Field

35. **INTERPRET** This problem is about the volume required for storing a given amount of electrostatic energy.

DEVELOP In a uniform field, Equation 23.8 can be written as

$$U = \frac{1}{2}\varepsilon_0 E^2 \times (\text{volume})$$

Knowing U and E allows us to deduce the volume of the region.

EVALUATE Substituting the values given in the problem statement, the volume is

$$\text{volume} = \frac{2U}{\varepsilon_0 E^2} = \frac{2(4\text{ MJ})}{(8.85\text{ pF/m})(30\text{ kV/m})^2} = 1.00 \times 10^9\text{ m}^3 = 1\text{ km}^3$$

ASSESS This is a very big volume occupied by a car battery. In reality, not all the energy stored goes to creating the field.

37. **INTERPRET** In this problem we are asked to find the electric energy stored in a proton by assuming it to be a uniformly charged sphere.

DEVELOP For this model of the proton, the field strength at the surface is $E = ke/R^2$ (from spherical symmetry and Gauss's law). Thus, the energy density in the surface electric field is

$$u = \frac{1}{2}\varepsilon_0 E^2 = \frac{1}{2}\frac{1}{4\pi k}\left(\frac{ke}{R^2}\right)^2 = \frac{ke^2}{8\pi R^4}$$

EVALUATE With $R = 1\text{ fm} = 1 \times 10^{-15}$ m, the energy density is

$$u = \frac{ke^2}{8\pi R^4} = \frac{(9 \times 10^9\text{ N}\cdot\text{m}^2\text{/C}^2)(1.6 \times 10^{-19}\text{ C})^2}{8\pi(1 \times 10^{-15}\text{ m})^4} = 9.17 \times 10^{30}\text{ J/m}^3 = 57.3\text{ keV/fm}^3$$

ASSESS The energy density is enormous, given the small size of the proton.

PROBLEMS

39. **INTERPRET** This problem is about the work done to create a certain charge distribution. The work is equal to the energy stored in the system.

DEVELOP When a charge q (assumed positive) is on the inner sphere, the potential difference between the spheres is

$$V = -\int_b^a E dr = -\int_b^a \frac{kqdr}{r^2} = kq\left(\frac{1}{a} - \frac{1}{b}\right)$$

To transfer an additional charge dq from the outer sphere requires work $dW = V\,dq$.

EVALUATE The total work required to transfer charge Q (leaving the spheres oppositely charged) is

$$W = \int_0^Q Vdq = \int_0^Q kq\left(\frac{1}{a} - \frac{1}{b}\right)dq = \frac{1}{2}kQ^2\left(\frac{1}{a} - \frac{1}{b}\right)$$

ASSESS Since $U = Q^2/2C$, this shows that the capacitance of this spherical capacitor is

$$C = \frac{1}{k(a^{-1} - b^{-1})} = \frac{ab}{k(b-a)}$$

Note that capacitance depends only on the geometry of the system, and is independent of V and Q.

41. **INTERPRET** We find the capacitance of a pair of coaxial cylinders, using the definition of capacitance.

DEVELOP From Example 22.4, we can see that the potential difference between two points distances a and b from a cylinder of charge per length λ is $V = \frac{\lambda}{2\pi\varepsilon_0}\ln(\frac{a}{b})$. The outer cylinder contributes nothing to the potential between the two surfaces, so we may set the potential at b to zero and use $V = \frac{\lambda}{2\pi\varepsilon_0}\ln(\frac{b}{a})$ as the magnitude of the voltage difference between the two conductors. The total charge on the inner cylinder is $Q = \lambda L$, and the definition of capacitance is $C = \frac{Q}{V}$.

EVALUATE

$$C = \frac{Q}{V} = \frac{\lambda L}{\frac{\lambda}{2\pi\varepsilon_0}\ln\left(\frac{b}{a}\right)} = \frac{2\pi\varepsilon_0 L}{\ln\left(\frac{b}{a}\right)}$$

ASSESS This capacitance depends only on the geometry, as we would expect.

43.   **INTERPRET**   This problem is about the capacitance of a charge configuration.

DEVELOP   When the third (middle) plate is positively charged, the electric field (not near an edge) is approximately uniform and away from the plate, with magnitude $E = \sigma/\varepsilon_0$. Since half of the total charge Q is on either side (by symmetry), the charge density is $\sigma = Q/2A$. The potential difference between the third plate and the outer two plates (which are both at the same potential and carry charges of $-Q/2$ on their inner surfaces) is

$$V = Ed = \frac{\sigma d}{\varepsilon_0} = \frac{Qd}{2\varepsilon_0 A}$$

EVALUATE   Using Equation 23.1, the capacitance is

$$C = \frac{Q}{V} = \frac{2\varepsilon_0 A}{d}.$$

ASSESS   As expected, capacitance only depends on the geometric factors (e.g., area, distance, . . .) of the configuration. Note that the above arrangement is like two capacitors in parallel.

45.   **INTERPRET**   In this problem, we are asked to compare the amount of energy stored in two different capacitors.

DEVELOP   The energy stored in a capacitor can be calculated using Equation 23.3: $U = CV^2/2$.

EVALUATE   The energies stored in the two capacitors are

$$U_1 = \frac{1}{2}C_1V_1^2 = \frac{1}{2}(1\ \mu F)(250\ V)^2 = 31.3\ mJ$$

$$U_2 = \frac{1}{2}C_2V_2^2 = \frac{1}{2}(470\ pF)(3\ kV)^2 = 2.12\ mJ$$

Thus, the energy stored in the second is about 14.8 times less than the first one.

ASSESS   The general expression for the ratio of the energies stored in two capacitors is

$$\frac{U_2}{U_1} = \frac{C_2V_2^2/2}{C_1V_1^2/2} = \left(\frac{C_2}{C_1}\right)\left(\frac{V_2}{V_1}\right)^2$$

47.   **INTERPRET**   In this problem we are asked to compare the amount of charge and energy stored in three different capacitors.

DEVELOP   The charge stored in a capacitor is $Q = CV$ (Equation 23.1), and the energy stored is $U = CV^2/2$ (Equation 23.3).

EVALUATE   **(a)** The stored charges are

$Q_1 = C_1V_1 = (0.01\ \mu F)(300\ V) = 3\ \mu C$ for the first capacitor

$Q_2 = C_2V_2 = (0.1\ \mu F)(100\ V) = 10\ \mu C$ for the second

$Q_3 = C_3V_3 = (30\ \mu F)(5\ V) = 150\ \mu C$ for the third

(b) The stored energies for the three capacitors are

$$U_1 = \frac{1}{2}C_1V_1^2 = \frac{1}{2}(3\ \mu C)(300\ V) = 450\ \mu J$$

$$U_2 = \frac{1}{2}C_2V_2^2 = \frac{1}{2}(0.1\ \mu F)(100\ V)^2 = 500\ \mu J$$

$$U_3 = \frac{1}{2}C_3V_3^2 = \frac{1}{2}(30\ \mu F)(5\ V)^2 = 375\ \mu J$$

(c) The cost effectiveness, measured in J/¢, is 18.0, 14.3, and 4.26 for these capacitors, respectively.

ASSESS   The first capacitor is the most cost effective of the three.

49.   **INTERPRET**   This problem is about the power consumption of a camera flashtube and the capacitor required for its operation.

DEVELOP   Power, $P = W/t$, as defined in Equation 6.15, is the rate at which energy is used. This is the equation we shall use for our calculations.

EVALUATE   **(a)** The power delivered to the tube during the flashing process is

$$P_{flash} = \frac{W}{t} = \frac{5.0\ J}{1\ ms} = 5\ kW$$

(b) The energy stored in a capacitor is $U = \frac{1}{2}CV^2$. Thus, the capacitance to be used to supply the flash energy is

$$C = \frac{2U}{V^2} = \frac{2(5\text{ J})}{(200\text{ V})^2} = 250\ \mu\text{F}$$

(c) The average power consumption during the 10-second interval is $P_{av} = \frac{5.0\text{ J}}{10\text{ s}} = 0.5$ W.

ASSESS The average power P_{av} is only about 10^{-4} times P_{flash}. This makes sense because the camera only flashes for 1 ms every 10 s.

51.   **INTERPRET** This problem is about the change in the capacitance and the stored energy due to an insertion of a conducting slab between the capacitor plates.

DEVELOP For part **(a)**, we first note that the charge on the plates remains the same, and so does the electric field ($E = \sigma/\varepsilon_0$) in the gaps between the plates and the slab. However, the separation (i.e., the thickness of the field region) between the plates has been changed to d'. Therefore, the capacitance becomes $C' = \varepsilon_0 A/d'$. For part **(b)**, when the charge is constant (no connections to anything isolates the system), the energy stored is inversely proportional to the capacitance, $U = Q^2/2C$.

EVALUATE   **(a)** With the plate separation reduced to 40% of its original value $d' = d_1 + d_2 = 0.4d$, the capacitance is increased to

$$C' = \frac{\varepsilon_0 A}{d'} = \frac{\varepsilon_0 A}{0.4d} = 2.5\ C$$

(b) The stored energy becomes

$$U' = \frac{Q^2}{2C'} = \frac{Q^2}{2(2.5C)} = 0.40\ U$$

or the energy decreases to 40% of its original value.

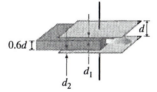

ASSESS With the slab inserted, there is less field region and less energy stored. While the slab is being inserted, work is done by electrical force to conserve energy. Note that the configuration behaves like a series combination of two parallel plate capacitors:

$$\frac{1}{C'} = \frac{1}{C_1} + \frac{1}{C_2} = \frac{d_1}{\varepsilon_0 A} + \frac{d_2}{\varepsilon_0 A} = \frac{d_1 + d_2}{\varepsilon_0 A} = \frac{0.4d}{\varepsilon_0 A} = \frac{1}{2.5C} \quad \rightarrow C' = 2.5C$$

53.   **INTERPRET** In this problem we want to connect capacitors of known capacitance and voltage rating to obtain the desired equivalent capacitance and voltage rating.

DEVELOP In parallel, the voltage across each element is the same, so to increase the voltage rating of a combination of equal capacitors, series connections must be considered. The general result of Problem 31 shows that for two equal capacitors in series, the voltage across each is one half the total, so the voltage rating of a series combination is doubled.

EVALUATE   **(a)** To obtain the desired voltage rating, we must use two capacitors in series so that the voltage becomes 50 V + 50 V = 100 V. However, the capacitance is now $C = (2\ \mu\text{F})(2\ \mu\text{F})/(2\ \mu\text{F} + 2\ \mu\text{F}) = 1\ \mu\text{F}$, and we need to increase the total capacitance to twice that of just two in series, without altering the voltage rating. This can be accomplished with a parallel combination of two pairs in series, i.e., a parallel combination of two 1 μF, 100 V series pairs. (Note that for equal capacitor elements, a parallel combination of two pairs in series has the same properties as a series combination of two pairs in parallel.)

(b) In this situation, four capacitors in series are required (rating 200 V, and capacitance $C^{-1} = 4(2\ \mu\text{F})^{-1}$ or $C = \frac{1}{2}\mu\text{F}$).

ASSESS   Schematically the connections described look like the following.

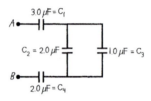

One may use Equations 23.5 and 23.6 to verify that the configurations indeed have the desired capacitance and voltage rating as specified in the problem statement.

55.   **INTERPRET**   This problem involves an assemblage of capacitors in a circuit, and we are interested in the energy stored in one of the capacitors.

DEVELOP   Number the capacitors as shown. Relative to points A and B, C_1, C_4, and the combination of C_2 and C_3 are in series. The energy of the $C_3 = 1\ \mu$F capacitor is $U_3 = \frac{1}{2}C_3V_3^2$, where the voltage drop across C_3 is $V_3 = V_{23}$, since C_2 and C_3 are in parallel. On the other hand, since C_1, C_4, and C_{23} are in series, we have $V = V_1 + V_4 + V_{23}$ and $Q = C_1V_1 = C_4V_4 = C_{23}V_{23}$. Once we know V_{23}, U_3 can be calculated.

EVALUATE   From the equation above, we find

$$V = V_{23}(1 + C_{23}/C_1 + C_{23}/C_4) = V_{23}\left(1 + \frac{3}{3} + \frac{3}{2}\right) = \frac{7}{2}V_{23} \quad \rightarrow \quad V_{23} = \frac{2}{7}V$$

This gives $U_3 = \frac{1}{2}C_3V_3^2 = \frac{1}{2}(1\ \mu\text{F})(2 \times 50\ \text{V}/7)^2 = 102\ \mu$J.

$$
\begin{array}{c}
3.0\ \mu\text{F} = C_1 \\
A \dashv\vdash \\
C_2 = 2.0\ \mu\text{F} \quad\quad 1.0\ \mu\text{F} = C_3 \\
B \dashv\vdash \\
2.0\ \mu\text{F} = C_4
\end{array}
$$

ASSESS   Since the combination of C_2 and C_3 are in series with C_1 and C_4, C_3 does not "feel" the full 50 V applied across AB, so its working voltage is lower. In fact we find it to be $V_3 = (2/7)V$.

57.   **INTERPRET**   This problem is about connecting two capacitors with different voltage ratings in series.

DEVELOP   In Problem 56, we have verified that the voltages across the individual capacitors are

$$V_1 = \frac{C_2V}{C_1 + C_2} \quad \text{and} \quad V_2 = \frac{C_1V}{C_1 + C_2}$$

where V is the voltage across the combination. The equations allow us to determine the maximum voltage that could be applied.

EVALUATE   The separate ratings require

$$V_1 = \frac{C_2}{C_1 + C_2}V = \frac{0.2\ \mu\text{F}}{0.1\ \mu\text{F} + 0.2\ \mu\text{F}}V = \frac{2}{3}V < 50\ \text{V}$$

Similarly, we find

$$V_2 = \frac{C_1}{C_1 + C_2}V = \frac{0.1\ \mu\text{F}}{0.1\ \mu\text{F} + 0.2\ \mu\text{F}}V = \frac{1}{3}V < 200\ \text{V}$$

The more stringent limit is $V \leq \frac{3}{2}(50\ \text{V}) = 75$ V.

ASSESS   We use the more stringent condition for V. If the less stringent one were used, then $V < 3(200\ \text{V}) = 600$ V and the voltage across C_1 would be 400 V, which exceeds the voltage rating of 50 V.

59.   **INTERPRET**   This problem involves inserting a piece of dielectric material in a capacitor.

DEVELOP   From Equation 23.4, one finds that the capacitance in the presence of dielectric is $C = \kappa C_0 = \frac{\kappa\varepsilon_0 A}{d}$. As for part **(b)**, the maximum voltage for this capacitor is $V_{\text{max}} = E_{\text{max}}d$.

EVALUATE   **(a)** With reference to Table 23.1, the thickness is

$$d = \frac{\kappa\varepsilon_0 A}{C} = \frac{(2.6)(8.85\ \text{pF/m})\pi(0.15\ \text{m})^2}{470\ \text{pF}} = 3.46\ \text{mm}$$

Since this is much less than the radius of the plates, the parallel plate approximation (plane symmetry) is a good one.

(b) The dielectric breakdown field for polystyrene is $E_{max} = 25$ kV/mm, so the maximum voltage is $V_{max} = E_{max}d =$ (25 kV/mm)(3.46 mm) = 86.5 kV.

ASSESS   In practice, the working voltage would be less than this by a comfortable safety margin.

61.   **INTERPRET**   This problem is about connecting two capacitors in parallel.

DEVELOP   The equivalent capacitance of two capacitors connected in parallel is (Equation 23.5)

$$C = C_1 + C_2$$

In this problem, $C_1 = 0.001\ \mu F = 1000$ pF, and C_2 has a range of 10–30 pF.

EVALUATE   The combination covers a range from 1010 to 1030 pF, or about

$$\frac{\pm 10}{1020} = 0.0098 = 0.98\%$$

from the central value.

ASSESS   We expect the variation to be very small since $C_1 \gg C_2$.

63.   **INTERPRET**   This problem involves electrostatic energy contained within a charged sphere. Our system has radial symmetry.

DEVELOP   From Example 21.1, we see that the radially symmetric field inside the sphere is $E_r = kQr/R^3$. so the energy density is

$$u(r) = \frac{1}{2}\varepsilon_0 E_r^2 = \frac{1}{2(4\pi k)}\left(\frac{kQr}{R^3}\right)^2 = \frac{kQ^2 r^2}{8\pi R^6}$$

The energy within the sphere can be found by integrating over the volume.

EVALUATE   With thin spherical shells of radius r for volume elements, $dV = 4\pi r^2 dr$, the integral for the energy is

$$U = \int_{\text{sphere}} u dV = \int_0^R \left(\frac{kQ^2 r^2}{8\pi R^6}\right) 4\pi r^2 dr = \frac{kQ^2}{2R^6}\int_0^R r^4 dr = \frac{kQ^2}{10R}$$

(This is just the energy stored inside the sphere. For the energy outside the sphere, and the total energy, see Problems 64 and 65.)

ASSESS   The result shows that U is inversely proportional to R. This means that the stored energy decreases if the same amount of charge Q is distributed over a greater volume. Our result can be compared to the situation (Problem 64) where the total charge Q is distributed over its *surface*. In that case, the total energy stored in its electric field is $U = kQ^2/2R$.

65.   **INTERPRET**   This problem is about the electrostatic energy of a charged sphere, both within and outside the sphere. Our system has radial symmetry.

DEVELOP   The field outside a spherically symmetric distribution of radius R is the same for the charge Q uniformly spread over the volume or the surface (thanks to Gauss's law). Thus, Problem 64 gives the energy in the electric field outside a uniformly charged spherical volume, while Problem 63 gives the energy inside. The total energy is

$$U = \frac{kQ^2}{2R} + \frac{kQ^2}{10R} = \frac{3kQ^2}{5R}$$

EVALUATE   Applied to a U^{235}-nucleus, the result gives

$$U = \frac{3kQ^2}{5R} = \frac{3(9\times 10^9\ \text{N}\cdot\text{m}^2/\text{C}^2)(92e)^2}{5(3.3\ \text{fm})} = 3.55\times 10^{-10}\ \text{J} = 2.22\ \text{GeV}$$

ASSESS   This Coulomb energy is about 1% of the mass energy of the U^{235}-nucleus, mc^2.

67. **INTERPRET** This problem involves electrostatic energy contained within a volume around a charged wire. Our system has line symmetry.

DEVELOP The electric field outside the wire (assumed to have line symmetry) is radially away from the axis with magnitude $E_r = 2k\lambda/r$ (see Equation 21.6). The energy density in a cylindrical shell of radius r, length L, and volume $dV = 2\pi rLdr$, is

$$u = \frac{1}{2}\varepsilon_0 E^2 = \frac{1}{2(4\pi k)}\left(\frac{2k\lambda}{r}\right)^2 = \frac{k\lambda^2}{2\pi r^2}$$

EVALUATE Thus, the energy in the space mentioned in this problem is

$$U = \int udV = \int_R^{3R}\left(\frac{k\lambda^2}{2\pi r^2}\right)2\pi rLdr = k\lambda^2 L \ln 3$$

$$= (9\times 10^9 \text{ N}\cdot\text{m}^2\text{/C}^2)(28 \text{ }\mu\text{C/m})^2(1 \text{ m})\ln 3 = 7.75 \text{ J}$$

ASSESS The energy density (energy/volume) decreases as $1/r^2$. This means that more energy is concentrated in the region of space closer to the wire.

69. **INTERPRET** Our object of interest is a spherical capacitor, and we want to explore the limit where $b - a \ll a$.

DEVELOP The capacitance for a spherical capacitor can be derived by noting that the potential difference between two concentric conducting spheres is

$$V = kQ\left(\frac{1}{a} - \frac{1}{b}\right) = \frac{kQ(b-a)}{ab} \rightarrow C = \frac{Q}{V} = \frac{ab}{k(b-a)}$$

The limit $d \equiv b - a \ll a$ can be taken to show that C reduces to that of a parallel-plate capacitor.

EVALUATE With $d \equiv b - a \ll a$, we find

$$C = \frac{ab}{k(b-a)} = \frac{4\pi\varepsilon_0 ab}{d} = \frac{4\pi\varepsilon_0 a(a+d)}{d} \approx \frac{4\pi\varepsilon_0 a^2}{d} = \frac{\varepsilon_0 A}{d}$$

which is the result of Equation 23.2, with $A = 4\pi a^2$ being the area of the spherical plates.

ASSESS The limit $d \equiv b - a \ll a$ means that the radius of the sphere is much greater than the separation between the two spheres. So locally we have effectively two plates that appear to be parallel.

71. **INTERPRET** This problem is about the size an electron has that makes its electrostatic energy equal to its mass energy.

DEVELOP The electrostatic energy stored in the field of a classical electron, whose charge e is distributed uniformly over the surface of a sphere of radius R, is $U = ke^2/2R$ (see Problem 64). The radius of the electron can then be obtained by setting this equal to the electron's mass energy, $U = m_e c^2$.

EVALUATE Equating the two energies gives $m_e c^2 = ke^2/2R$, or

$$R = \frac{ke^2}{2m_e c^2} = 1.41 \text{ fm}$$

where constants from the inside front cover were used.

ASSESS The "classical radius of the electron," based on a consideration of the scattering of electromagnetic waves from free electrons, called Thomson scattering, is actually equal to $r_e = 2R = ke^2/m_e c^2$.

73. **INTERPRET** This problem is about the effect of inserting a dielectric material into a parallel-plate capacitor, which is connected to a battery.

DEVELOP We first note that the capacitance depends on the configuration and electrical properties of the plates and insulating materials, not on the external connections, so

$$C = C_0 \frac{(\kappa x + L - x)}{L}$$

as in the preceding problem. If the capacitor remains connected to a battery, the voltage is constant, $V = V_0$.

EVALUATE **(b)** The energy is $U = \frac{1}{2}CV_0^2 = \frac{1}{2}C_0 V_0^2 \frac{(\kappa x + L - x)}{L}$. For $x = L/2$, we get $U = C_0 V_0^2(\kappa + 1)/4$. Note that this is different from the preceding problem, because the battery does work.

(c) When the capacitor is connected to a battery, Equation 6.8 ($F_x = -dU/dx$) for the force does not apply. However, for particular values of charge and voltage on the capacitor, the force on the slab considered here is the same, regardless of the external connections. In the preceding problem we found that $F_x = \frac{1}{2}C_0V^2(\kappa - 1)/L$, where V was the particular voltage (and, because of the special form of the capacitance, $C(x)$, the particular charge q did not appear). Since $V = V_0$ in this problem,

$$F_x = \frac{1}{2}\frac{C_0V_0^2(\kappa - 1)}{L}$$

ASSESS   The force on the slab is to the right, drawing the dielectric between the plates.

75.   **INTERPRET**   We find the energy per length stored in the electric field of a uniformly charged rod. We will use Gauss's law to determine the electric field within the rod, and then integrate the energy density.

DEVELOP   We use Gauss's law, with cylindrical symmetry, to find the electric field using $\int \vec{E} \cdot d\vec{a} = \frac{q_{enclosed}}{\varepsilon_0}$. Once we have this field, we integrate the energy density $u_E = \frac{1}{2}\varepsilon_0 E^2$ over the cylinder to find the total energy $U = \int u_E dV$.

EVALUATE

$$\int \vec{E} \cdot d\vec{a} = \frac{q_{enclosed}}{\varepsilon_0} \rightarrow E(2\pi rL) = \frac{\pi r^2 L\rho}{\varepsilon_0} \rightarrow E = \frac{r\rho}{2\varepsilon_0}$$

$$U = \frac{1}{2}\varepsilon_0 \int_0^L \int_0^{2\pi} \int_0^R \left(\frac{r\rho}{2\varepsilon_0}\right)^2 rdrd\phi dz = \frac{\rho^2}{4\varepsilon_0}2\pi L \int_0^R r^3 dr$$

$$\rightarrow \frac{U}{L} = \frac{\pi\rho^2}{2\varepsilon_0}\frac{1}{4}R^4 = \frac{\pi\rho^2 R^4}{8\varepsilon_0}$$

ASSESS   Increasing the radius increases the energy per length dramatically.

77.   **INTERPRET**   Find the electrostatic energy stored between two parallel plates of a parallel-plate capacitor, and then differentiate to find the force between the plates.

DEVELOP   We find the electric field between the plates using $E = \frac{\sigma}{\varepsilon_0} = \frac{Q}{A\varepsilon_0}$. This is constant, so total energy stored in this field is then $U = u_E V$. We can find the force by using $F_x = -\frac{dU}{dx}$.

EVALUATE

(a)

$$U = u_E V = \left(\frac{1}{2}\varepsilon_0 E^2\right)(Ax) = \frac{Q^2}{2A\varepsilon_0}x$$

(b)

$$F_x = -\frac{dU}{dx} = -\frac{Q^2}{2A\varepsilon_0}$$

This is half the value you would obtain by multiplying the charge on one plate by the field between the plates.

ASSESS   The answer we get for **(b)** is half the field times the charge on one plate: but we must remember that the field between the plates is created by *both* charged plates. A charge is not affected by the field it creates. Only the field created by the *other* plate causes a force on each plate, and the other plate creates half the field.

79.   **INTERPRET**   Given the electric field as a function of radius inside a cylindrical capacitor, we find the energy density and the total energy stored, and we show that this total energy is consistent with the energy stored in a cylindrical capacitor.

DEVELOP   The electric field is given in the problem as

$$\vec{E} = \left(\frac{\lambda}{2\pi\varepsilon_0 r}\right)\hat{r}$$

The energy density is $u_E = \frac{1}{2}\varepsilon_0 E^2$, and the capacitance for a given length L (from Problem 23.41) is

$$C = \frac{2\pi\varepsilon_0 L}{\ln\left(\frac{b}{a}\right)}$$

EVALUATE First we find the energy density as a function of radius:

$$u_E = \frac{1}{2}\frac{\lambda^2}{4\pi^2\varepsilon_0 r^2}$$

Next we integrate this over the volume of the capacitor to find the total energy U:

$$U = \int_0^L \int_0^{2\pi} \int_a^b \frac{\lambda^2}{8\pi^2\varepsilon_0 r^2} r\,dr\,d\phi\,dz = \frac{2\pi L\lambda^2}{8\pi^2\varepsilon_0}\int_a^b \frac{1}{r}dr = \frac{L\lambda^2}{4\pi\varepsilon_0}\ln\left(\frac{b}{a}\right)$$

From Problem 23.41, we have

$$V = \frac{\lambda}{2\pi\varepsilon_0}\ln\left(\frac{b}{a}\right) \text{ so } \frac{1}{2}CV^2 = \frac{1}{2}\left(\frac{2\pi\varepsilon_0 L}{\ln\left(\frac{b}{a}\right)}\right)\left(\frac{\lambda}{2\pi\varepsilon_0}\ln\left(\frac{b}{a}\right)\right)^2 = \frac{1}{2}\frac{L\lambda^2}{2\pi\varepsilon_0}\ln\left(\frac{b}{a}\right) = U$$

ASSESS It is always a relief, in a complicated problem like this, to discover that things are consistent with what we already know. Checking that we get the same value for U by multiple methods is a good method of verification.

81. **INTERPRET** We find the capacitance of a "twin lead" antenna with given dimensions, using the potential given and the definition of capacitance

DEVELOP We are given that

$$\Delta V = \frac{\lambda}{\pi\varepsilon_0}\ln\left(\frac{b-a}{a}\right)$$

The characteristics of the cable are $a = 0.5$ mm, $b = 12$ mm, and $L = 30$ m. The capacitance is $C = \frac{Q}{V}$.

EVALUATE

$$C = \frac{Q}{V} = \frac{\lambda L}{\left[\frac{\lambda}{\pi\varepsilon_0}\ln\left(\frac{b-a}{a}\right)\right]} = \frac{\pi\varepsilon_0 L}{\ln\left(\frac{b-a}{a}\right)} = 266 \text{ pF}$$

ASSESS This is less than the maximum allowable capacitance, so our design is ok.

ELECTRIC CURRENT

EXERCISES

Section 24.1 Electric Current

15. **INTERPRET** This problem is about the number of charges involved in setting up an electric current.

    DEVELOP For a steady current, the amount of charge crossing a given area in time Δt can be found by using Equation 24.1a, $I = \Delta Q / \Delta t$.

    EVALUATE A battery rated at $80 \text{ A} \cdot \text{h}$ can supply a net charge of

    $$\Delta Q = I \Delta t = (80 \text{ C/s})(3600 \text{ s}) = 2.88 \times 10^5 \text{C}$$

    ASSESS By definition, $1 \text{ A} = 1 \text{C/s}$, or $1 \text{ C} = 1 \text{ A} \cdot \text{s}$. So the result makes sense.

17. **INTERPRET** In this problem we are asked to find the current density, given the electric current and the cross section.

    DEVELOP The current density J, is defined as the current per unit area, or $J = I/A$. The area of the cross section is $A = \pi R^2 = \pi d^2/4$.

    EVALUATE The cross section of a wire is uniform, so the density is

    $$J = \frac{I}{A} = \frac{I}{\pi d^2/4} = \frac{10 \text{ A}}{\pi (1.29 \times 10^{-3} \text{m})^2/4} = 7.65 \times 10^6 \text{ A/m}^2$$

    ASSESS If the current I is kept fixed, the smaller the cross-sectional area A, the greater the current density J.

Section 24.2 Conduction Mechanisms

19. **INTERPRET** In this problem we are asked to calculate the electric field in a current-carrying conductor.

    DEVELOP To find the electric field, we make use of Ohm's law (which applies to silver), $J = E/\rho$, and the definition of current density, which is assumed to be uniform in the wire.

    EVALUATE Using Table 24.1, we find the electric field to be

    $$E = \rho J = \frac{\rho I}{\pi d^2/4} = \frac{(1.59 \times 10^{-8} \ \Omega \cdot \text{m})(7.5 \text{ A})}{\pi (9.5 \times 10^{-4} \ \text{m})^2/4} = 0.168 \text{ V/m}$$

    ASSESS The value is a lot smaller than the electric field we discussed in electrostatic situations. Since silver is such a good conductor, a small field can drive a substantial current.

21. **INTERPRET** This problem is about applying Ohm's law to find the resistivity of a rod.

    DEVELOP If the rod has a uniform current density and obeys Ohm's law (Equations 24.4a and 4b), then its resistivity is

    $$\rho = \frac{E}{J} = \frac{E}{I/A} = \frac{E(\pi d^2/4)}{I}$$

    EVALUATE Substituting the values given in the problem, we find the resistivity of the rod to be

    $$\rho = \frac{E(\pi d^2/4)}{I} = \frac{(1.4 \text{ V/m})\pi (10^{-2} \ \text{m})^2/4}{50 \text{ A}} = 2.20 \times 10^{-6} \ \Omega \cdot \text{m}$$

    ASSESS With reference to Table 24.1, we see that the value is within the range of resistivity of a metallic conductor.

Section 24.3 Resistance and Ohm's Law

23.   **INTERPRET**   This problem involves using Ohm's law to calculate the resistance of a heating coil.

DEVELOP   The macroscopic form the Ohm's law is probably applicable to the heating coil, which is typically a coil of wire. Equation 24.5 gives $R = V/I$.

EVALUATE   Substituting the values given in the problem statement, we find the resistance to be

$$R = \frac{V}{I} = \frac{120 \text{ V}}{4.8 \text{ A}} = 25 \text{ }\Omega$$

ASSESS   This is a fairly large resistance. The resistance of the coil used for heating is usually quite high.

25.   **INTERPRET**   In this problem we want to apply Ohm's law to calculate the current across a resistor.

DEVELOP   Ohm's law, $I = V/R$, given in Equation 24.5 provides the connection between current, resistance, and voltage. This equation allows us to compute the current I.

EVALUATE   Substituting the values given in the problem statement, we find the current to be

$$I = \frac{V}{R} = \frac{110 \text{ V}}{47 \text{ k}\Omega} = 2.34 \text{ mA}$$

ASSESS   From Ohm's law, we see that current is inversely proportional to resistance when V is kept fixed. In our case, a large resistance yields a small current.

27.   **INTERPRET**   This problem is about applying Ohm's law to calculate the current across a resistor.

DEVELOP   Ohm's law, $I = V/R$, given in Equation 24.5 provides the connection between current, resistance, and voltage. This equation allows us to compute the current I.

EVALUATE   Substituting the values given in the problem statement, we find the current to be

$$I = \frac{V}{R} = \frac{45 \text{ V}}{1.8 \text{ k}\Omega} = 25 \text{ mA}$$

ASSESS   From Ohm's law, we see that current is inversely proportional to resistance when V is kept fixed. In our case, a large resistance yields a small current.

Section 24.4 Electric Power

29.   **INTERPRET**   This problem is about electric power of a motor, given the current drawn and the voltage across the terminals.

DEVELOP   Equation 24.7, $P = IV$, provides the connection between electric current, voltage and electric power. This equation is what we shall use to compute the power P.

EVALUATE   Substituting the values given in the problem, we find the power consumption to be
$$P = IV = (125 \text{ A})(11 \text{ V}) = 1.38 \text{ kW}$$

ASSESS   A large electric power is needed to start a car. This is why a very high current is required in the starter motor to crank the engine.

31.   **INTERPRET**   This problem is about the current drawn from a battery, given its voltage and power rating.

DEVELOP   Equation 24.7, $P = IV$, provides the connection between electric current, voltage, and electric power. This equation is what we shall use to compute the current I.

EVALUATE   Rearranging Equation 24.7, we find

$$I = \frac{P}{V} = \frac{240 \text{ }\mu\text{W}}{1.5 \text{ V}} = 160 \text{ }\mu\text{A}$$

ASSESS   We expect the current drawn from the battery to be very small since little power is needed to operate a watch.

33.   **INTERPRET**   In this problem we are asked to calculate the resistance of a light bulb.

DEVELOP   Since voltage is fixed, we use Equation 24.8b, $P = V^2/R$, to find the bulb's resistance.

EVALUATE   Substituting the values given in the problem, we obtain

$$R = \frac{V^2}{P} = \frac{(120 \text{ V})^2}{60 \text{ W}} = 240 \text{ }\Omega$$

ASSESS   This is the resistance of the light bulb at its operating temperature. Light bulbs actually are non-Ohmic because their resistance varies with temperature.

Section 24.5 Electrical Safety

35.   **INTERPRET**   We use Ohm's law to find the resistance necessary for a given voltage to drive a specified current.
      DEVELOP   Ohm's law is $V = IR$. We are given $V = 120$ V and $I = 2.5$ mA, and need to find R.
      EVALUATE

$$R = \frac{V}{I} = \frac{30 \text{ V}}{2.5 \text{ mA}} = 12 \text{ k}\Omega$$

ASSESS   If your skin was wet, your resistance would be much lower and this voltage would pose a serious threat.

PROBLEMS

37.   **INTERPRET**   This problem is about the average current in a given time interval and the number of ions in the channel.
      DEVELOP   The average current is a time-weighted average over sub-intervals

$$I_{av} = \frac{\Delta Q}{\Delta t} = \frac{\sum I_i \Delta t_i}{\Delta t} = \frac{\sum I_i \Delta t_i}{\sum \Delta t_i}$$

The number of singly-charged ions passing through the channel while current is flowing is $N = \Delta Q/e$.
      EVALUATE   (a) The average current is

$$I_{av} = \frac{\sum I_i \Delta t_i}{\sum \Delta t_i} = (20\%)(2.4 \text{ pA}) + (80\%)(0) = 0.48 \text{ pA}$$

(b) The number of ions is

$$N = \frac{\Delta Q}{e} = \frac{I \Delta t}{e} = \frac{(2.4 \text{ pA})(1.0 \text{ ms})}{(1.60 \times 10^{-19} \text{ C})} = 1.5 \times 10^4$$

ASSESS   This is a very small current. Nevertheless, there exists instrumentation that can measure current down to the pico-ampere range.

39.   **INTERPRET**   This problem is about finding the total current in a film, given the current density and the linear dimensions of the film.
      DEVELOP   Since current density J is current per unit area, assuming the current density is perpendicular to the cross-sectional area of the film, the current is $I = JA$.
      EVALUATE   Substituting the values given in the problem statement, we find the current to be

$$I = JA = (6.8 \times 10^5 \text{ A/m}^2)(2.5 \text{ }\mu\text{m} \times 0.18 \text{ mm}) = 306 \text{ }\mu\text{A}$$

ASSESS   The current density 0.68 MA/m² is approximately the maximum safe current density in typical household wiring.

41.   **INTERPRET**   In this problem we are given the microscopic parameters, and asked to calculate the drift speed of charge carriers in different regions.
      DEVELOP   The drift speed of a charge can be calculated using either Equation 24.2 or Equation 24.3:

$$v_d = \frac{I}{nAq} = \frac{J}{nq}$$

EVALUATE   In the copper wire, the drift speed is

$$v_d = \frac{I}{neA} = \frac{10^{-1} \text{ A}}{(1.1 \times 10^{29} \text{ m}^{-3})(1.6 \times 10^{-19} \text{ C})(\frac{1}{4}\pi \times 10^{-8} \text{ m}^2)} = 0.723 \text{ mm/s}$$

In the solution, the positive and negative ions have equal charge magnitudes, number, and current densities, by hypothesis, so $J = n_+(2e)v_d + n_-(-2e)(-v_d) = 4nev_d$. Thus, with uniform current density, $J = I/A$, the drift speed is

$$v_d = \frac{J}{4ne} = \frac{I}{4neA} = \frac{10^{-1}\ \text{A}}{4(6.1 \times 10^{23}\ \text{m}^{-3})(1.6 \times 10^{-19}\ \text{C})(\tfrac{1}{4}\pi \times 10^{-4}\ \text{m}^2)} = 3.26\ \text{mm/s}$$

Similarly, in the vacuum tube, we have

$$v_d = \frac{I}{neA} = \frac{10^{-1}\ \text{A}}{(2.2 \times 10^{16}\ \text{m}^{-3})(1.6 \times 10^{-19}\ \text{C})(\tfrac{1}{4}\pi \times 10^{-6}\ \text{m}^2)} = 3.62 \times 10^7\ \text{m/s}$$

ASSESS   Since the drift speed v_d is inversely proportional to the number density n, it is greatest in the vacuum tube, but smallest in the copper wire.

43.   **INTERPRET**   This problem involves using Ohm's law to find the current density and current, given the potential difference between two ends of a wire.

DEVELOP   Equation 24.4b, $J = E/\rho$, can be used to find the current density. The corresponding current is simply equal to $I = JA$.

EVALUATE   **(a)** Substituting the values given in the problem statement, we find the current density to be

$$J = \frac{E}{\rho} = \frac{V/L}{\rho} = \frac{(2.5\ \text{V})/(6\ \text{m})}{9.71 \times 10^{-8}\ \Omega \cdot \text{m}} = 4.29 \times 10^6\ \text{A/m}^2$$

(b) The current is $I = JA = J(\pi d^2/4) = (4.29 \times 10^6\ \text{A/m}^2)(\tfrac{1}{4}\pi \times 10^{-6}\ \text{m}^2) = 3.37\ \text{A}$.

ASSESS   Since the potential difference between the ends of the wire is only 2.5 V, we expect the resistance to be small as well. The resistance of the wire is $R = \frac{V}{I} = \frac{2.5\ \text{V}}{3.37\ \text{A}} = 0.74\ \Omega$. We can also find R by using Equation 24.6: $R = \rho L/A$.

45.   **INTERPRET**   In this problem we are asked to compare the voltages across wires made up of silver and iron.

DEVELOP   The resistance of a uniform wire of Ohmic material is given by Equation 24.6: $R = \rho L/A$. Therefore, for equal L, A, and I, the ratio of the voltages is

$$\frac{V_{\text{Fe}}}{V_{\text{Ag}}} = \frac{IR_{\text{Fe}}}{IR_{\text{Ag}}} = \frac{\rho_{\text{Fe}}}{\rho_{\text{Ag}}}$$

EVALUATE   Using Table 24.1 for resistivities, we obtain

$$\frac{V_{\text{Fe}}}{V_{\text{Ag}}} = \frac{\rho_{\text{Fe}}}{\rho_{\text{Ag}}} = \frac{9.71 \times 10^{-8}\ \Omega \cdot \text{m}}{1.59 \times 10^{-8}\ \Omega \cdot \text{m}} = 6.11$$

ASSESS   Since iron has higher resistivity than silver, for the same length, area, and current, the voltage across the iron wire would be higher.

47.   **INTERPRET**   This problem compares resistances of wires made up of different materials. Ohm's law is involved.

DEVELOP   The resistance of a uniform wire of Ohmic material is given by Equation 24.6: $R = \rho L/A$. Therefore, the resistance per unit length is

$$\frac{R}{L} = \frac{\rho}{A} = \frac{\rho}{\pi d^2/4}$$

EVALUATE   Equal values of R/L for copper and aluminum wires imply that

$$\frac{\rho_{\text{Cu}}}{d_{\text{Cu}}^2} = \frac{\rho_{\text{Al}}}{d_{\text{Al}}^2} \quad \rightarrow \quad \frac{d_{\text{Al}}}{d_{\text{Cu}}} = \sqrt{\frac{\rho_{\text{Al}}}{\rho_{\text{Cu}}}} = \sqrt{\frac{2.65 \times 10^{-8}\Omega \cdot \text{m}}{1.68 \times 10^{-8}\Omega \cdot \text{m}}} = 1.26$$

where we have used Table 24.1 for resistivities.

ASSESS   Since $d \sim \rho^{1/2}$, the higher the resistivity, the greater the diameter of the wire.

49.   **INTERPRET**   We need to find the diameter of copper and of aluminum wire that would provide the requested resistance. We will use the resistivity of the two materials to find the resistance of the wires. We will also find the mass and cost for each wire, using the price per kilogram and the density.

DEVELOP The resistance of a cylinder is $R = \frac{\rho L}{A}$. The resistivity of copper, from Table 24.1, is $\rho_c = 1.68 \times 10^{-8}$ Ωm and the resistivity of aluminum is $\rho_a = 2.65 \times 10^{-8}$ Ωm. For parts (a) and (b), we will use $L = 1.0$ km, and solve for the radius of each wire such that $R = 50$ mΩ. For part (c), we will find the mass of each 1-km wire, multiply by the cost per kg for each type, and determine which is cheaper. $c_c = \$4.65/\text{kg}$, $c_a = \$2.30/\text{kg}$, $density_c = 8900$ kg/m^3, and $density_a = 2700$ kg/m^3.

EVALUATE

$$R = \frac{\rho L}{A} = \frac{\rho L}{\pi r^2} \rightarrow r = \sqrt{\frac{\rho L}{\pi R}}$$

(a)
$$d_c = 2r_c = \sqrt{\frac{\rho_c L}{\pi R}} = 21 \text{ mm}$$

(b)
$$d_a = 2r_a = \sqrt{\frac{\rho_a L}{\pi R}} = 26 \text{ mm}$$

(c) The cost per meter of each wire is $(\text{cost/kg}) \times (density) \times \pi r^2$. For copper, the cost/meter is \$13.90, and for aluminum the cost/meter is \$3.29.

ASSESS Although copper is a better conductor, and less is required, in this case aluminum is more economical due to its lower density and lower cost per kg.

51. **INTERPRET** In this problem we are asked to calculate resistance, given the voltage drop and the current. Ohm's law is involved.

    DEVELOP The resistance at the battery terminal can be found by using Ohm's law, $I = V/R$.

    EVALUATE From Ohm's law, we have

    $$R = \frac{V}{I} = \frac{4.2 \text{ V}}{125 \text{ A}} = 33.6 \text{ m}\Omega$$

    ASSESS The resistance of the battery cable in Example 24.4 was $R = 0.60$ mΩ, so most of the resistance just calculated was in the connection.

53. **INTERPRET** This problem is about the dependence of resistance and power dissipation on the geometry of the material.

    DEVELOP Equation 24.6, $R = \rho L/A$, relates the resistance of a material to its geometry (length and cross-sectional area). When voltage is fixed, the power dissipated is given by Equation 24.8b, $P = V^2/R$. Thus, we see that at the same voltage, the ratio of the power dissipated is the inverse of the ratio of the resistances, which in turn, goes as the inverse of the square of the ratio of the diameters.

    $$\frac{P_1}{P_2} = \frac{V^2/R_1}{V^2/R_2} = \frac{R_2}{R_1} = \frac{\rho L/A_2}{\rho L/A_1} = \frac{A_1}{A_2} = \frac{\pi d_1^2/4}{\pi d_2^2/4} = \left(\frac{d_1}{d_2}\right)^2$$

    EVALUATE From the equation above, we see that if $P_1 = 2P_2$, then $d_1 = \sqrt{2}d_2$.

    ASSESS Our result shows that power dissipated increases with the area, or the square of the diameter, $P \sim A \sim d^2$.

55. **INTERPRET** This problem is about using electric power to do mechanical work.

    DEVELOP If there are no losses, the electrical power supplied to the motor, $P_{in} = IV$, equals the mechanical power expended lifting the weight, $P_{out} = Fv$.

    EVALUATE With $P_{in} = P_{out}$, we find the current to be

    $$I = \frac{Fv}{V} = \frac{(15 \text{ N})(0.25 \text{ m/s})}{6 \text{ V}} = 0.625 \text{ A}$$

    ASSESS In reality, the motor will not be 100% efficient. So the current drawn will be somewhat higher than 0.625 A.

57. **INTERPRET** This problem is about doing mechanical work with electric power.

    DEVELOP The electrical power supplied to the motor is $P_{in} = IV$. With 90% efficiency, the mechanical power expended lifting the weight is $P_{out} = Fv = 0.90P_{in} = 0.90IV$.

EVALUATE Thus, the current drawn is

$$I = \frac{Fv}{0.90\,V} = \frac{(200\text{ N})(3.1\text{ m/s})}{0.90(240\text{ V})} = 2.87\text{ A}$$

ASSESS The less efficient the motor operates, the more the current that must be drawn to supply the required power.

59. **INTERPRET** This problem is about using electric power to boil water. Electric energy is converted to thermal energy.

DEVELOP The power required to bring the water to its boiling point is

$$P = \frac{\Delta Q}{\Delta t} = \frac{m_w c_w \Delta T}{\Delta t}$$

EVALUATE **(a)** Substituting the values given in the problem, we find the power to be

$$P = \frac{m_w c_w \Delta T}{\Delta t} = \frac{(250\text{ ml})(1\text{ g/ml})(4.184\text{ J/g}\cdot\text{k})(100°\text{C} - 10°\text{C})}{85\text{ s}} = 1.11\text{ kW}$$

(b) With $P = V^2/R$ (appropriate to an Ohmic device), one finds the heater's resistance to be

$$R = \frac{V^2}{P} = \frac{(120\text{ V})^2}{1.11\text{ kW}} = 13.0\ \Omega$$

ASSESS Besides the power supplied to heat up water, additional power is also needed to compensate for any power lost to the surroundings or cup, plus that used by the heater itself. All these factors have been neglected in the problem.

61. **INTERPRET** In this problem we are asked to calculate the drift speed of electrons in aluminum.

DEVELOP Using Equation 24.2, the drift speed of electrons in the wire is

$$v_d = \frac{I}{neA} = \frac{I}{ne(\pi d^2/4)}$$

where n is the number density of conduction electrons.

EVALUATE For aluminum, the number density is

$$n = \frac{(3.5\text{ electrons/ion})(2702\text{ kg/m}^3)}{(26.98\text{ u/ion} \times 1.66 \times 10^{-27}\text{ kg/u})} = 2.11 \times 10^{29}\text{ electrons/m}^3$$

Thus, the drift speed is

$$v_d = \frac{I}{ne(\pi d^2/4)} = \frac{20\text{ A}}{(2.11 \times 10^{29}\text{ m}^{-3})(1.6 \times 10^{-19}\text{ C})\pi(0.21\text{ cm})^2/4} = 0.171\text{ mm/s}$$

ASSESS The drift speed is very small. With this speed, it would take about 100 minutes for an electron to travel 1 m. However, as explained in Example 24.1, electrons inside the conducting wire all get their "marching orders" from the electric field that's established almost instantaneously. Consequently, when you flip the switch, electrons throughout the wire start to move almost instantaneously and light comes on immediately.

63. **INTERPRET** We find the radius of a conical wire. We do this by integrating over each thin disk that makes up the wire, as shown in Figure 24.21. We also show that this resistance is the same as for a wire with elliptical cross section.

DEVELOP The resistance of each disk dx that makes up this cone is

$$dR = \frac{\rho dL}{A} = \frac{\rho dx}{\pi r^2}$$

The radius of each disk changes linearly from a to b as we move down the wire, so

$$r = a + \frac{(b-a)}{L}x$$

We find the total resistance by integrating dR from $x = 0$ to $x = L$. For comparison at the end of the problem, we will use the area of an ellipse,

$$A = \pi ab, \quad \text{and} \quad R = \frac{\rho L}{A}$$

EVALUATE

$$R = \int_0^L \frac{\rho}{\pi\left(a + \frac{b-a}{L}x\right)^2} dx = \frac{\rho L}{(a-b)\pi\left(a + \frac{(b-a)x}{L}\right)}\Bigg|_0^L = \frac{L\rho}{\pi ab}$$

ASSESS   This is the same result as for a wire with elliptical cross-section and area $A = \pi ab$, since $R = \frac{\rho L}{A} = \frac{\rho L}{\pi ab}$.

65.   **INTERPRET**   We find the resistance of a cylinder which has varying resistivity. We integrate over the length of the cylinder, using the equation for resistance

$$dR = \frac{\rho(x)dx}{A}$$

DEVELOP   We are given an equation for the resistivity:

$$\rho(x) = \rho_0\left(1 + \frac{x}{L}\right)e^{x/L}$$

where $\rho_0 = 2.41\times10^{-3}$ Ωm. The radius of the cylinder is $r = 0.0025$ m, and the length is $L = 0.015$ m. We integrate

$$dR = \frac{\rho(x)dx}{A}$$

from $x = 0$ to $x = L$.

EVALUATE

$$dR = \frac{\rho(x)dx}{A} \rightarrow R = \int_0^L \frac{\rho_0\left(1 + \frac{x}{L}\right)e^{x/L}}{\pi r^2}dx = \frac{\rho_0}{\pi r^2}\int_0^L\left(1 + \frac{x}{L}\right)e^{x/L}dx$$

$$\rightarrow R = \frac{\rho_0}{\pi r^2}[xe^{x/L}]_0^L = \frac{\rho_0 Le}{\pi r^2}$$

ASSESS   The resistivity increases with length, and this value of R is larger than the constant-resistivity value $\frac{\rho_0 L}{\pi r^2}$. That makes sense.

67.   **INTERPRET**   We use the dimensions of a resistor, and the desired resistance, to find the necessary resistivity of the material. We use the relationship between resistivity and resistance to solve the problem.

DEVELOP $R = \frac{\rho L}{A}$. We are told that the length of the resistor is $L = 10$ μm, the width is $w = 1.4$ μm, and the thickness is $t = 0.85$ μm. The desired resistance is $R = 470$ Ω.

EVALUATE

$$R = \frac{\rho L}{wt} \rightarrow \rho = \frac{Rwt}{L} = 5.59\times10^{-5}\ \Omega\text{m}$$

ASSESS   With the initial resistivity, the resistance would have $R = \frac{\rho L}{wt} = 71$ Ω, which is different enough that it would probably cause trouble.

69.   **INTERPRET**   We find the speed at which an electric vehicle can climb a given slope at a required speed, given the voltage, the amp-hour rating of the batteries, the mass of the vehicle, and the efficiency of the motor. We will use the equation for electrical power.

DEVELOP   We will first use energy methods to find the power required, then the current needed to supply that power at a voltage of $V = 312$ V, using $P = IV$. Only 85% of the electrical power is converted to mechanical power, so we will have to take that into consideration as well. The batteries can supply 100 amp-hours, so once we know the current we can find the time.

The slope is $\theta = 10°$, the mass of the car is $m = 1500$ kg, and the desired speed is $v = 45$ km/h $= 12.5$ m/s.

EVALUATE The power necessary is $P = Fv = (mg\sin\theta)v$. The electrical power converted to mechanical power is $P = (85\%)IV$, sos

$$mgv\sin\theta = 0.85IV \rightarrow I = \frac{mgv\sin\theta}{0.85V} = 120 \text{ A}$$

The batteries can supply current such that

$$It = 100 \text{ Amp}\cdot\text{hour, so } t = \frac{100 \text{ A}\cdot\text{hour}}{120 \text{ A}} = 50 \text{ minutes}$$

ASSESS Unfortunately, this number is about right. Improving battery technology is one of the research areas critical for further EV and hybrid vehicle development.

EXERCISES

Section 25.1 Circuits, Symbols, and Electromotive Force

15. **INTERPRET** This problem is about how various circuit elements can be connected to form a closed series circuit.
    DEVELOP In a series circuit, the same current must flow through all elements.
    EVALUATE One possibility is shown below. The order of elements and the polarity of the battery connections are not specified.

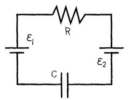

    ASSESS An important feature about a series circuit is that the current through all the components must be the same. With two batteries, the direction of the current flow is determined by the polarity of the larger of the two voltage ratings.

17. **INTERPRET** This problem explores the connection between the emf of a battery and the energy it delivers.
    DEVELOP Electromotive force, or emf, is defined as work per unit charge, $\varepsilon = W/q$.
    EVALUATE Substituting the values given in the problem statement, we find the emf to be

$$\varepsilon = \frac{W}{q} = \frac{27 \text{ J}}{3 \text{ C}} = 9 \text{ V}$$

    ASSESS For an ideal battery with zero internal resistance, the emf is equal to the terminal voltage (potential difference across the battery terminals).

19. **INTERPRET** This problem is about the chemical energy used up in the battery for the work done.
    DEVELOP The power delivered by an emf is $P = I\varepsilon$. Therefore, if the voltage and current remain constant, then the energy converted would be $W = Pt = I\varepsilon t$.
    EVALUATE Substituting the values given, the energy used in

$$W = I\varepsilon t = (5 \text{ A})(12 \text{ V})(3600 \text{ s}) = 216 \text{ kJ}$$

    ASSESS The result makes sense; the energy used up is proportional to the current drawn, the emf, and the duration the headlights were left on.

Section 25.2 Series and Parallel Circuits

21. **INTERPRET** This problem is about connecting two resistors in parallel.
    DEVELOP The equivalent resistance of two resistors connected in parallel can be found by Equation 25.3a;

$$\frac{1}{R_{\text{parallel}}} = \frac{1}{R_1} + \frac{1}{R_2}$$

The equation allows us to determine R_2 when R_{parallel} and R_1 are known.

EVALUATE   The solution for R_2 in Equation 25.3a is

$$R_2 = \frac{R_1 R_{\text{parallel}}}{R_1 - R_{\text{parallel}}} = \frac{(56 \text{ k}\Omega)(45 \text{ k}\Omega)}{56 \text{ k}\Omega - 45 \text{ k}\Omega} = 229 \text{ k}\Omega$$

ASSESS   Our result shows that $R_2 > R_{\text{parallel}}$. This is consistent with the fact that the equivalent resistance R_{parallel} is smaller than R_1 and R_2.

23.   **INTERPRET**   This problem is about the internal resistance of the battery in Exercise 22.

DEVELOP   The starter circuit contains all the resistances in series, as in Figure 25.9. (We assume R_L includes the resistance of the cables, connections, etc., as well as that of the motor.) With the defective starter, the terminal voltage is

$$V_T = \varepsilon - IR_{\text{int}} \quad \rightarrow \quad 6 \text{ V} = 12 \text{ V} - (300 \text{ A})R_{\text{int}}$$

EVALUATE   From the equation above, we find the internal resistance to be

$$R_{\text{int}} = \frac{\varepsilon - V_T}{I} = \frac{(12 \text{ V} - 6 \text{ V})}{300 \text{ A}} = 0.02 \text{ }\Omega$$

ASSESS   The terminal voltage $V_T = 6.0$ V is substantially less than the battery's emf $\varepsilon = 12$ V. The two are equal only in the ideal case where the internal resistance vanishes.

25.   **INTERPRET**   In this problem we are asked to find all possible values of equivalent resistance that could be obtained with three resistors.

DEVELOP   Since each resistor can be placed either in parallel or in series, there are eight combinations using all three resistors.

EVALUATE   Let $R_1 = 1.0 \Omega, R_2 = 2.0 \text{ }\Omega$, and $R_3 = 3.0 \text{ }\Omega$. The possible results are (a) one in series with two in parallel:

$$R_3 + \frac{R_1 R_2}{R_1 + R_2} = 3 \text{ }\Omega + \frac{(1 \text{ }\Omega)(2 \text{ }\Omega)}{1 \text{ }\Omega + 2 \text{ }\Omega} = \frac{11}{3} \text{ }\Omega$$

$$R_2 + \frac{R_1 R_3}{R_1 + R_3} = 2 \text{ }\Omega + \frac{(1 \text{ }\Omega)(3 \text{ }\Omega)}{1 \text{ }\Omega + 3 \text{ }\Omega} = \frac{11}{4} \text{ }\Omega$$

$$R_1 + \frac{R_2 R_3}{R_2 + R_3} = 1 \text{ }\Omega + \frac{(2 \text{ }\Omega)(3 \text{ }\Omega)}{2 \text{ }\Omega + 3 \text{ }\Omega} = \frac{11}{5} \text{ }\Omega$$

(b) one in parallel with two in series:

$$\frac{R_3(R_1 + R_2)}{R_1 + R_2 + R_3} = + \frac{(3 \text{ }\Omega)(1 \text{ }\Omega + 2 \text{ }\Omega)}{1 \text{ }\Omega + 2 \text{ }\Omega + 3 \text{ }\Omega} = \frac{9}{6} \text{ }\Omega = \frac{3}{2} \text{ }\Omega$$

$$\frac{R_2(R_1 + R_3)}{R_1 + R_2 + R_3} = + \frac{(2 \text{ }\Omega)(1 \text{ }\Omega + 3 \text{ }\Omega)}{1 \text{ }\Omega + 2 \text{ }\Omega + 3 \text{ }\Omega} = \frac{8}{6} \text{ }\Omega = \frac{4}{3} \text{ }\Omega$$

$$\frac{R_1(R_2 + R_3)}{R_1 + R_2 + R_3} = + \frac{(1 \text{ }\Omega)(2 \text{ }\Omega + 3 \text{ }\Omega)}{1 \text{ }\Omega + 2 \text{ }\Omega + 3 \text{ }\Omega} = \frac{5}{6} \text{ }\Omega$$

(c) three in series: $R_1 + R_2 + R_3 = 1 \text{ }\Omega + 2 \text{ }\Omega + 3 \text{ }\Omega = 6 \text{ }\Omega$.

(d) three in parallel:

$$\left(R_1^{-1} + R_2^{-1} + R_3^{-1}\right)^{-1} = \frac{R_1 R_2 R_3}{R_1 R_2 + R_1 R_3 + R_2 R_3} = \frac{(1 \text{ }\Omega)(2 \text{ }\Omega)(3 \text{ }\Omega)}{(1 \text{ }\Omega)(2 \text{ }\Omega) + (1 \text{ }\Omega)(3 \text{ }\Omega) + (2 \text{ }\Omega)(3 \text{ }\Omega)} = \frac{6}{11} \text{ }\Omega$$

ASSESS   The equivalent resistance is a maximum when all three are connected in series, as in (c), and a minimum when all are connected in parallel, as in (d).

Section 25.3 Kirchoff's Laws and Multiloop Circuits

27.   **INTERPRET**   This problem asks for the currents in a multi-loop circuit.

DEVELOP   The general solution of the two loop equations and one node equation given in Example 25.4 can be found using determinants (or I_1 and I_2 can be found in terms of I_3, as in Example 25.4). The equations and the solution are:

$$I_1 R_1 + 0 + I_3 R_3 = \varepsilon_1 \quad \text{(loop 1)}$$
$$0 - I_2 R_2 + I_3 R_3 = \varepsilon_2 \quad \text{(loop 2)}$$
$$I_1 - I_2 - I_3 = 0 \quad \text{(node A)}$$

$$\Delta \equiv \begin{vmatrix} R_1 & 0 & R_3 \\ 0 & -R_2 & R_3 \\ 1 & -1 & -1 \end{vmatrix} = R_1 R_2 + R_2 R_3 + R_3 R_1, \quad I_1 = \frac{1}{\Delta} \begin{vmatrix} \varepsilon_1 & 0 & R_3 \\ \varepsilon_2 & -R_2 & R_3 \\ 0 & -1 & -1 \end{vmatrix} = \frac{\varepsilon_1 (R_2 + R_3) - \varepsilon_2 R_3}{\Delta}$$

$$I_2 = \frac{1}{\Delta} \begin{vmatrix} R_1 & \varepsilon_1 & R_3 \\ 0 & \varepsilon_2 & R_3 \\ 1 & 0 & -1 \end{vmatrix} = \frac{\varepsilon_1 R_3 - \varepsilon_2 (R_1 + R_3)}{\Delta}, \quad I_3 = \frac{1}{\Delta} \begin{vmatrix} R_1 & 0 & \varepsilon_1 \\ 0 & -R_2 & \varepsilon_2 \\ 1 & -1 & 0 \end{vmatrix} = \frac{\varepsilon_2 R_1 + \varepsilon_1 R_2}{\Delta}$$

EVALUATE   With the particular values of emf's and resistors in this problem, we have

$$\Delta = R_1 R_2 + R_2 R_3 + R_3 R_1 = (2\,\Omega)(4\,\Omega) + (4\,\Omega)(1\,\Omega) + (1\,\Omega)(2\,\Omega) = 14\,\Omega^2$$

and the currents are

$$I_1 = [(R_2 + R_3)\varepsilon_1 - R_3 \varepsilon_2]\Delta^{-1} = [(4\,\Omega + 1\,\Omega)(6\,\text{V}) - (1\,\Omega)(1\,\text{V})]/(14\,\Omega^2) = 2.07\,\text{A}$$
$$I_2 = [R_3 \varepsilon_1 - (R_1 + R_3)\varepsilon_2]\Delta^{-1} = [(1\,\Omega)(6\,\text{V}) - (2\,\Omega + 1\,\Omega)(1\,\text{V})]/(14\,\Omega^2) = 0.214\,\text{A}$$
$$I_3 = (R_2 \varepsilon_1 + R_1 \varepsilon_2)\Delta^{-1} = [(4\,\Omega)(6\,\text{V}) + (2\,\Omega)(1\,\text{V})]/(14\,\Omega^2) = 1.86\,\text{A}$$

ASSESS   The same results could be obtained by retracing the reasoning of Example 25.4, with $\varepsilon_2 = 1.0$ V replacing the original value in loop 2. Then, everything is the same until the equation for loop 2: $1.0 + 4I_2 - I_3 = 0$.

29. **INTERPRET**   We find the current through a resisitor in a given circuit, using Kirchhoff's laws. We will use the loops and nodes drawn in Example 25.4.

DEVELOP   The circuit is given to us in Figure 25.14, with one change: $\varepsilon_2 = 2.0$V. We will use node A and loops 1 and 2. These will give us three equations, which we will use to solve the three unknown currents. At node A, $-I_1 + I_2 + I_3 = 0$. For loop 1, $\varepsilon_1 - I_1 R_1 - I_3 R_3 = 0$. For loop 2, $\varepsilon_2 + I_2 R_2 - I_3 R_3 = 0$.

EVALUATE   We want current I_2, so eliminate the other two currents. The node equation gives us $I_1 = I_2 + I_3$. Substitute this into the equation for loop 1 and solve for I_3:

$$\varepsilon_1 - (I_2 + I_3)R_1 - I_3 R_3 = 0 \rightarrow I_3 = \frac{\varepsilon_1 - I_2 R_1}{R_1 + R_3}$$

Now substitute this value into the equation for loop 2 and solve for I_2.

$$\varepsilon_2 + I_2 R_2 - \frac{\varepsilon_1 - I_2 R_1}{R_1 + R_3} R_3 = 0 \rightarrow \varepsilon_2 (R_1 + R_3) + I_2 R_2 (R_1 + R_3) - \varepsilon_1 R_3 + I_2 R_1 R_3 = 0$$

$$\rightarrow \varepsilon_2 (R_1 + R_3) + I_2 (R_1 R_2 + R_2 R_3 + R_1 R_3) - \varepsilon_1 R_3 = 0$$

$$\rightarrow I_2 = \frac{\varepsilon_1 R_3 - \varepsilon_2 (R_1 + R_3)}{R_1 R_2 + R_2 R_3 + R_1 R_3} = \frac{(6\text{V})(1\Omega) - (2\text{V})(3\Omega)}{8\Omega^2 + 4\Omega^2 + 2\Omega^2} = 0 \text{ V}/\Omega = 0 \text{ A}$$

ASSESS   The current through resistor R_2 is zero! Looking back at the original diagram, we can see that this would mean that battery 2 is supplying no current and the voltage drops through resistors 1 and 3 equal the voltage supplied by battery 1. This is a somewhat unexpected solution, but it is consistent.

Section 25.4 Electrical Measurements

31. **INTERPRET**   This problem is about the measurement error caused by the non-zero resistance of an ammeter.

DEVELOP   The current in the circuit of Fig. 25.29 is

$$I = \frac{V}{R_1 + R_2} = \frac{150\,\text{V}}{5\,\text{k}\Omega + 10\,\text{k}\Omega} = 10\,\text{mA}$$

With the ammeter inserted (in series with the resistors), the resistance is increased by $R_A = 100\,\Omega$.

EVALUATE   The resulting current after including R_A is

$$I' = \frac{V}{R_1 + R_2 + R_A} = \frac{150\,\text{V}}{5\,\text{k}\Omega + 10\,\text{k}\Omega + 0.10\,\text{k}\Omega} = 9.93\,\text{mA}$$

The value is about 0.662% lower than I.

ASSESS   The current reading by the ammeter is lower due to its internal resistance.

Section 25.5 Capacitors in Circuits

33.   **INTERPRET**   In this problem we are asked to show that the quantity RC, the product of resistance and capacitance, has the units of time.

DEVELOP   The SI units for R and C are W and F, respectively. The units can be rewritten as

$$1\,\Omega = 1\frac{V}{A} = 1\frac{V}{C/s} = 1\frac{V \cdot s}{C}, \quad 1\,F = 1\frac{C}{V}$$

EVALUATE   From the expressions above, the SI units for the time constant, RC, are

$$1\,\Omega \cdot F = 1\left(\frac{V \cdot s}{C}\right)\left(\frac{C}{V}\right) = 1\,s$$

as stated.

ASSESS   The quantity RC is the characteristic time for changes to occur in an RC circuit.

35.   **INTERPRET**   This problem is about the time dependence of the capacitor voltage in a charging RC circuit.

DEVELOP   The capacitor voltage as a function of time is given by Equation 25.6:

$$V_C = \varepsilon(1 - e^{-t/RC})$$

EVALUATE   After five time constants, $t = 5RC$, the equation above gives a voltage of

$$\frac{V_C}{\varepsilon} = 1 - e^{-5} = 1 - 6.74 \times 10^{-3} \approx 99.3\%$$

of the applied voltage.

ASSESS   As time goes on and after many more time constants, we find essentially no current flowing to the capacitor, and the capacitor could be considered as being fully charged for all practical purposes.

37.   **INTERPRET**   We find the voltage across a capacitor in a circuit, given that the capacitor is "fully charged." We will take this to mean that the current through the capacitor is zero, and use the results of Example 25.7b, with Ohm's law, to find the voltage required.

DEVELOP   In Figure 25.24a, we see the circuit in question. If the capacitor is fully charged, then no current flows through it and the circuit is equivalent to the circuit shown in 25.24c. So we find the current through resistor R_2 in Figure 25.24c and then determine the voltage across resistor R_2, which is the same as the voltage across the capacitor.

EVALUATE   The current through resistor R_2 is given in Example 25.7 as $I = \frac{\varepsilon}{R_1 + R_2}$. The voltage is given by Ohm's law as $V = IR = \frac{\varepsilon}{R_1 + R_2}R_2 = \varepsilon\left(\frac{R_2}{R_1 + R_2}\right)$.

ASSESS   In the limit of long charging times, this circuit behaves like a voltage divider.

PROBLEMS

39.   **INTERPRET**   The problem asks for the current in a resistor which is part of a more complex circuit.

DEVELOP   The circuit in Fig. 25.30, with a battery connected across points A and B, is similar to the circuit analyzed in Example 25.3. In this case, we have one 1.0 W resistor in parallel with two 1.0 W resistors in series:

$$\frac{1}{R_\parallel} = \frac{1}{1\,\Omega} + \frac{1}{1\,\Omega + 1\,\Omega} = \frac{3}{2\,\Omega} \quad \rightarrow \quad R_\parallel = \frac{2}{3}\,\Omega$$

and the total resistance is $R_{tot} = 1\,\Omega + 1\,\Omega + \frac{2}{3}\,\Omega = \frac{8}{3}\,\Omega$. The total current (that through the battery) is

$I_{tot} = \frac{\varepsilon}{R_{tot}} = \frac{6V}{8\,\Omega/3} = \frac{9}{4}\,A = 2.25\,A.$

EVALUATE   The voltage across the parallel combination is

$$V_\parallel = I_{tot}R_\parallel = \left(\frac{9}{4}\,A\right)\left(\frac{2}{3}\,\Omega\right) = \frac{3}{2}\,V$$

which is the voltage across the vertical $R_v = 1\,\Omega$ resistor. Thus, the current through this resistor is then

$I_v = \frac{V_\parallel}{R_v} = \frac{3V/2}{1\,\Omega} = 1.5\,A.$

ASSESS   We have a total of 2.25 A of current flowing around the circuit. At the vertex of the triangular loop, it is split into $I_v = 1.5\,A$ and $I' = I_{tot} - I_v = 0.75A$. The voltage drop across the vertical resistor ($V_\parallel = 1.5\,V$) is the same as that going through point C and the two 1.0-W resistors: $V' = (0.75\,A)(1\,\Omega + 1\,\Omega) = 1.5V$.

41.   **INTERPRET**   The circuit has two batteries connected in series. We apply Kirchhoff's law to solve for the current that flows through the discharged battery.

DEVELOP   Terminals of like polarity are connected with jumpers of negligible resistance. Kirchhoff's voltage law gives

$$\varepsilon_1 - \varepsilon_2 - IR_1 - IR_2 = 0$$

EVALUATE   Solving the equation above, we obtain

$$I = \frac{\varepsilon_1 - \varepsilon_2}{R_1 + R_2} = \frac{12\text{ V} - 9\text{V}}{0.02\ \Omega + 0.08\ \Omega} = 30\text{ A}$$

ASSESS   When you try to jump start a car, you connect positive to positive and negative to negative terminals. The connection is what was illustrated in the figure.

43.   **INTERPRET**   This problem is about rate of energy dissipation in the resistor.

DEVELOP   For a short-circuited battery, the current is $I = \varepsilon/R_{\text{int}}$, so the dissipated power is $P = I^2 R_{\text{int}} = \frac{\varepsilon^2}{R_{\text{int}}}$.

EVALUATE   Substituting the values given in the problem, we find the rate of energy dissipation to be

$$P = \frac{\varepsilon^2}{R_{\text{int}}} = \frac{(6\text{ V})^2}{2.5\ \Omega} = 14.4\text{ W}$$

ASSESS   With ε held fixed at 6 V, we see that the power dissipated is inversely proportional to the internal resistance R_{int}.

45.   **INTERPRET**   The circuit in this problem contains a battery – the emf source, and two resistors in series.

DEVELOP   Kirchhoff's loop law gives $\varepsilon - IR_1 - IR_2 = 0$, or $I = \varepsilon/(R_1 + R_2)$. Therefore, the voltage across R_2 is

$$V_2 = IR_2 = \frac{\varepsilon R_2}{R_1 + R_2}$$

This is the equation we shall use to solve for R_2. Once R_2 is known, the power dissipated is simply equal to $P_2 = V_2^2/R_2$.

EVALUATE   **(a)** The equation above gives

$$R_2 = \frac{R_1 V_2}{\varepsilon - V_2} = \frac{(270\ \Omega)(4.5\text{ V})}{12\text{ V} - 4.5\text{ V}} = 162\ \Omega$$

(b) The power dissipated is $P_2 = \frac{V_2^2}{R_2} = \frac{(4.5\text{ V})^2}{162\Omega} = 125$ mW.

ASSESS   For completeness, let's calculate the power dissipated in R_1 and the total power. The voltage across R_1 is $V_1 = \varepsilon - V_2 = 12\text{V} - 4.5\text{V} = 7.5\text{V}$, and the power dissipated in R_1 is $P_1 = \frac{V_1^2}{R_1} = \frac{(7.5\text{ V})^2}{270\ \Omega} = 208$ mW. The total power dissipated in the circuit is

$$P_{\text{tot}} = \frac{\varepsilon^2}{R_1 + R_2} = \frac{(12\text{V})^2}{270\ \Omega + 162\ \Omega} = 333\text{ mW}$$

which indeed is equal to the sum of P_1 and P_2.

47.   **INTERPRET**   The circuit in this problem contains a battery—the emf source, and three resistors. We want to analyze the voltage across the one which is a variable resistor.

DEVELOP   The resistors in parallel have an equivalent resistance of $R_\parallel = RR_1/(R + R_1)$. The other R, and $R_\parallel$, is a voltage divider in series with voltage ε.

EVALUATE   **(a)** Using Equation 25.2, we find the voltage across R_1 to be

$$V_\parallel = \frac{\varepsilon R_\parallel}{R + R_\parallel} = \frac{\varepsilon R_1}{R + 2R_1}$$

(b) The sketch of $V_{\parallel}$ as a function of R_1 is shown on the right.

(c) If $R_1 = 0$, then $V_{\parallel} = 0$. On the other hand, if $R_1 = \infty$, then $V_{\parallel} = \varepsilon/2$ (the value when R_1 is removed). If $R_1 = 10R$, $V_{\parallel} = (10/21)\varepsilon$.

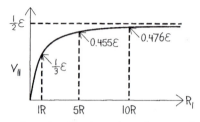

ASSESS   The limit $R_1 = 0$ corresponds to the situation where the second resistor is shorted out. The limit $R_1 = \infty$ is an open circuit with no current going through it.

49. **INTERPRET**   This problem asks for the power dissipated in a resistor which is part of a more complex circuit.
    DEVELOP   The three resistors in parallel have an effective resistance of

    $$\frac{1}{R_{\parallel}} = \frac{1}{2\,\Omega} + \frac{1}{4\,\Omega} + \frac{1}{6\,\Omega} = \frac{11}{12\,\Omega} \;\rightarrow\; R_{\parallel} = \frac{12}{11}\,\Omega$$

    The equivalent resistance of the circuit is $R_{\text{tot}} = R_1 + R_{\parallel} = 1\Omega + \frac{12}{11}\Omega = \frac{23}{11}\Omega$. Equation 25.2 gives the voltage across them as

    $$V_{\parallel} = \frac{\varepsilon R_{\parallel}}{R_{\text{tot}}} = \frac{(6\text{V})(12\Omega/11)}{23\Omega/11} = \frac{72}{23}\;\text{V}$$

    EVALUATE   Thus, the power dissipated in the 4Ω resistor is

    $$P_4 = \frac{V_{\parallel}^2}{R_4} = \frac{(72\text{V}/23)^2}{4\Omega} = 2.45\;\text{W}$$

    ASSESS   With ε held fixed at 6 V, we see that the power dissipated is inversely proportional to the resistance.

51. **INTERPRET**   The problem asks for the equivalent resistance between two points in a complex circuit.
    DEVELOP   The equivalent resistance is determined by the current which would flow through a pure emf if it were connected between A and B: $R_{AB} = \varepsilon/I$. Since I is but one of six branch currents, the direct solution of Kirchhoff's circuit laws is tedious $(6 \times 6$ determinants). However, because of the special values of the resistors in Fig. 25.34, a symmetry argument greatly simplifies the calculation.
    The equality of the resistors on opposite sides of the square implies that the potential difference between A and C equals that between D and B, i.e.,

    $$V_A - V_C = V_D - V_B$$

    Equivalently, $V_A - V_D = V_C - V_B$. Since $V_A - V_C = I_1 R$, $V_A - V_D = I_2(2R)$, etc., the symmetry argument requires that both R-resistors on the perimeter carry the same current, I_1, and both $2R$-resistors carry current I_2. Then Kirchhoff's current law implies that the current through E is $I_1 + I_2$, and the current through the central resistor is $I_1 - I_2$ (as added to Figure 25.34). Now there are only two independent branch currents, which can be found from Kirchhoff's voltage law, applied, for example,

    $$\varepsilon - I_1 R - I_2(2R) = 0 \quad (\text{loop }ACBA)$$
    $$-I_1 R - (I_1 - I_2)R + I_2(2R) = 0 \quad (\text{loop }ACDA)$$

    These equations may be rewritten as

    $$I_1 + 2I_2 = \frac{\varepsilon}{R}$$
    $$-2I_1 + 3I_2 = 0$$

    with solution $I_1 = 3\varepsilon/7R$ and $I_2 = 2\varepsilon/7R$.

EVALUATE The sum of the two currents gives $I = I_1 + I_2 = 5\mathcal{E}/7R$, which leads to

$$R_{AB} = \frac{\mathcal{E}}{I} = \frac{7R}{5}$$

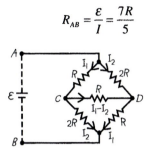

ASSESS The configuration of resistors in Figure 25.34 is called a Wheatstone bridge.

53. **INTERPRET** This problem asks for the current in a resistor which is part of a more complex circuit. The solution requires analyzing a circuit with series and parallel components.

DEVELOP Let us choose the positive sense for each of the three branch currents in Figure 25.35 as upward through their respective emf's (at least one must be negative, of course), and consider the two smaller loops shown. Kirchhoff's circuit laws give:

$$I_a + I_b + I_c = 0 \qquad \text{(top node)}$$
$$(R_1 + R_2)I_a - R_3 I_b = \mathcal{E}_1 - \mathcal{E}_2 \qquad \text{(left loop)}$$
$$-R_3 I_b + R_4 I_c = \mathcal{E}_3 - \mathcal{E}_2 \qquad \text{(right loop)}$$

Solve for I_a and I_c from the loop equations and substitute into the node equation:

$$\frac{(\mathcal{E}_1 - \mathcal{E}_2) + R_3 I_b}{R_1 + R_2} + I_b + \frac{(\mathcal{E}_3 - \mathcal{E}_2) + R_3 I_b}{R_4} = 0$$

The current in R_3 is I_b.

EVALUATE Solving for I_b, we find

$$I_b = -\frac{(\mathcal{E}_1 - \mathcal{E}_2)R_4 + (\mathcal{E}_3 - \mathcal{E}_2)(R_1 + R_2)}{R_3 R_4 + (R_1 + R_2)(R_3 + R_4)} = -\frac{(6 \text{ V} - 1.5 \text{ V})(820 \ \Omega) + (4.5 \text{ V} - 1.5 \text{ V})(420 \ \Omega)}{(560 \ \Omega)(820 \ \Omega) + (420 \ \Omega)(1380 \ \Omega)}$$
$$= -4.77 \text{ mA}$$

A negative current is downward through $\mathcal{E}_2$ in Figure 25.35.

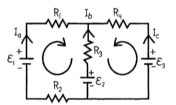

ASSESS Substituting $I_b = -4.77$ mA into the equations above, we find $I_a = 4.35$ mA and $I_c = 0.42$ mA. One can readily verify that the solutions satisfy all the equations.

55. **INTERPRET** In this problem the voltage across a given resistor is measured using a voltmeter which behaves like a resistance.

DEVELOP With a meter of resistance R_m connected as indicated, the circuit reduces to two pairs of parallel resistors in series. The total resistance is

$$R_{\text{tot}} = \frac{(30 \text{ k}\Omega)R_m}{30 \text{ k}\Omega + R_m} + \frac{40 \text{ k}\Omega}{2}$$

The voltage reading is

$$V_m = R_m I_m = \frac{R_m (30 \text{ k}\Omega) I_{\text{tot}}}{30 \text{ k}\Omega + R_m}$$

where $I_{\text{tot}} = (100\text{V})/R_{\text{tot}}$ (the expression for V_m follows from Equation 25.2, with R_1 and R_2 as the above pairs, or from I_m as a fraction of I_{tot}).

EVALUATE   For the three voltmeters specified, $I_{\text{tot}} = 2.58$ mA, 2.14 mA, and 2.00 mA, while $V_m = 48.4$ V, 57.3 V, and 59.9 V, respectively. (After checking the calculations, round off to two figures.)

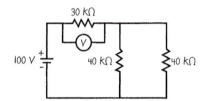

ASSESS   Of course, 60 V is the ideal voltmeter reading. This reading corresponds to an ideal voltmeter that has infinite resistance.

57.   **INTERPRET**   In this problem an ammeter is used to measure the current in a circuit. The ammeter is connected in series with the resistor.

    DEVELOP   The internal resistance of an ideal battery is zero, so the resistor has a value of $R = \varepsilon/I = (12\text{V})/(1\text{A}) = 12\,\Omega$. With the ammeter in place, the current would be

$$I = \frac{V}{R + R_m}$$

    EVALUATE   **(a)** Substituting the values given in the problem statement, we find the current to be

$$I = \frac{V}{R + R_m} = \frac{12\text{ V}}{12.1\,\Omega} = 0.992\text{ A}$$

    (b) If the resistance of the ammeter were neglected in the calculation, one would obtain

$$R' = \frac{12\text{ V}}{0.992\text{ A}} = 12.1\,\Omega$$

    ASSESS   This value R' differs from the actual value of $R = 12\,\Omega$ by 0.83%. However, subtraction of R_m is a correction that could be included easily.

59.   **INTERPRET**   This problem is about a discharging capacitor in an RC circuit, and we want to find the time to reach a given voltage.

    DEVELOP   A capacitor discharging through a resistor is described by exponential decay, with time constant RC (Equation 25.8):

$$V(t) = V(0)e^{-t/RC}$$

The energy in the capacitor is

$$U_C(t) = \frac{1}{2}CV(t)^2 = \frac{1}{2}CV(0)^2 e^{-2t/RC} = U_C(0)e^{-2t/RC}$$

    EVALUATE   **(a)** The time it takes to reach $V(t) = 5$ V is

$$t = RC\ln\left(\frac{V(0)}{V(t)}\right) = (500\text{ k}\Omega)(1\mu\text{F})\ln 2 = 347\text{ ms}$$

    (b) Similarly, the time it takes for the energy to decrease to half its initial value is

$$t = \frac{1}{2}RC\ln\left(\frac{U_C(0)}{U_C(t)}\right) = \frac{1}{2}(500\text{ k}\Omega)(1\mu\text{F})\ln 2 = 173\text{ ms}$$

    ASSESS   The time constant in this problem is $RC = (500\text{ k}\Omega)(1\,\mu\text{F}) = 500$ ms. When $t = RC$, the voltage is reduced by a factor of e, or $V(t)/V(0) = e^{-1} = 0.368$. Therefore, it takes less than one time constant for the value of $V(t)$ to be halved from its initial value.

61.   **INTERPRET**   This problem involves energy dissipation in an RC circuit. The object of interest is the capacitance.

    DEVELOP   A capacitor discharging through a resistor is described by exponential decay, with time constant RC (Equation 25.8):

$$V(t) = V(0)e^{-t/RC}$$

The energy in the capacitor is

$$U_C(t) = \frac{1}{2}CV(t)^2 = \frac{1}{2}CV(0)^2 e^{-2t/RC} = U_C(0)e^{-2t/RC}$$

EVALUATE If 2 J is dissipated in time t, the energy stored in the capacitor drops from $U_C(0) = 5$ J to $U_C(t) = 3$ J (assuming there are no losses due to radiation, etc.). From the equation above, we find the capacitance to be

$$C = \frac{2t}{R\ln(U_C(0)/U_C(t))} = \frac{2(8.6 \text{ ms})}{(10 \text{ k}\Omega)\ln(5 \text{ J}/3 \text{ J})} = 3.37\,\mu\text{F}$$

ASSESS In this problem the time constant is $RC = (10 \text{ k}\Omega)(3.37\,\mu\text{F}) = 33.7$ ms. Therefore, at 8.6 ms (about $0.255\, RC$), the energy decreases by a factor $e^{-2(0.255)} \approx 0.6$. This is precisely what we found.

63. **INTERPRET** This problem is about the long-term and short-term values of current and voltage of an RC circuit.
     DEVELOP In addition to the explanation in Example 25.7, we note that when the switch is in the closed position, Kirchhoff's voltage law applied to the loop containing both resistors yields $\mathcal{E} = I_1 R_1 + I_2 R_2$, and to the loop containing just R_2 and C, $V_C = I_2 R_2$.
     EVALUATE (a) If the switch is closed at $t = 0$, Example 25.7 shows that $V_C(0) = 0$, $I_2(0) = 0$, and

$$I_1(0) = \frac{\mathcal{E}}{R_1} = \frac{100 \text{ V}}{4 \text{ k}\Omega} = 25 \text{ mA}$$

(b) After a long time, $t = \infty$, Example 25.7 also shows that

$$I_1(\infty) = I_2(\infty) = \frac{\mathcal{E}}{R_1 + R_2} = \frac{100 \text{ V}}{10 \text{ k}\Omega} = 10 \text{ mA}$$

and $V_C(\infty) = I_2(\infty)R_2 = (10\text{mA})(6\text{k}\Omega) = 60\text{V}$.
(c) Under the conditions stated, the fully charged capacitor ($V_C = 60$ V) simply discharges through R_2. (R_1 is in an open-circuit branch, so $I_1 = 0$ for the entire discharging process.) The initial discharging current is

$$I_2 = \frac{V_C}{R_2} = \frac{60 \text{ V}}{6 \text{ k}\Omega} = 10 \text{ mA}$$

(d) After a very long time, I_2 and V_C decay exponentially to zero.
ASSESS We deduced the short-term and long-term behavior of the RC circuit without having to solve a complicated differential equation. A long time after the circuit has been closed, the capacitor becomes fully charged with no current flowing to it. When the circuit is reopened, the capacitor starts to discharge and eventually loses all its stored energy.

65. **INTERPRET** In this problem we are asked to find the voltage and internal resistance of a battery using the measured voltage values of two voltmeters.
     DEVELOP The internal resistance of the battery (R_i) and the resistance of the voltmeter (R_m) are in series with the battery's emf, so the current is $I = \mathcal{E}/(R_i + R_m)$. The potential drop across the meter (its reading) is

$$V_m = IR_m = \frac{\mathcal{E}R_m}{R_i + R_m}$$

From the given data, we can write

$$4.36 \text{ V} = \frac{\mathcal{E}(1 \text{ k}\Omega)}{R_i + 1 \text{ k}\Omega} \quad \text{and} \quad 4.41 \text{ V} = \frac{\mathcal{E}(1.5 \text{ k}\Omega)}{R_i + 1.5 \text{ k}\Omega}$$

or $R_i + 1\text{k}\Omega = \mathcal{E}(1\text{k}\Omega/4.36\text{V})$ and $R_i + 1.5\text{k}\Omega = \mathcal{E}(1.5 \text{ k}\Omega/4.41 \text{ V})$.
EVALUATE Solving the simultaneous equations for $\mathcal{E}$ and R_i gives

$$\mathcal{E} = (1.5 \text{ k}\Omega - 1 \text{ k}\Omega)\left(\frac{1.5 \text{ k}\Omega}{4.41 \text{ V}} - \frac{1 \text{ k}\Omega}{4.36 \text{ V}}\right)^{-1} = 4.51 \text{ V}$$

and $R_i = (4.51\text{V})(1 \text{ k}\Omega/4.36\text{V}) - 1 \text{ k}\Omega = 35.2\Omega$.
ASSESS An ideal voltmeter has infinite resistance. Thus, when we let $R_m \to \infty$, its reading approaches the battery voltage $\mathcal{E}$.

67.   **INTERPRET**   The electric field at the node increases due to charge accumulation and eventually reaches the breakdown field strength.

DEVELOP   The charge on the node (whether positive or negative) accumulates at a rate of $I = dq/dt = 1\mu A = 1\mu C/s$, so $|q(t)| = (1\mu A)t$ (where we assume that $q(0) = 0$). If the node is treated approximately as an isolated sphere, the electric field strength at its surface is

$$E = \frac{k\,|q|}{r^2} = \frac{kIt}{r^2}$$

Electric breakdown occurs when $E = E_b = 3$ MV/m .

EVALUATE   The time when the breakdown happens is

$$t = \frac{E_b r^2}{kI} = \frac{(3\text{ MV/m})(0.5\text{ mm})^2}{(9\times10^9\text{ m/F})(1\;\mu A)} = 83.3\;\mu s$$

ASSESS   This problem shows that Kirchhoff's node law must hold, or else there would be a charge buildup at the node which quickly leads to an electric breakdown.

69.   **INTERPRET**   This problem is about energy stored in the capacitor that's part of the RC circuit. We are asked to show that it only stores half the energy the battery supplies.

DEVELOP   The power supplied by the battery charging a capacitor (initially uncharged) in an RC circuit is

$$P = I\varepsilon = \frac{\varepsilon^2}{R}e^{-t/RC}$$

where the current is given by Equation 25.5: $I = (\varepsilon/R)e^{-t/RC}$. The total energy supplied is

$$U_{\text{battery}} = \int_0^\infty P\,dt = \frac{\varepsilon^2}{R}\int_0^\infty e^{-t/RC}\,dt = C\varepsilon^2(e^0 - e^{-\infty}) = C\varepsilon^2$$

EVALUATE   The energy stored in the fully charged capacitor is

$$U_C(\infty) = \frac{1}{2}CV(\infty)^2 = \frac{1}{2}C\varepsilon^2 = \frac{1}{2}U_{\text{battery}}$$

Thus, we see that the energy stored in the capacitor is only half of that supplied by the battery.

ASSESS   The other half of the energy supplied by the battery is dissipated in the resistor:

$$U_R = \int_0^\infty I^2R\,dt = \frac{\varepsilon^2}{R}\int_0^\infty e^{-2t/RC}\,dt = \frac{1}{2}C\varepsilon^2$$

71.   **INTERPRET**   This problem is about finding the voltage and internal resistance of a battery. We are given the current values when the battery is connected to resistors of known resistance.

DEVELOP   The circuit diagram is like Fig. 25.10, and Kirchhoff's voltage law is

$$\varepsilon - IR_{\text{int}} - IR_L = 0$$

For the two cases given, this may be written as

$$\varepsilon - (26\text{ mA})R_{\text{int}} = (26\text{ mA})(50\;\Omega) = 1.3\text{ V}$$
$$\varepsilon - (43\text{ mA})R_{\text{int}} = (43\text{ mA})(22\;\Omega) = 0.946\text{ V}$$

EVALUATE   Solving for ε and R_{int}, we find

$$R_{\text{int}} = \frac{1.3\text{ V} - 0.946\text{ V}}{43\text{ mA} - 26\text{ mA}}\Omega = 20.8\;\Omega \quad\text{and}\quad \varepsilon = (26\text{ mA})(50\;\Omega + 20.8\;\Omega) = 1.84\text{ V}$$

ASSESS   The terminal voltage of the battery is $V = \varepsilon - IR_{\text{int}} = 1.84\text{ V} - I(20.8\;\Omega)$, which is lower than ε. When the battery is connected to a resistor of resistance R, the current in the circuit is $I = \varepsilon/(R + R_{\text{int}})$.

73.   **INTERPRET**   Our circuit consists of an array of resistors of infinite extent, and we're asked to find the equivalent resistance.

DEVELOP   Since the circuit line is infinite, the addition or deletion of one more element leaves the equivalent resistance unchanged. This can be represented diagrammatically as

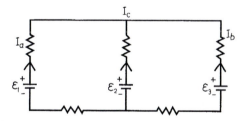

The right-hand picture represents R in series with the parallel combination R and R_{eq}. Thus,

$$R_{eq} = R + \frac{RR_{eq}}{R + R_{eq}}$$

EVALUATE   Solving for R_{eq}, one finds $R_{eq}^2 - RR_{eq} - R^2 = 0$, or

$$R_{eq} = (1 + \sqrt{5})R/2 = 1.62R$$

Note that only the positive root is physically meaningful for a resistance.

ASSESS   Let's see how this limiting value is reached. With only two resistors, the equivalent resistance is $R_1 = R + R = 2R$. Next, consider four resistors (the four on the left of Fig. 25.41). The equivalent resistance is

$$R_2 = R + \frac{1}{1/R + 1/2R} = R + \frac{2R}{3} = 1.67R$$

Continuing the same line of reasoning leads to the quadratic equation which we solved to obtain $R_{eq} = (1 + \sqrt{5})R/2 = 1.62R$.

75.   **INTERPRET**   This problem asks for the current through an emf source which is part of a more complex circuit. The solution requires analyzing a circuit with series and parallel components.

DEVELOP   Let us choose the positive sense for each of the three branch currents in Fig. 25.43 as upward through their respective emf's (at least one must be negative, of course), and consider the right loop and the big loop. Kirchhoff's circuit laws give:

$$
\begin{aligned}
I_a + I_b + I_c &= 0 && \text{(top node)} \\
I_a(2R) - I_b(2R) &= \varepsilon_1 - \varepsilon_3 && \text{(big loop)} \\
I_b(2R) - I_c R &= \varepsilon_3 - \varepsilon_2 && \text{(right loop)}
\end{aligned}
$$

Solve for I_a and I_c from the loop equations and substitute into the node equation:

$$\frac{(\varepsilon_1 - \varepsilon_3) + 2RI_b}{2R} + I_b + \frac{2RI_b - (\varepsilon_3 - \varepsilon_2)}{R} = 0$$

The current in ε_3 is I_b.

EVALUATE   Solving For I_b, we find

$$I_b = \frac{(3\varepsilon_3 - 2\varepsilon_2 - \varepsilon_1)}{8R} = \frac{(60 \text{ mV} - 90 \text{ mV} - 75 \text{ mV})}{8(1.5 \text{ M}\Omega)} = -8.75 \text{ nA}$$

The negative sign means that the direction of I_b is opposite of what was shown in the diagram.

ASSESS   The negative sign in I_b can be easily understood by noting that ε_3 is smaller than ε_1 and ε_2.

77.   **INTERPRET**   We represent a "leaky" capacitor with an equivalent circuit diagram, and determine the time constant for this circuit. We will also show that the time constant does not depend on the geometry of the capacitor, but only on its material properties.

DEVELOP   For part (a), see the figure later. For part (b), we will use the resistance of the insulation material, $R = \rho \frac{d}{A}$, where d is the thickness of the material and A is the area of the plates. We will also use the equation for parallel-plate capacitance, $C = \kappa \varepsilon_0 \frac{A}{d}$, where $\kappa = 5.6$ is the dielectric constant of glass. The time constant we are seeking is $\tau = RC$.

EVALUATE

(b) $\tau = RC = \rho \frac{d}{A} \kappa \varepsilon_0 \frac{A}{d} = \rho \kappa \varepsilon_0$. This is independent of the geometrical terms d and A, and depends only on the material properties. $\tau = \rho \kappa \varepsilon_0 = 595$ s.

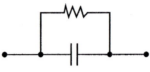

ASSESS This is actually pretty good for a capacitor. Materials with high resistivity and high dielectric constant will make capacitors with longer leakage time constants.

79. **INTERPRET** We use Kirchhoff's laws to write a system of equations for the circuit shown in Figure 25.39, and from the resulting equations we will determine the current through resistor R_2. We will need 4 equations.

DEVELOP First we make a diagram of the circuit, as shown in the figure later. Nodes A and B give us duplicate information, so we will use only node A, along with the three loops.

Node A: $I_1 - I_2 - I_3 - I_4 = 0$
Loop 1: $\varepsilon - I_1 R - I_2 R = 0$
Loop 2: $I_2 R - \frac{Q_3}{C} = 0$
Loop 3: $\frac{Q_3}{C} - \frac{Q_4}{C} - I_4 R = 0$

We will solve for I_2 as a function of time.

EVALUATE

Node A: $I_1 = I_2 + I_3 + I_4$.

Loop 1:

$$\varepsilon - (I_2 + I_3 + I_4)R - I_2 R = 0 \rightarrow I_4 = \frac{\varepsilon}{R} - 2I_2 - I_3$$

$$\rightarrow \frac{dI_4}{dt} = -2\frac{dI_2}{dt} - \frac{dI_3}{dt}$$

Loop 2:

$$\frac{Q_3}{C} = I_2 R \rightarrow \frac{I_3}{C} = R\frac{dI_2}{dt} \rightarrow \frac{dI_3}{dt} = RC\frac{d^2 I_2}{dt^2}$$

Loop 3:

$$\frac{I_3}{C} - \frac{I_4}{C} - R\frac{dI_4}{dt} = 0 \rightarrow R\frac{dI_2}{dt} - \frac{1}{C}\left(\frac{\varepsilon}{R} - 2I_2 - I_3\right) - R\left(-2\frac{dI_2}{dt} - \frac{dI_3}{dt}\right) = 0$$

$$\rightarrow \frac{dI_2}{dt} - \frac{1}{RC}\frac{\varepsilon}{R} + \frac{2}{RC}I_2 + \frac{dI_2}{dt} + 2\frac{dI_2}{dt} + RC\frac{d^2 I_2}{dt^2} = 0$$

$$\rightarrow \frac{d^2 I_2}{dt^2} + \frac{4}{RC}\frac{dI_2}{dt} + \frac{2}{(RC)^2}I_2 = \frac{\varepsilon}{R(RC)^2}$$

This is a second-order linear differential equation with constant coefficients. We can solve the homogenous equation using the characteristic equation:

$$\lambda^2 + \frac{4}{RC}\lambda + \frac{2}{(RC)^2} = 0$$

$$\rightarrow \lambda = \frac{1}{2}\left[-\frac{4}{RC} \pm \sqrt{\frac{16}{(RC)^2} - \frac{8}{(RC)^2}}\right] = \frac{-2}{RC} \pm \frac{2}{RC}\sqrt{2}$$

$$\rightarrow \lambda = \left\{-\frac{2}{RC}(1+\sqrt{2}), -\frac{2}{RC}(1-\sqrt{2})\right\}$$

So the solution to the homogenous equation is

$$I_2(t) = A_1 e^{-\frac{2}{RC}(1+\sqrt{2})t} + A_2 e^{-\frac{2}{RC}(1-\sqrt{2})t}$$

and the solution to the inhomogenous equation is

$$I_2(t) = A_1 e^{-\frac{2}{RC}(1+\sqrt{2})t} + A_2 e^{-\frac{2}{RC}(1-\sqrt{2})t} + \frac{\varepsilon}{2R}$$

Now we need the initial condition on I_2 and $\frac{dI_2}{dt}$. Since both capacitors are initially uncharged and essentially short circuits, $I_1(0) = I_3(0) = \frac{dQ_3}{dt}\big|_0 = \frac{\varepsilon}{R}$ and the *initial* voltage across the central capacitor is given by $V_C = \varepsilon(1 - e^{-t/RC})$. This voltage creates a current through R_2 of

$$I_2\big|_{t=0} = \frac{V_C}{R}\bigg|_{t=0} = \frac{\varepsilon}{R}(1 - e^{-t/RC})\bigg|_{t=0}$$

so

$$I_2(0) = 0 \text{ and } \frac{dI_2}{dt}\bigg|_{t=0} = -\frac{\varepsilon}{R}\left(-\frac{1}{RC}\right)e^{-t/RC}\bigg|_{t=0} = \frac{\varepsilon}{R^2C}$$

Applying the boundary condition $I_2(0) = 0$ to the solution obtained previously gives us $0 = A_1 + A_2 + \dfrac{\varepsilon}{2R}$, and applying

$$\frac{dI_2}{dt}\bigg|_{t=0} = \frac{\varepsilon}{R^2C} \text{ gives us } \frac{\varepsilon}{R^2C} = A_1\left(-\frac{2}{RC}(1+\sqrt{2})\right) + A_2\left(-\frac{2}{RC}(1-\sqrt{2})\right)$$

from which we can determine that $A_1 = A_2 = -\dfrac{\varepsilon}{4R}$.

So, our final solution is

$$I_2(t) = \frac{\varepsilon}{4R}\left[2 - e^{-\frac{2}{RC}(1+\sqrt{2})t} - e^{-\frac{2}{RC}(1-\sqrt{2})t}\right]$$

ASSESS This was a difficult problem, but the technique used to set it up is the same as for an easier one: Kirchhoff's laws.

81. **INTERPRET** We need to create a resistance value using three different resistors. We can combine the resistors in series and/or parallel. We will find the appropriate combination by making some educated guesses and seeing what works.

DEVELOP The three resistors are $R_1 = 3.30 \text{ k}\Omega$, $R_2 = 4.70 \text{ k}\Omega$, and $R_3 = 1.50 \text{ k}\Omega$. Resistors add in series, and in parallel their reciprocals add.

One guess would be to place R_1 and R_2 in parallel so that the combination was smaller than either, then adding R_3 in series to this combination. Try it!

EVALUATE $\frac{1}{R_p} = \frac{1}{R_1} + \frac{1}{R_2} \rightarrow R_p = 1.94\text{k}\Omega$, and $R_p + R_3 = 3.44 \text{ k}\Omega$ which is just what we wanted.

ASSESS Another reasonable guess might be to put R_1 and R_3 in series, then put R_2 in parallel across that combination: but that method gives us only $2.37 \text{ k}\Omega$.

EXERCISES

Section 26.2 Magnetic Force and Field

17. **INTERPRET** This problem is about the magnetic force exerted on a moving electron.

DEVELOP The magnetic force on a charge q moving with velocity $\vec{v}$ is given by Equation 26.1: $\vec{F}_B = q\vec{v} \times \vec{B}$. The magnitude of $\vec{F}_B$ is

$$F_B = |\vec{F}_B| = |q\vec{v} \times \vec{B}| = |q|vB\sin\theta$$

EVALUATE **(a)** The magnetic field is a minimum when $\sin\theta = 1$ (the magnetic field perpendicular to the velocity). Thus,

$$B_{min} = \frac{F_B}{|q|v} = \frac{5.4 \times 10^{-15} \text{ N}}{(1.6 \times 10^{-19} \text{ C})(2.1 \times 10^{7} \text{ m/s})} = 1.61 \times 10^{-3} \text{ T} = 16.1 \text{ G}$$

(b) For $\theta = 45°$, the magnetic field is

$$B = \frac{F_B}{|q|v\sin\theta} = \frac{5.4 \times 10^{-15} \text{ N}}{(1.6 \times 10^{-19} \text{ C})(2.1 \times 10^{7} \text{ m/s})\sin 45°} = \sqrt{2} \, B_{min} = 22.7 \text{ G}$$

ASSESS The magnetic force on the electron is very tiny. The magnetic field required to produce this force can be compared to the Earth's magnetic field, which is about 1 G.

19. **INTERPRET** In this problem we are asked to find the magnetic force exerted on a moving proton.

DEVELOP The magnetic force on a charge q moving with velocity $\vec{v}$ is given by Equation 26.1: $\vec{F}_B = q\vec{v} \times \vec{B}$. The magnitude of $\vec{F}_B$ is

$$F_B = |\vec{F}_B| = |q\vec{v} \times \vec{B}| = |q|vB\sin\theta$$

The charge of the proton is $q = 1.6 \times 10^{-19}$ C.

EVALUATE **(a)** When $\theta = 90°$, the magnitude of the magnetic force is

$$F_B = qvB\sin 90° = (1.6 \times 10^{-19} \text{ C})(2.5 \times 10^{5} \text{ m/s})(0.5 \text{ T}) = 2.0 \times 10^{-14} \text{ N}$$

(b) When $\theta = 30°$, the force is

$$F_B = qvB\sin 30° = (1.6 \times 10^{-19} \text{ C})(2.5 \times 10^{5} \text{ m/s})(0.5 \text{ T})\sin 30° = 1.0 \times 10^{-14} \text{ N}$$

(c) When $\theta = 0°$, the force is $F_B = qvB\sin 0° = 0$.

ASSESS The magnetic force is a maximum ($F_{B,max} = |q|vB$) when $\theta = 90°$ and a minimum ($F_{B,min} = 0$) when $\theta = 0°$.

21. **INTERPRET** This problem is about the speed of a given charge if it is to pass through the velocity selector undeflected.

DEVELOP In the presence of both electric and magnetic fields, the force on a moving charge is (see Equation 26.2):

$$\vec{F} = \vec{F}_E + \vec{F}_B = q(\vec{E} + \vec{v} \times \vec{B})$$

The condition for a charged particle to pass undeflected through the velocity selector is $\vec{F}_E = -\vec{F}_B$, or $v = E/B$.

EVALUATE Substituting the values given in the problem statement, we obtain

$$v = \frac{E}{B} = \frac{24 \text{ kN/C}}{0.06 \text{ T}} = 400 \text{ km/s}$$

ASSESS Only particles with this speed would pass undeflected through the mutually perpendicular fields; at any other speed, particles would be deflected.

Section 26.3 Charged Particles in Magnetic Fields

23.   **INTERPRET**   This problem is about an electron undergoing circular motion in a uniform magnetic field. We want to know its period, or the time it takes to complete one revolution.

DEVELOP   Using Equation 26.3, the radius of the circular motion is $r = mv/|e| B$. Therefore, the period of the motion is

$$T = \frac{2\pi r}{v} = \frac{2\pi}{v} \frac{mv}{|e|B} = \frac{2\pi m}{|e|B}$$

EVALUATE   Substituting the values given in the problem statement, we find the period to be

$$T = \frac{2\pi m}{|e|B} = \frac{2\pi(9.11 \times 10^{-31} \text{ kg})}{(1.6 \times 10^{-19} \text{ C})(10^{-4} \text{ T})} = 358 \text{ ns}$$

ASSESS   The period is independent of the electron's speed and orbital radius. However, it is inversely proportional to the magnetic field strength.

25.   **INTERPRET**   In this problem electrons and protons have the same kinetic energy, and they are undergoing circular motion in the same magnetic field. We want to compare the radii of their orbits.

DEVELOP   From Equation 26.3, the radius of the circular motion is $r = mv/qB$. For a non-relativistic particle, $K = \frac{1}{2}mv^2$, or $v = \sqrt{2K/m}$. Therefore,

$$r = \frac{mv}{qB} = \frac{m}{qB}\sqrt{\frac{2K}{m}} = \frac{\sqrt{2Km}}{qB}$$

EVALUATE   For protons and electrons having the same kinetic energy and in the same magnetic field, the ratio of their radii is

$$\frac{r_p}{r_e} = \sqrt{\frac{m_p}{m_e}} = \sqrt{1836} \approx 43$$

ASSESS   The fact that $r \sim m^{1/2}$ implies that heavier particles are more difficult to bend.

27.   **INTERPRET**   This problem is about two protons undergoing circular motion and colliding head-on.

DEVELOP   In an elastic head-on collision between particles of equal mass, the particles exchange velocities ($v_{1f} = v_{2i}$ and $v_{2f} = v_{1i}$). Moving in a plane perpendicular to the magnetic field, each proton describes a different circle of radius $r = mv/eB$, with period (which is independent of r and v) of $T = 2\pi m/eB$.

EVALUATE   Substituting the values given, we obtain

$$T = \frac{2\pi m}{eB} = \frac{2\pi(1.67 \times 10^{-27} \text{ kg})}{(1.6 \times 10^{-19} \text{ C})(5 \times 10^{-2} \text{ T})} = 1.31 \text{ } \mu s$$

After one period, each proton would be back at the site of the collision, and could collide again.

ASSESS   In solving the problem, we have ignored Coulomb repulsion between the two protons.

Section 26.4 The Magnetic Force on a Current

29.   **INTERPRET**   In this problem we are asked about the magnetic field strength of a straight current-carrying wire.

DEVELOP   Equation 26.5 gives the magnetic force on a straight current carrying wire in a uniform magnetic field, $\vec{F} = I\vec{L} \times \vec{B}$. The magnitude of the force is $F = ILB\sin\theta$.

EVALUATE   (a) From the magnitude of $\vec{F}$ and the given data, we find the magnetic field strength to be

$$B = \frac{F}{IL\sin\theta} = \frac{0.31 \text{ N/m}}{(15 \text{ A})\sin 25°} = 48.9 \text{ mT}$$

(b) By placing the wire perpendicular to the field ($\sin\theta = 1$) a maximum force per unit length of

$$\frac{F}{L} = IB = (15 \text{ A})(48.9 \text{ mT}) = 0.734 \text{ N/m}$$

could be attained.

ASSESS   From the definition of cross product between two vectors, we see that the magnetic force $\vec{F}$ is perpendicular to both the current direction $\vec{L}$ and the magnetic field $\vec{B}$, and the magnitude of $\vec{F}$ is a maximum when $\vec{L} \perp \vec{B}$.

31.   **INTERPRET**   Two forces are involved in this problem: the magnetic force and the gravitational force. We want to find the magnetic field strength such that the two forces are equal in magnitude.

DEVELOP   A magnetic force equal in magnitude to the weight of the wire requires that

$$F_B = F_g \quad \rightarrow \quad ILB = mg$$

since the wire is perpendicular to the field.

EVALUATE   The equation above implies that the field strength is

$$B = \frac{mg}{IL} = \frac{(m/L)g}{IL} = \frac{(75 \text{ g/m})(9.8 \text{ m/s}^2)}{6.2 \text{ A}} = 0.119 \text{ T}$$

ASSESS   This field strength is much greater than the typical value of 0.01 T produced by a bar magnet.

Section 26.5 Origins of the Magnetic Field

33.   **INTERPRET**   This problem is about the magnetic field produced by a current-carrying loop.

DEVELOP   As shown in Example 26.4, the magnetic field at a point P on the axis of a circular loop of radius a carrying current I is (Equation 26.9):

$$B = \frac{\mu_0 I a^2}{2(x^2 + a^2)^{3/2}}$$

EVALUATE   (a) At the center, $x = 0$, and the field strength is

$$B = \frac{\mu_0 I}{2a} = \frac{(4\pi \times 10^{-7} \text{ N/A}^2)(650 \text{ mA})}{2(1 \text{ cm})} = 40.8 \ \mu\text{T}$$

(b) At $x = 20$ cm on the axis, we have

$$B = \frac{\mu_0 I a^2}{2(x^2 + a^2)^{3/2}} = \frac{(4\pi \times 10^{-7} \text{ N/A}^2)(650 \text{ mA})(1 \text{ cm})^2}{2[(20 \text{ cm})^2 + (1 \text{ cm})^2]^{3/2}} = 5.09 \text{ nT}$$

ASSESS   The direction of the field is along the axis. The field strength is greatest at the center of the loop since the point is closest to the current distribution.

35.   **INTERPRET**   This problem is about the magnetic field produced by a current-carrying wire.

DEVELOP   Equation 26.10 gives the magnetic field strength of an infinitely long straight wire:

$$B = \frac{\mu_0 I}{2\pi r}$$

The expression is applicable if r is much smaller compared to the length of the wire.

EVALUATE   Using the equation above, we find the current to be

$$I = \frac{2\pi r B}{\mu_0} = \frac{2\pi (1.2 \text{ cm})(67 \ \mu\text{T})}{4\pi \times 10^{-7} \text{ N/A}^2} = 4.02 \text{ A}$$

ASSESS   The current is proportional to the magnetic field strength. Note that the magnetic field lines are concentric circles, as illustrated in Fig. 26.18.

Section 26.6 Magnetic Dipoles

37.   **INTERPRET**   This problem is about the magnetic field strength produced by a magnetic dipole.

DEVELOP   As discussed in Section 26.6, the magnetic field strength at a distance x along the axis of a magnetic dipole moment μ is

$$B = \frac{\mu_0}{2\pi} \frac{\mu}{x^3}$$

The magnetic poles are two such points on the Earth's surface.

EVALUATE   Substituting the values given, we find the field strength to be

$$B = \frac{\mu_0}{2\pi} \frac{\mu}{R_E^3} = \left(2 \times 10^{-7} \frac{\text{N}}{\text{A}^2}\right) \frac{(8 \times 10^{22} \text{A} \cdot \text{m}^2)}{(6.37 \times 10^6 \text{ m})^3} = 6.20 \times 10^{-5} \text{ T} = 0.62 \text{ G}$$

ASSESS   The main component of the Earth's magnetic field is a dipole. The magnetic field near the surface of the Earth is about 0.5 G.

39.   **INTERPRET**   This problem is about an electric motor. We are asked to find the magnetic field strength, given the torque and the current.

DEVELOP   The maximum torque on a plane circular coil follows from Equations 26.14:

$$\tau_{max} = \mu B = NI\pi R^2 B$$

EVALUATE   Solving for B using the above equation, we obtain

$$B = \frac{\tau_{max}}{\mu} = \frac{\tau_{max}}{NI\pi R^2} = \frac{1.2 \text{ N} \cdot \text{m}}{(250)(3.3 \text{ A})\pi(3.1 \text{ cm})^2} = 482 \text{ mT}$$

ASSESS   This field strength is rather high, but reasonable for producing the torque needed to rotate the motor.

Section 26.8 Ampère's Law

41.   **INTERPRET**   This problem involves application of Ampère's law since current is encircled by a loop.

DEVELOP   Applying Ampère's law (Equation 26.16) to the loop shown in Fig. 26.42 (going clockwise) gives

$$\oint \vec{B} \cdot d\vec{r} = \mu_0 I_{encircled} \rightarrow B(2r) = (75 \text{ }\mu\text{T})(2 \times 0.2 \text{ m}) = (4\pi \times 10^{-7} \text{ N/A}^2)I_{encircled}$$

Note that the sides of the loop perpendicular to $\vec{B}$ give no contribution to the line integral.

EVALUATE   Thus, the encircled current is

$$I_{encircled} = \frac{(75 \text{ }\mu\text{T})(2 \times 0.2 \text{ m})}{4\pi \times 10^{-7} \text{ N/A}^2} = 23.9 \text{ A}$$

ASSESS   As explained in the text, the current flows along the boundary surface between the regions of oppositely directed $\vec{B}$, positive into the page in Fig. 26.42, for clockwise circulation around the loop.

43.   **INTERPRET**   This problem is about the magnetic field of a long current-carrying wire of radius R. We want to show that both Equations 26.17 and 26.18 lead to the same result when $r = R$.

DEVELOP   Equation 26.17, $B = \mu_0 I/2\pi r$, holds for $r \geq R$ while Equation 26.18, $B = \mu_0 Ir/2\pi R^2$, holds for $r \leq R$.

EVALUATE   When $r = R$, both equations give $B = \mu_0 I/2\pi R$.

ASSESS   We expect both equations to give the same result for the magnetic field since the encircled current at $r = R$ is $I_{encircled} = I$ in both cases.

PROBLEMS

45.   **INTERPRET**   This problem is about magnetic force exerted on a moving charged particle.

DEVELOP   The magnetic force on a moving charge can be calculated using Equation 26.1: $\vec{F} = q\vec{v} \times \vec{B}$. The force is a cross product of $\vec{v}$ and $\vec{B}$.

EVALUATE   (a) From Equation 26.1, we find the force to be

$$\vec{F} = q\vec{v} \times \vec{B} = (50 \text{ }\mu\text{C})[(5 \text{ m/s})\hat{i} + (3.2 \text{ m/s})\hat{k}] \times [(9.4 \text{ T})\hat{i} + (6.7 \text{ T})\hat{j}]$$
$$= (50 \times 10^{-6} \text{ N})(5 \times 6.7\hat{k} + 3.2 \times 9.4\hat{j} - 3.2 \times 6.7\hat{i})$$
$$= (-1.072\hat{i} + 1.504\hat{j} + 1.675\hat{k}) \times 10^{-3} \text{ N}$$

The magnitude and direction can be found from the components, if desired.

(b) The dot products $\vec{F} \cdot \vec{v}$ and $\vec{F} \cdot \vec{B}$ are proportional to $(-1.072)(5) + (1.675)(3.2) = 0$ and $(-1.072)(9.4) + (1.504)(6.7) = 0$, respectively, since the cross product of two vectors is perpendicular to each factor. (We did not round off the components of $\vec{F}$, so that the vanishing of the dot products could be exactly confirmed.)

ASSESS The fact that the product $\vec{F} \cdot \vec{v}$ vanishes can also be shown as follows:

$$\vec{F} \cdot \vec{v} = (q\vec{v} \times \vec{B}) \cdot \vec{v} = q(\vec{v} \times \vec{v}) \cdot \vec{B} = 0$$

where we have used the vector identity: $(\vec{A} \times \vec{B}) \cdot \vec{C} = (\vec{C} \times \vec{A}) \cdot \vec{B}$.

47. **INTERPRET** This problem is about magnetic force exerted on a moving charged particle. We are interested in the angle between $\vec{v}$ and $\vec{B}$.

    DEVELOP The magnetic force on a moving charge can be calculated using Equation 26.1: $\vec{F} = q\vec{v} \times \vec{B}$. The magnitude of the force is

    $$F = |\vec{F}| = |q\vec{v} \times \vec{B}| = qvB\sin\theta$$

    where θ is the angle between $\vec{v}$ and $\vec{B}$.

    EVALUATE Equation 26.1 gives

    $$\sin\theta = \frac{F}{qvB} = \frac{\sqrt{(2.5^2) + (7.0)^2}\ \mu N}{(1.4\ \mu C)(185\ m/s)\sqrt{(42^2) + (-15)^2}\ mT} = 0.644$$

    Then $\theta = 40.1°$ or $140°$. Both values are possible since $\sin\theta = \sin(180° - \theta)$.

    ASSESS The result is reasonable since $|\sin\theta| \leq 1$.

49. **INTERPRET** This problem is about a charged particle undergoing circular motion in a magnetic field, and we want to express the radius of the orbit in terms of its charge, mass, kinetic energy, and the magnetic field strength.

    DEVELOP From Equation 26.3, the radius of the circular motion is $r = mv/qB$. For a non-relativistic particle, $K = \frac{1}{2}mv^2$, or $v = \sqrt{2K/m}$.

    EVALUATE Therefore, the radius of the orbit is

    $$r = \frac{mv}{qB} = \frac{m}{qB}\sqrt{\frac{2K}{m}} = \frac{\sqrt{2Km}}{qB}$$

    ASSESS Our result indicates that the radius is proportional to $\sqrt{K}$, or v. Thus, the greater the kinetic energy of the particle, the larger its radius.

51. **INTERPRET** In this problem an electron is moving in a magnetic field with a velocity that has both parallel and perpendicular components to the magnetic field. The path is a spiral.

    DEVELOP The radius depends only on the perpendicular velocity component, $r = \frac{mv_{\perp}}{eB}$. On the other hand, the distance moved parallel to the field is $d = v_{\parallel}T$, where T is the cyclotron period.

    EVALUATE (a) The radius of the spiral path is

    $$r = \frac{mv_{\perp}}{eB} = \frac{(9.11 \times 10^{-31}\ kg)(3.1 \times 10^6\ m/s)}{(1.6 \times 10^{-19}\ C)(0.25\ T)} = 70.6\ \mu m$$

    (b) Since $v_{\parallel} = v_{\perp}$, the distance moved parallel to the field is

    $$d = v_{\parallel}T = v_{\perp}\left(\frac{2\pi m}{eB}\right) = 2\pi\left(\frac{mv_{\perp}}{eB}\right) = 2\pi r = 2\pi(70.6\ \mu m) = 444\ \mu m$$

    ASSESS Since motion parallel to the field is not affected by the magnetic force, with $v_{\parallel} = v_{\perp}$, the distance traveled in $t = T$ along the direction of the field is simply $d = 2\pi r$.

53. **INTERPRET** This problem is about the current that's needed to produce a magnetic force to balance the current-carrying rod against gravity.

    DEVELOP An upward magnetic force on the rod equal (in magnitude) to its weight is the minimum force necessary. Since the rod is perpendicular to the magnetic field, the magnetic force is $\vec{F} = I\vec{L} \times \vec{B}$, or $F = ILB\sin\theta = ILB$.

    EVALUATE (a) The minimum current is obtained by setting $ILB = mg$, or

    $$I = \frac{mg}{LB} = \frac{(0.018\ kg)(9.8\ m/s^2)}{(0.2\ m)(0.15\ T)} = 5.88\ A$$

    (b) The force is upward for current flowing from A to B, consistent with the right-hand rule for the cross product.

ASSESS   A current of 5.88 A sounds reasonable. The weight of the rod is about $F_g = mg = 0.176$ N. To support the weight with an upward magnetic force, we need a strong enough magnetic field and big enough current such that $ILB \geq mg$.

55.   **INTERPRET**   In this problem the magnetic field exerts a torque on a current-carrying loop, causing the normal of the loop to make an angle with the field.

DEVELOP   The magnetic torque exerted on a dipole moment $\vec{\mu}$ by the magnetic field $\vec{B}$ is given by Equation 26.14: $\vec{\tau} = \vec{\mu} \times \vec{B}$. The magnitude of $\vec{\tau}$ is

$$\tau = |\vec{\tau}| = \mu B \sin\theta$$

where $\vec{\mu} = IA\hat{n}$, with $A = \pi R^2$ being the area of the loop, and $\hat{n}$ the unit vector in the normal direction of the plane of the loop.

EVALUATE   Substituting the values given in the problem, we find the field strength to be

$$B = \frac{\tau}{\mu\sin\theta} = \frac{\tau}{I\pi R^2 \sin\theta} = \frac{0.015 \text{ N}\cdot\text{m}}{(12 \text{ A})\pi(5 \text{ cm})^2 \sin 25°} = 377 \text{ mT}$$

ASSESS   The torque tends to align the magnetic dipole moment with the magnetic field. It is at a maximum, $\tau_{max} = \mu B$, when $\theta = 90°$.

57.   **INTERPRET**   This problem is about the change in potential energy of a magnetic dipole moment.

DEVELOP   From Equation 26.15, the potential energy of a magnetic dipole in a magnetic field is $U = -\vec{\mu}\cdot\vec{B}$. Therefore, the energy required to reverse the orientation of a proton's magnetic moment from parallel to antiparallel to the applied magnetic field is $\Delta U = 2\mu B$.

EVALUATE   Substituting the value given, we find the energy to be

$$\Delta U = 2\mu B = 2(1.41\times 10^{-26} \text{ A}\cdot\text{m}^2)(7.0 \text{ T}) = 1.97\times 10^{-25} \text{ J} = 1.23\times 10^{-6} \text{ eV}$$

ASSESS   The potential energy of a dipole moment is a minimum $(U = -\mu B)$ when it is parallel to the magnetic field, but a maximum $(U = +\mu B)$ when it is antiparallel to the field. Positive work must be done to flip the dipole.

59.   **INTERPRET**   This problem is about the magnetic field at the center of a current-carrying coil. We are interested in the number of turns the coil has.

DEVELOP   Equation 26.9 (see Example 26.4) can be modified for N turns of wire so at the center of a flat circular coil, the magnetic field is $B = N\mu_0 I/2a$. This equation can be used to solve for N.

EVALUATE   The number of turns in the coil is

$$N = \frac{2aB}{\mu_0 I} = \frac{(0.2 \text{ m})(2.3 \text{ mT})}{(4\pi\times 10^{-7} \text{ N/A}^2)(0.5 \text{ A})} = 732$$

ASSESS   This is quite a lot of turns. The same field strength is obtained with one turn of coil carrying a current $NI = (732)(0.5 \text{ A}) = 366$ A.

61.   **INTERPRET**   This problem asks for the direction of the magnetic compass points when it is placed directly underneath a current-carrying cable and subject to the influence of the Earth's magnetic field.

DEVELOP   A compass needle (small dipole magnet) is free to rotate in a horizontal plane until it is aligned with the direction of the total horizontal magnetic field (see Equation 26.14). On the other hand, a long, straight wire (the power line) carrying a current of 500 A parallel to the ground, in the direction of magnetic north, produces a magnetic field, at a distance 10 m below, to the west, with magnitude given by Equation 26.17 (see diagram):

$$B_y = \frac{\mu_0 I}{2\pi r} = \frac{(4\pi\times 10^{-7} \text{ N/A}^2)(500 \text{ A})}{2\pi(10 \text{ m})} = 10^{-5} \text{ T} = 0.1 \text{ G}$$

The horizontal component of the Earth's magnetic field is $B_x = 0.24$ G.

EVALUATE   The compass needle will point

$$\theta = \tan^{-1}\left(\frac{B_y}{B_x}\right) = \tan^{-1}\left(\frac{0.1}{0.24}\right) = 22.6°$$

west of magnetic north.

ASSESS   The compass needle points in the direction of the magnetic field, which in this case, is the vector sum of the Earth's magnetic field and the field of a current-carrying power line.

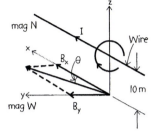

63.   **INTERPRET**   Our system consists of three parallel wires carrying current in the same direction. We are interested in the force on each wire.

DEVELOP   Since $L = 4.6$ m $\gg 3.5$ cm $= d$, the magnitude of the attractive force between each pair of wires is approximated by Equation 26.11: $F = \mu_0 I^2 L/2\pi d$.

EVALUATE   The forces on a given wire, due to the other two, make an angle of $60°$, as shown, so the net force is

$$F = |\vec{F_1} + \vec{F_2}| = 2F\cos 30° = \frac{\sqrt{3}\mu_0 I^2 L}{2\pi d} = \frac{\sqrt{3}(4\pi \times 10^{-7} \text{ N/A}^2)(20 \text{ A})^2(4.6 \text{ m})}{2\pi(3.5 \text{ cm})} = 18.2 \text{ mN}$$

ASSESS   The net force exerted on one wire by the other two wires is attractive since the currents flow in the same direction.

65.   **INTERPRET**   The system is a long conducting rod having a non-uniform current density. We are interested in the magnetic field strength both inside and outside the rod. The problem involves Ampere's law.

DEVELOP   The magnetic field of a long conducting rod is approximately cylindrically symmetric, as discussed in Section 26.8. The magnetic field can be found by using Ampere's law:

$$\oint \vec{B} \cdot d\vec{r} = \mu_0 I_{\text{encircled}}$$

EVALUATE   (a) Inside the rod, Ampère's law can be used to find the field, as in Example 26.8, by integrating the current density over a smaller cross-sectional area, corresponding to $I_{\text{encircled}}$ for an amperian loop with $r \le R$. Then,

$$I_{\text{encircled}} = \int_0^r J_0(r/R)2\pi r\, dr = \frac{2\pi J_0 r^3}{3R}$$

Here, area elements were chosen to be circular rings of radius r, thickness dr, and area $dA = 2\pi r\, dr$. Ampère's law gives $2\pi r B = \mu_0 I_{\text{encircled}}$, or $B = \mu_0 J_0 r^2/3R$, for $r \le R$.

(b) The field outside ($r \ge R$) is given by Equation 26.17, and has direction circling the rod according to the right-hand rule. The total current can be related to the current density by integrating over the cross-sectional area of the rod,

$$I = \int \vec{J} \cdot d\vec{A} = \int_0^R J_0(r/R)2\pi r\, dr = \frac{2\pi J_0 R^2}{3}$$

Equation 26.17 can then be written as $B = \frac{\mu_0 I}{2\pi r} = \frac{\mu_0 J_0 R^2}{3r}$ for $r \ge R$.

ASSESS   The magnetic field increases as r^2 inside the rod, but decreases as $1/r$ outside the rod. At $r = R$, both expressions give the same result: $B(r = R) = \mu_0 J_0 R/3$.

67.   **INTERPRET**   The system consists of two large current-carrying plates. The current distributions have plane symmetry, and we apply the superposition principle.

DEVELOP   Equation 26.19, $B_1 = \mu_0 J_s/2$, and Fig. 26.32 give the magnitude and direction of the individual fields. The total field is the superposition of fields due to two (approximately infinite) flat parallel current sheets.

EVALUATE   (a) Between the plates, both fields are in the negative y direction. Thus,

$$\vec{B}_{\mathrm{btw}} = -\frac{1}{2}\mu_0 J_{s,1}\hat{j} - \frac{1}{2}\mu_0 J_{s,2}\hat{j} = -\mu_0 J_s \hat{j}$$

(b) Outside the plates, the fields are in opposite directions, and thus cancel, $\vec{B}_{\mathrm{out}} = 0$.

ASSESS   The situation here is analogous to the electric field of two infinite planes carrying opposite charges (a capacitor). The field is non-zero in between the plates but vanishes outside.

69.   **INTERPRET**   This problem deals with the magnetic field of a current-carrying solenoid. We are interested in the number of turns the solenoid has and the power it dissipates.

DEVELOP   A length-diameter ratio of 10 to 1 is large enough for Equation 26.20, $B = \mu_0 nI$, to be a good approximation to the field near the solenoid's center. This is the equation we shall use to calculate the number of turns. On the other hand, the power the solenoid dissipates is given by $P = I^2 R$.

EVALUATE   (a) Using Equation 26.20, we find the number of turns per unit length to be

$$n = \frac{B}{\mu_0 I} = \frac{10^{-1}\ \mathrm{T}}{(4\pi \times 10^{-7}\ \mathrm{N/A^2})(35\ \mathrm{A})} = 2.27 \times 10^3\ \mathrm{m^{-1}}$$

This implies that the total number of turns is $N = nL = 2.27 \times 10^3$.

(b) A direct current is used in the solenoid, so the power dissipated (Joule heat) is

$$P = I^2 R = (35\ \mathrm{A})^2 (2.7\ \Omega) = 3.31\ \mathrm{kW}$$

ASSESS   That's a lot of turns in one meter. So the solenoid is very tightly wound to produce such a strong field at its center.

71.   **INTERPRET**   In this problem we are asked to derive to expression for the magnetic field of a solenoid by treating it as being made up of a large number of current loops.

DEVELOP   Consider a small length of solenoid, dx, to be like a coil of radius R and current $nI\,dx$. Using Equation 26.9, the axial field is

$$dB = \frac{\mu_0 (nI\,dx) R^2}{2(x^2 + R^2)^{3/2}}$$

with direction along the axis according to the right-hand rule. For a very long solenoid, we can integrate this from $x = -\infty$ to $x = +\infty$ to find the total field.

EVALUATE   Integrating over dx from $x = -\infty$ to $x = +\infty$, we find the magnetic field to be

$$B_{\mathrm{sol}} = \frac{\mu_0 nI R^2}{2}\int_{-\infty}^{\infty}\frac{dx}{(x^2 + R^2)^{3/2}} = \frac{\mu_0 nI R^2}{2}\left.\frac{x}{R^2\sqrt{x^2 + R^2}}\right|_{-\infty}^{\infty} = \mu_0 nI$$

This is the expression given in Equation 26.20.

ASSESS   For a finite solenoid, a similar integral gives the field at any point on the axis only, for example, at the center of a solenoid of length L,

$$B(0) = \frac{\mu_0 nIL}{\sqrt{L^2 + 4R^2}}$$

73.   **INTERPRET**   This problem is about the magnetic field of a coaxial cable. Ampère's law can be applied since the current distribution has line symmetry.

DEVELOP   For a long, straight cable, the magnetic field can be found from Ampère's law. The field lines are cylindrically symmetric and form closed loops, hence must be concentric circles, which we also choose as amperian loops. Take positive circulation counterclockwise so that positive current is out of the page. Then

$$\oint \vec{B} \cdot d\vec{r} = 2\pi r B = \mu_0 I_{\mathrm{encircled}}$$

Assume that the current density in each conductor is uniform; i.e., the current is proportional to the cross-sectional area. We may calculate $I_{encircled}$ in four regions of space.

EVALUATE   (a) For $r \le R_a$, the encircled current is

$$I_{encircled} = I\left(\frac{\pi r^2}{\pi R_a^2}\right) = I\left(\frac{r^2}{R_a^2}\right)$$

so $B = \mu_0 Ir/2\pi R_a^2$. Numerically, $R_a = 0.5$ mm, $R_b = 5$ mm, $R_c = 5.2$ mm, and $I = 0.1$ A, so inside the inner conductor when $r = 0.1$ mm $\le R_a$, the magnetic field is

$$B = \frac{\mu_0 Ir}{2\pi R_a^2} = \frac{(4\pi \times 10^{-7}\ T\cdot m/A)(0.1\ A)(0.1\ mm)}{2\pi (0.5\ mm)^2} = 8\ \mu T$$

(b) For $R_a \le r \le R_b$, the current encircled is $I_{encircled} = I$, so $B = \mu_0 I/2\pi r$. Therefore, for $R_a \le r = 5$ mm $\le R_b$, the magnetic field is

$$B = \frac{\mu_0 I}{2\pi r} = \frac{(4\pi \times 10^{-7}\ T\cdot m/A)(0.1\ A)}{2\pi (5\ mm)} = 4\ \mu T$$

(c) For $r \ge R_c$, $I_{encircled} = 0$, so the magnetic field is $B = 0$. Thus, at $r = 2$ cm $\ge R_c$, $B = 0$.

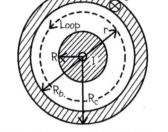

ASSESS   For completeness, let's calculate the magnetic field in the region $R_b \le r \le R_c$. The current encircled by the amperian loop is

$$I_{encircled} = I - I\frac{\pi(r^2 - R_b^2)}{\pi(R_c^2 - R_b^2)} = I\left(\frac{R_c^2 - r^2}{R_c^2 - R_b^2}\right)$$

and the magnetic field is

$$B(r) = \frac{\mu_0 I}{2\pi(R_c^2 - R_b^2)}\left(\frac{R_c^2}{r} - r\right)$$

The outer radius R_c is the inner radius plus the thickness of the outer conductor. For $r = R_b$ and $r = R_c$, the above equation gives $B = \mu_0 I/2\pi R_b$ and $B = 0$, respectively, in agreement with the results obtained before.

75.   **INTERPRET**   This problem is about the magnetic field of a current-carrying conducting bar. Symmetry holds approximately in certain limits.

DEVELOP   Very near the conductor, but far from any edge, the field is like that due to a large current sheet. On the other hand, very far from the conductor, the field is like that due to a long, straight wire.

EVALUATE   (a) Approximating the bar by a large current sheet with $J_s = I/w$, Equation 26.19 gives $B \approx \frac{\mu_0 I}{2w}$.
(b) Approximating the bar by a long, straight wire. Equation 26.17 gives $B \approx \mu_0 I/2\pi r$.

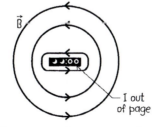

I out of page

ASSESS   The conductor exhibits different approximate symmetries, depending on where the field point is chosen.

77.   **INTERPRET**   The system is a solid conducting wire having a non-uniform current density. We are interested in the magnetic field strength both inside and outside the wire. The problem involves Ampère's law.

DEVELOP   The total current in the wire can be obtained by integrating the current density over the cross sectional area. The magnetic field of a long conducting wire is approximately cylindrically symmetric, as discussed in Section 26.8. The magnetic field can be found by using Ampère's law:

$$\oint \vec{B} \cdot d\vec{r} = \mu_0 I_{encircled}$$

EVALUATE   (a) Using thin rings as the area elements with $dA = 2\pi r dr$, the total current in the wire (z axis out of the page) is

$$I = \int_0^R J \, dA = \int_0^R J_0\left(1 - \frac{r}{R}\right)2\pi r \, dr = 2\pi J_0 \left(\frac{r^2}{2} - \frac{r^3}{3R}\right)\bigg|_0^R = \frac{1}{3}\pi R^2 J_0$$

(b) A concentric amperian loop outside the wire encircles the total current, so Ampère's law gives

$$2\pi r B = \mu_0 I = \mu_0 \left(\frac{1}{3}\pi R^2 J_0\right)$$

or $B = \mu_0 J_0 R^2/6r$.

(c) Inside the wire, Ampère's law gives $2\pi r B = \mu_0 I_{encircled}$. The calculation in part **(a)** shows that within a loop of radius $r < R$,

$$I_{encircled} = \int_0^r J dA = 2\pi J_0\left(\frac{r^2}{2} - \frac{r^3}{3R}\right)\bigg|_0^r = \pi J_0 r^2\left(1 - \frac{2r}{3R}\right)$$

Therefore, $B = \frac{1}{2}\mu_0 J_0 r(1 - \frac{2r}{3R})$.

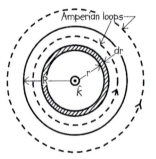

ASSESS   At $r = R$, both equations give $B = \mu_0 J_0 R/6$. Since from part **(a)** $J_0 = 3I/\pi R^2$, the magnetic field can also be written as $B = \mu_0 I/2\pi R$. The form is the same as that shown in Equation 26.17.

79.   **INTERPRET**   The problem is about magnetic dipole moment formed by a current-carrying wire of length L.

DEVELOP   The number of turns of radius r that can be formed from a wire of length L is

$$N = \frac{L}{2\pi r} \quad \rightarrow \quad r = \frac{L}{2\pi N}$$

The magnitude of the magnetic dipole moment of such a coil is $\mu = NIA = NI\pi r^2$.

EVALUATE   (a) Substituting the expression for r into the equation for μ, we find

$$\mu = NI\pi\left(\frac{L}{2\pi N}\right)^2 = \frac{IL^2}{4\pi N}$$

(b) The magnetic dipole moment is clearly a maximum when N is a minimum, and the smallest value of N is, of course, one.

ASSESS   Since the radius is proportional to $1/N$, the area of the circular coil is proportional to $1/N^2$. This makes sense because with the length L kept fixed, the greater the number of turns, the smaller the area of the coil. Thus, with $\mu = NIA = NI\pi r^2$, we see that $\mu \sim 1/N$.

81. **INTERPRET** Two forces are involved in this problem: gravitational and magnetic. At equilibrium, the two forces cancel exactly.

DEVELOP If the height h is small compared to the length of the rods, we can use Equation 26.11 for the repulsive magnetic force between the horizontal rods (upward on the top rod)

$$F_B = \frac{\mu_0 I^2 L}{2\pi h}$$

The rod is in equilibrium when this equals its weight, $F_g = mg$.

EVALUATE The equilibrium condition $F_B = F_g$ gives

$$h = \frac{\mu_0 I^2 L}{2\pi mg} = \frac{(4\pi \times 10^{-7} \text{ N/A}^2)(66 \text{ A})^2 (0.95 \text{ m})}{2\pi(0.022 \text{ kg})(9.8 \text{ m/s}^2)} = 3.84 \text{ mm}$$

ASSESS The height h is indeed small compared to 95 cm. So our assumption is justified.

83. **INTERPRET** We find the spacing between two current loops at which the second derivative of the magnetic field becomes zero. We will use the results of Example 26.4, which give us the field for a single loop, along the axis of the loop, and use superposition to extend this to two loops.

DEVELOP The field along the axis of a single loop is

$$B = \frac{\mu_0 I a^2}{2(x^2 + a^2)^{3/2}}$$

(Equation 26.9.) The field due to two such loops each at a distance $x = \frac{b}{2}$ from the center is then

$$B = \frac{\mu_0 I a^2}{((x - \frac{b}{2})^2 + a^2)^{3/2}}$$

From this equation for the field at the center of these two loops, we will find the second derivative of B and then find the value of b for which the second derivative is zero.

EVALUATE

$$\frac{d^2 B}{dx^2} = \frac{15\mu_0 I a^2 \left(x - \frac{b}{2}\right)^2}{\left(a^2 + \left(x - \frac{b}{2}\right)^2\right)^{7/2}} - \frac{3\mu_0 I a^2}{\left(a^2 + \left(x - \frac{b}{2}\right)^2\right)^{5/2}} = -\frac{384\mu_0 I a^2 (a^2 - (b - 2x)^2)}{(4a^2 + (b - 2x)^2)^{7/2}}$$

At $x = 0$,

$$\frac{d^2 B}{dx^2}\bigg|_{x=0} = \frac{384\mu_0 I a^2 (a^2 - b^2)}{(4a^2 + b^2)^{7/2}} = 0 \rightarrow a^2 = b^2$$

ASSESS The spacing between the coils must equal the radius of the coils. This is a standard method of producing a uniform magnetic field in a region.

85. **INTERPRET** We find the net force on a wire due to three other wires. All wires are parallel and carry the same current, although the direction of the currents varies. We will use the equation for force between two parallel wires to find the force due to each wire, then add these force vectors.

DEVELOP The force per length between two parallel wires is $\frac{F}{L} = \frac{\mu_0 I_1 I_2}{2\pi d}$. This force is attractive if the currents flow in the same direction, so the forces on the wire at the lower left corner of the square are as shown in the figure later. We will add these three force vectors. The currents are all $I = 2.5$ A, and the distance along the edge of the square is $d = 0.15$ m.

EVALUATE

$$\frac{\vec{F_1}}{L} = -\frac{\mu_0 I^2}{2\pi d}\hat{j}, \quad \frac{\vec{F_2}}{L} = -\frac{\mu_0 I^2}{2\pi(d\sqrt{2})}\left(\frac{1}{2}\hat{i} + \frac{1}{2}\hat{j}\right) \quad \text{and} \quad \frac{\vec{F_3}}{L} = \frac{\mu_0 I^2}{2\pi d}\hat{i}$$

so the net force per length is

$$\frac{\vec{F}}{L} = \frac{\mu_0 I^2}{2\pi d}\left[\left(1 - \frac{1}{2\sqrt{2}}\right)\hat{i} - \left(1 + \frac{1}{2\sqrt{2}}\right)\hat{j}\right] = (5.39 \times 10^{-6}\text{N})\hat{i} - (1.13 \times 10^{-5}\text{N})\hat{j} \text{ or } \frac{\vec{F}}{L} = 1.25 \times 10^{-5}\text{N}$$

at an angle of 64.5° below the +x axis.

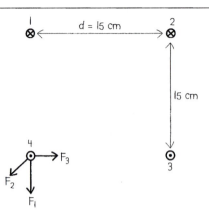

ASSESS   This is a very small force, even with this relatively large current.

87.   **INTERPRET**   We use the result from Problem 86 to find the magnetic field at the center of a square current loop. Most of the work has been done already—we just need to replace x_0 and L with values appropriate to this problem, for each side, then add the fields due to each side.

DEVELOP   We'll find the magnitude of the field due to one side of the square, then multiply by 4 since all sides contribute the same field in the same direction. The field due to one side is obtained by substituting $x_0 = -\frac{a}{2}$, $L = \frac{a}{2}$, and $y = \frac{a}{2}$ in the results from Problem 86,

$$B = \frac{\mu_0 I}{4\pi y}\left[\frac{x_0 + L}{\sqrt{(x_0 + L)^2 + y^2}} - \frac{x_0}{\sqrt{x_0^2 + y^2}}\right]$$

EVALUATE   For one side,

$$B = \frac{\mu_0 I}{4\pi \frac{a}{2}}\left[\frac{\frac{a}{2}}{\sqrt{(-\frac{a}{2}) + (\frac{a}{2})^2}}\right] = \frac{\mu_0 I}{4\pi}\left[\frac{1}{\sqrt{\frac{a}{2}}}\right] = \frac{\mu_0 I \sqrt{2}}{4\pi a}$$

The net field at the center is then $B = \frac{\mu_0 I \sqrt{2}}{\pi a}$.

ASSESS   We could extend this method to find the magnetic field due to *any* configuration of short wire segments—which is what integrating the Biot-Savart law does for us in the first place!

89.   **INTERPRET**   We find the force on a magnetic dipole located on the axis of a current loop by differentiating the potential energy. We will use the magnetic field equation developed in Example 26.4.

DEVELOP   The potential energy of a dipole in a magnetic field is given by $U = -\vec{\mu} \cdot \vec{B}$, where in this case

$$\vec{B} = \frac{\mu_0 I a^2}{2(x^2 + a^2)^{3/2}}\hat{i}$$

and $\vec{\mu} = \mu\hat{i}$. The force is $F_x = -\frac{dU}{dx}$.

EVALUATE

$$U = -\frac{\mu_0 I a^2 \mu}{2(x^2 + a^2)^{3/2}}$$

so

$$F = \frac{\mu_0 I a^2 \mu}{2}\frac{-3x}{(x^2 + a^2)^{5/2}}$$

At $x = a$,

$$F = \frac{\mu_0 I a^2 \mu}{2}\frac{-3a}{(2a^2)^{5/2}} = -\frac{3\mu_0 I \mu}{2}\frac{a^3}{4\sqrt{2}a^5} = -\frac{3\mu_0 I \mu}{8\sqrt{2}a^2}$$

ASSESS   This force is opposite the direction of x, so it is an attractive force in this case.

91. **INTERPRET** What current would it take to generate a planet-sized magnetic field? Here we approximate the Earth's geodynamo with a single loop, and find the current necessary. We use the results of Example 26.4 to find the magnetic field along the axis of a loop.

DEVELOP We will use

$$B = \frac{\mu_0 I a^2}{2(x^2 + a^2)^{3/2}}$$

where the radius of our loop is $a = 3 \times 10^6$ m and $B = 62 \times 10^{-6}$ T. We solve for I, given that $x = R_E = 6.38 \times 10^6$ m.

EVALUATE

$$I = \frac{2B(x^2 + a^2)^{3/2}}{\mu_0 a^2} = 3.8 \times 10^9 \text{ A}$$

ASSESS This seems like a very high current—but remember that's the current flowing through an entire planet, so the actual current density in this simplified model is quite low.

93. **INTERPRET** We find the magnetic field of a sheet of current by integrating a row of line currents, instead of by using Ampère's law.

DEVELOP We start with a sketch, as shown in the figure later. Since it is an infinite sheet of current, there is an equal amount of current to the right and to the left of the point P, and the vertical components of B cancel.

The field due to the line current dI is

$$dB = \frac{\mu_0 dI}{2\pi r} = \frac{\mu_0 J_s dx}{2\pi (x^2 + y^2)^{1/2}}$$

We need only the horizontal component of this field,

$$dB_x = dB \sin\theta = \frac{\mu_0 J_s dx}{2\pi (x^2 + y^2)^{1/2}} \left(\frac{y}{(x^2 + y^2)^{1/2}} \right)$$

We integrate this over all x to find the field due to an infinite sheet.

EVALUATE

$$dB_x = \frac{\mu_0 J_s y}{2\pi (x^2 + y^2)} dx \rightarrow B_x = \frac{\mu_0 J_s y}{2\pi} \int_{-\infty}^{\infty} \frac{dx}{x^2 + y^2} = \frac{\mu_0 J_s y}{2\pi} \left[\frac{1}{y} \tan^{-1}\left(\frac{x}{y} \right) \right]_{-\infty}^{\infty}$$

$$\rightarrow B_x = \frac{\mu_0 J_s}{2}$$

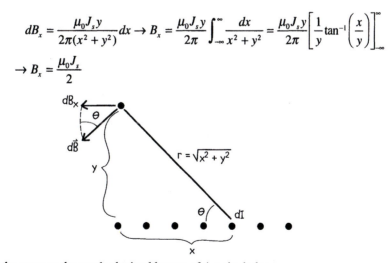

ASSESS This is the same as the result obtained by use of Ampère's law.

27 ELECTROMAGNETIC INDUCTION

EXERCISES

Sections 27.2 Faraday's Law and 27.3 Induction and Energy

15. **INTERPRET** In this problem we are asked to verify that the SI unit of the rate of change of magnetic flux is volt.

DEVELOP We first note that the left-hand-side of Equation 27.2, $\varepsilon = -d\Phi_B/dt$, represents the induced emf which has units of volt. From the definition of magnetic flux given in Equation 27.1a, we see that it has SI units of $T \cdot m^2$.

EVALUATE The reasoning above shows that the units of $d\Phi_B/dt$ are

$$T \cdot m^2/s = (N/A \cdot m)(m^2/s) = (N \cdot m/A \cdot s) = J/C = V$$

ASSESS Faraday's law relates the induced emf to the change in flux. It is the rate of change of flux, and not the flux or the magnetic field that gives rise to an induced emf.

17. **INTERPRET** This problem is about the rate of change of magnetic flux through a loop due to a changing magnetic field.

DEVELOP For a stationary plane loop in a uniform magnetic field, the magnetic flux is given by Equation 27.1b, $\Phi_B = \vec{B} \cdot \vec{A} = BA\cos\theta$. Note that the SI unit of flux, $T \cdot m^2$, is also called a weber, Wb. The rate of change of magnetic flux is $d\Phi_B/dt = \Delta\Phi_B/\Delta t$.

EVALUATE (a) The magnetic field at the beginning ($t_1 = 0$) is

$$\Phi_1 = B_1 A = \frac{1}{4}\pi d^2 B_1 = \frac{1}{4}\pi(40\text{ cm})^2(5\text{ mT}) = 6.28 \times 10^{-4}\text{ Wb}$$

(b) The magnetic field at $t_2 = 25$ ms is

$$\Phi_2 = B_2 A = \frac{1}{4}\pi d^2 B_2 = \frac{1}{4}\pi(40\text{ cm})^2(55\text{ mT}) = 6.91 \times 10^{-3}\text{ Wb}$$

(c) Since the field increases linearly, the rate of change of magnetic flux is

$$\frac{d\Phi_B}{dt} = \frac{\Delta\Phi_B}{\Delta t} = \frac{\Phi_2 - \Phi_1}{t_2 - t_1} = \frac{6.91\times10^{-3}\text{ Wb} - 0.628\times10^{-3}\text{ Wb}}{25\text{ ms}} = 0.251\text{ V}$$

From Faraday's law, this is equal to the magnitude of the induced emf, which causes a current

$$I = \frac{|\varepsilon|}{R} = \frac{0.251\text{ V}}{100\ \Omega} = 2.51\text{ mA}$$

in the loop.

$$|\varepsilon| = \left|\frac{d\Phi_B}{dt}\right| = \frac{d}{dt}(N_{coil}B_{sol}A_{sol}) = N_{coil}A_{sol}\left|\frac{dB_{sol}}{dt}\right|$$

(d) The direction must oppose the increase of the external field downward, hence the induced field is upward and I is CCW when viewed from above the loop.

ASSESS Since $\Delta\Phi_B/\Delta t = (\Delta B/\Delta t)A$ with the area of the loop kept fixed, the induced emf and hence the current scale linearly with the value $\Delta B/\Delta t$.

19. **INTERPRET** The problem asks for the number of turns the coil must have in order to produce a given emf when it is placed in a time-varying magnetic field.

$$|\varepsilon| = \left|\frac{d\Phi_B}{dt}\right| = \frac{d}{dt}(N_{coil}\, B_{sol}\, A_{sol}) = N_{coil}\, A_{sol}\left|\frac{dB_{sol}}{dt}\right|$$

DEVELOP When the coil is wrapped around the solenoid, all of the flux in the solenoid ($B_{sol}\, A_{sol}$, for a long thin solenoid) goes through each of the N_{coil} turns of the coil. Then the induced emf in the coil is

$$|\varepsilon| = \left|\frac{d\Phi_B}{dt}\right| = \frac{d}{dt}(N_{coil}\, B_{sol}\, A_{sol}) = N_{coil}\, A_{sol}\left|\frac{dB_{sol}}{dt}\right|$$

EVALUATE Substituting the values given in the problem statement, the number of turns in the coil is

$$N_{coil} = \frac{|\varepsilon|}{|dB_{sol}/dt|\, A_{sol}} = \frac{15\text{ V}}{(2.4\text{ T/s})\pi(0.1\text{ m})^2} = 199 \text{ turns}$$

ASSESS The number of turns is proportional to the induced emf, but inversely proportional to the rate of change of magnetic field.

Section 27.4 Inductance

21. **INTERPRET** This problem asks for the self-inductance of an inductor.
DEVELOP The induced emf in an inductor is given by Equation 27.5: $\varepsilon_L = -L\, dI/dt$. With ε_L and dI/dt given, we can use the equation to compute the self-inductance L.
EVALUATE From Equation 27.5, we find the self-inductance to be

$$L = \left|-\frac{\varepsilon_L}{dI/dt}\right| = \frac{40\text{ V}}{100\text{ A/s}} = 0.4 \text{ H}$$

ASSESS Our value of self-inductance is reasonable; inductances in common electronic circuits usually range from micro-henrys to several henrys.

23. **INTERPRET** The problem asks for the number of turns in the inductor which in our case is a long solenoid.
DEVELOP The inductance of a long solenoid is given by Equation 27.4 (in Example 27.6):

$$L = \mu_0 n^2 A l = \frac{\mu_0 N^2 A}{l}$$

The equation allows us to determine N, the number of turns.
EVALUATE From the equation above, we find the value of N to be

$$N = \sqrt{\frac{Ll}{\mu_0 A}} = \sqrt{\frac{(5.8\text{ mH})(15\text{ cm})}{(4\pi \times 10^{-7}\text{ H/m})\pi(1.1\text{ cm})^2}} = 1.35 \times 10^3$$

ASSESS This corresponds to $n = N/l = 1.35 \times 10^3/(15\text{cm}) = 90/\text{cm}$. That's a lot of turns in one centimeter. So the inductor is very tightly wound.

25. **INTERPRET** This problem is about the resistance in a series RL circuit.
DEVELOP The buildup of current in an RL circuit with a battery is given by Equation 27.7:

$$I(t) = I_\infty(1 - e^{-Rt/L})$$

where $I_\infty = \varepsilon_0/R$ is the final current. This is the equation we employ to solve for R.
EVALUATE Substituting the values given, one finds the resistance to be

$$R = -\frac{L}{t}\ln\left(1 - \frac{I}{I_\infty}\right) = -\frac{1.8\text{ mH}}{3.1\ \mu\text{s}}\ln(1 - 0.20) = 130\ \Omega$$

ASSESS We find R to be inversely proportional to t. This means that the greater the value of R, the shorter the time it takes for the current to increase to 20% of its final value.

Section 27.5 Magnetic Energy

27. **INTERPRET** This problem is about the magnetic energy stored in an inductor. The object of interest is the current.
DEVELOP The amount of energy stored in an inductor is given by Equation 27.9: $U = LI^2/2$. This equation allows us to determine the current I.

EVALUATE From the equation above, we find the current to be

$$I = \sqrt{\frac{2U}{L}} = \sqrt{\frac{2(50 \ \mu J)}{10 \ mH}} = 0.1 \ A$$

ASSESS Since the energy stored in the inductor is small, we expect the current (which is proportional to $\sqrt{U}$) to be a small one as well.

29. **INTERPRET** This problem is about the magnetic energy stored in a solenoid.

    DEVELOP We first note that the inductance of a solenoid is given by Equation 27.4: $L = \mu_0 n^2 Al = \mu_0 N^2 A/l$. Equation 27.9, $U = LI^2/2$, can then be used to find the amount of energy stored in the solenoid.

    EVALUATE Combining Equations 27.4 and 27.9, we find the stored energy to be

    $$U = \frac{1}{2} LI^2 = \frac{1}{2} \frac{\mu_0 N^2 AI^2}{l} = \frac{(4\pi \times 10^{-7} \ N/A^2)(500)^2 \pi (1.5 \ cm/2)^2 (65 \ mA)^2}{2(23 \ cm)} = 0.510 \ \mu J$$

    ASSESS This inductance of the solenoid is about $240 \mu H$. The stored energy is typical for a small inductor with a small current.

31. **INTERPRET** We are given the magnetic field strength and asked to compute the magnetic-energy density.

    DEVELOP Equation 27.10, $u_B = B^2/2\mu_0$, provides the connection between the magnetic energy density and the magnetic field strength.

    EVALUATE From Equation 27.10, the energy density is

    $$u_B = \frac{B^2}{2\mu_0} = \frac{(50 \ T)^2}{2(4\pi \times 10^{-7} \ N/A^2)} = 995 \ MJ/m^3$$

    ASSESS This is about 2.8% of the energy density content of gasoline, $u_{gasoline} = 35{,}000 \ MJ/m^3$.

Section 27.6 Induced Electric Fields

33. **INTERPRET** We're given the induced electric field and asked to find the rate of change of the magnetic field.

    DEVELOP The connection between the induced electric field and the rate of change of the magnetic field is given by Equation 27.11:

    $$\oint \vec{E} \cdot d\vec{r} = -\frac{d\Phi_B}{dt}$$

    The geometry of the induced electric field from the solenoid is described in Example 27.11, where

    $$2\pi r|E| = \left| -\frac{d(\pi R^2 B)}{dt} \right| = \pi R^2 \left| \frac{dB}{dt} \right|$$

    EVALUATE The rate of change of the magnetic field is

    $$\left| \frac{dB}{dt} \right| = \frac{2r \ |E|}{R^2} = \frac{2(12 \ cm)(45 \ V/m)}{(10 \ cm)^2} = 1.08 \times 10^3 \ T/s = 1.08 \ T/ms$$

    ASSESS The magnetic field is changing at a very fast rate. Note that the sign of dB/dt and the direction of the induced electric field are related by Lenz's law.

PROBLEMS

35. [NOTE: The third sentence of the problem statement should read: "Subsequently, the current increases according to $I = bt^2$, where b is a constant with the units A/s^2."]

    INTERPRET This problem is about the rate of change of magnetic field strength, given the induced current as a function of time.

    DEVELOP The flux through a stationary loop perpendicular to a magnetic field is $\Phi_B = BA$, so Faraday's law (Equation 27.2) and Ohm's law (Equation 24.5) relate this to the magnitude of the induced current:

    $$I = \frac{|\varepsilon|}{R} = \frac{|d\Phi_B/dt|}{R} = \frac{A \ |dB/dt|}{R}$$

EVALUATE   From the above expression, we find the rate of change of the magnetic field to be

$$\frac{dB}{dt} = \frac{IR}{A} = \frac{bR}{R}t^2$$

Integration yields $B(t) = (bR/3A)\, t^3$, where $B(0) = 0$ was specified.

ASSESS   Since the current increases as t^2, we expect the magnetic field strength to increase as t^3.

37.   INTERPRET   This problem is about the work done by an external agent to move a loop at a constant speed across a region with uniform magnetic field.

DEVELOP   The loop can be treated analogously to the situation analyzed in Section 27.3, under the heading "Motional EMF and Lenz's Law"; instead of exiting, the loop is entering the field region at constant velocity. All quantities have the same magnitudes, except the current in the loop is CCW instead of CW, as in Fig. 27.13.

EVALUATE   Since the applied force acts over a displacement equal to the side-length of the loop, the work done can be calculated directly:

$$W_{app} = \vec{F}_{app} \cdot \vec{l} = (IlB)l = Il^2 B$$

Since the induced current is

$$I = \frac{|\varepsilon|}{R} = \frac{|d\Phi_B/dt|}{R} = \frac{1}{R}\frac{d}{dt}(Blx) = \frac{Blv}{R}$$

the work applied by the external agent is

$$W_{app} = Il^2 B = \left(\frac{Blv}{R}\right)l^2 B = \frac{B^2 l^3 v}{R}$$

ASSESS   Alternatively, the work can be calculated from the conservation of energy:

$$P_{diss} = I^2 R = \frac{(Blv)^2}{R} \quad \rightarrow \quad W_{app} = P_{diss}t = \frac{(Blv)^2}{R}\frac{l}{v} = \frac{B^2 l^3 v}{R}$$

39.   INTERPRET   This problem is about the magnetic flux through a square due to a non-uniform field.

DEVELOP   We take elements of area to be $d\vec{A} = (2x_0 dx)\hat{k}$, which are rectangular strips parallel to the y axis. The flux through the loop is then equal to $\Phi_B = \int_{square} \vec{B} \cdot d\vec{A}$.

EVALUATE   The integral gives

$$\Phi_B = \int_{square} \vec{B} \cdot d\vec{A} = \left(\frac{B_0}{x_0^2}\right)2x_0 \int_0^{2x_0} x^2 dx = \frac{2B_0}{x_0}\frac{(2x_0)^3}{3} = \frac{16B_0 x_0^2}{3}$$

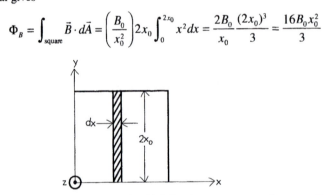

ASSESS   The quantity $16B_0 x_0^2/3$ has units of $T \cdot m^2$, which is what we expect of a magnetic flux.

41.   INTERPRET   This problem is about the current induced in a rectangular loop due to a nearby time-varying current source.

DEVELOP   The normal to the loop in Fig. 27.7 was taken to be in the direction of the magnetic field of the wire, or into the page, so the positive sense of circulation around the loop is clockwise (from the right-hand rule). Faraday's and Ohm's laws give an induced current in the loop of

$$I = \frac{|\varepsilon|}{R} = \frac{|d\Phi_B/dt|}{R}$$

The magnetic flux has been calculated in Example 27.2: $\Phi_B = \frac{\mu_0 I l}{2\pi} \ln\left(\frac{a+w}{a}\right)$.

EVALUATE   Combining the above expressions gives

$$I = \frac{|\varepsilon|}{R} = \frac{|d\Phi_B/dt|}{R} = \frac{\mu_0 l(dI/dt)}{2\pi R} \ln\left(\frac{a+w}{a}\right) = \frac{(4\pi \times 10^{-7}\ \text{N/A}^2)(6\ \text{cm})(25\ \text{A/s})}{2\pi(50\ \text{m}\Omega)} \ln\left(\frac{4.5\ \text{cm}}{1.0\ \text{cm}}\right)$$
$$= 9.02\mu A$$

By Lenz's law, the induced current is counterclockwise in the loop.

ASSESS   As the inward flux increases (because the current in the long wire is increasing), by Lenz's law, the induced current must flow in the counterclockwise direction to produce an outward flux to oppose the change.

43.   **INTERPRET**   This problem is about the current induced in a loop due to a time-varying magnetic field.

DEVELOP   Take the normal to the loop along the z axis, so that the positive sense of circulation for the induced emf and current is CCW (looking down on the x-y plane) as shown. The flux through the loop is

$$\Phi_B = \vec{B} \cdot \vec{A} = (B_0 \sin\omega t\ \hat{k}) \cdot (A\hat{k}) = B_0 A \sin\omega t$$

so the induced current is

$$I = \frac{\varepsilon}{R} = -\frac{d\Phi_B/dt}{R} = -\frac{A\,dB/dt}{R} = -\frac{AB_0\omega\cos\omega t}{R}$$

EVALUATE   (a) Using the expression above, the current at $t = 0$ is

$$I(0) = -\frac{\omega B_0 A}{R} = -\frac{(10\ \text{s}^{-1})(2\ \text{T})(150\ \text{cm}^2)}{5\ \Omega} = -60\ \text{mA}$$

The negative sign implies that the induced current is clockwise.

(b) Similarly, at $t = 0.1$ s, the induced current is

$$I(0.1\ \text{s}) = -\frac{\omega B_0 A}{R}\cos\omega t = -\frac{(10\ \text{s}^{-1})(2\ \text{T})(150\ \text{cm}^2)}{5\ \Omega}\cos[(10\ \text{s}^{-1})(0.1\ \text{s})]$$
$$= -60\ \text{mA}\cos(1\ \text{radian}) = -32.4\ \text{mA}$$

ASSESS   During the time interval in which the outward flux through the loop increases ($|\cos\omega t| > 0$, as viewed from above the loop), by Lenz's law, the induced current must flow in the clockwise direction to produce an inward flux to oppose the change.

45.   **INTERPRET**   We have a conducting coil rotating in a fixed magnetic field. The induction is due to the change in the orientation of the loop.

DEVELOP   The magnetic flux through one turn of coil is (see Example 27.5)

$$\Phi_1 = \vec{B} \cdot \vec{A} = BA\cos\theta = BA\cos(2\pi ft)$$

Therefore, using Faraday's law, the induced emf is

$$\varepsilon = -\frac{d\Phi_B}{dt} = -NBA[-2\pi f\sin(2\pi ft)]$$

The peak output voltage of an electric generator is $\varepsilon_{peak} = 2\pi fNBA$, where $A = \frac{1}{4}\pi d^2$ is the loop area in this case.

EVALUATE   Substituting the values given, we find

$$\varepsilon_{peak} = 2\pi fNBA = 2\pi(1000/60\ \text{s})(250)(0.1\ \text{T})\pi(0.1\ \text{m}/2)^2 = 20.6\ \text{V}$$

ASSESS   The value is typical of a car alternator, and this is attained when $|\sin(2\pi ft)| = 1$.

47.   **INTERPRET**   We have a circuit formed by the rails, the resistance, and the conducting bar. As the bar slides along the rails the circuit area increases and a current is induced. We are interested in the rate of work done by an external agent to move the conducting bar.

DEVELOP   To find the direction of the current in the loop, we note that since the area enclosed by the circuit, and the magnetic flux through it, are increasing, Lenz's law requires that the induced current opposes this with an upward induced magnetic field.

To answer part (b), we make use of the result obtained in Example 27.4, which analyzed the same situation. Therefore, the current in the bar was found to be $I = |\varepsilon|/R = Blv/R$.

Since this is perpendicular to the magnetic field, the magnetic force on the bar is $F_{mag} = IlB$ (to the left in Fig. 27.39). The agent pulling the bar at constant velocity must exert an equal force in the direction of v.

EVALUATE (a) From the right-hand rule, the induced current must circulate CCW. (Take the positive sense of circulation around the circuit CW, so that the normal to the area is in the direction of B, into the page.) An alternative way to determine the direction of the current is to note that the force on a (hypothetical) positive charge carrier in the bar, $\vec{F} = q\vec{v} \times \vec{B}$, is upward in Fig. 27.39, so current will circulate CCW around the loop containing the bar, the resistor, and the rails (i.e., downward in the resistor). (The force per unit positive charge is the motional emf in the bar.)

(b) The rate of work done by the external agent is $P = \vec{F} \cdot \vec{v} = IlBv = \frac{(Blv)^2}{R}$.

ASSESS The conservation of energy requires that the work done by the agent be equal to the rate energy is dissipated in the resistor (we neglected the resistance of the bar and the rails), $I^2 R = (Blv/R)^2 R = (Blv)^2/R$.

49. INTERPRET The circuit consists of the rails, the resistance, the conducting bar, and the battery. As the bar slides along the rails the circuit area increases and a current is induced. We are interested in the subsequent motion of the conducting bar.

DEVELOP To analyze the subsequent motion of the bar, we first note that the battery causes a CW current (downward in the bar) to flow in the circuit composed of the bar, resistor, and rails. (For positive circulation CW, the right-hand rule gives a positive normal to the area bounded by the circuit into the page, so that the flux $\Phi_B = \vec{B} \cdot \vec{A} = Blx$ is positive. The length of the circuit is x, as in Example 27.4.) Thus, there is a magnetic force $\vec{F}_{mag} = I\vec{l} \times \vec{B} = IlB$ to the right, which accelerates the bar in that direction. (Any other forces on the bar are assumed to cancel, or be negligible.)

EVALUATE (a) As in Example 27.4, there is an induced emf $\varepsilon_i = -d\Phi_B/dt = -Blv$, opposing the battery (with the negative sign indicating a CCW sense in the circuit). The instantaneous current is

$$I(t) = \frac{\varepsilon + \varepsilon_i}{R} = \frac{\varepsilon - Blv}{R}$$

Thus, as the speed v increases, the induced current I (and the accelerating force) decreases.

(b) Eventually $(t \to \infty)$, $I(\infty) = 0$, and the magnetic force $F_{mag} = I(\infty)lB = 0$. So the velocity v_∞ stays constant.

(c) When $I(\infty) = 0$, $\varepsilon - Blv_\infty = 0$, which implies that $v_\infty = \varepsilon/Bl$.

Although v_∞ doesn't depend on the resistance, the value of R does affect how rapidly v approaches v_∞. For large R, I charges slowly and v takes a long time to reach v_∞.

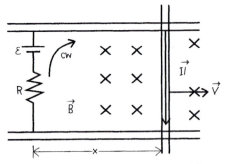

ASSESS In this problem, the equation of motion of the bar (mass m) is

$$m\frac{dv}{dt} = IlB = \frac{(\varepsilon - Blv)lB}{R} = \frac{(v_\infty - v)l^2 B^2}{R}$$

which can be rewritten as

$$\frac{dv}{v_\infty - v} = \frac{l^2 B^2}{mR} dt$$

For $v_0 = 0$, this integrates to $\ln(1 - v/v_\infty) = -l^2 B^2 t/mR$, or $v(t) = v_\infty(1 - e^{-l^2 B^2 t/mR})$. The time constant, $\tau = mR/l^2 B^2$, depends on the resistance.

51.   **INTERPRET**   The problem involves a changing magnetic field that induces an electric field. Both fields exert a force on a proton.

DEVELOP   The induced electric field inside a long thin cylindrical solenoid, whose magnetic field is increasing with time as $\vec{B} = bt\hat{k}$, (take $\hat{k}$ into the page as in Fig. 27.30), can be found by modifying the argument in Example 27.11 for radius $r < R$. The magnitude of $\vec{E}$ is

$$E = \left| -\frac{1}{2}r\frac{dB}{dt} \right| = \frac{rb}{2}$$

The induced electric field circulates CCW around the direction of $\vec{B}$, or $\vec{E} = -(bx/2)\hat{j}$ for the given point on the x axis ($x = 5$ cm, $y = z = 0$) inside the solenoid (whose axis we assume is the z axis). At $t = 0.4$ μs, the uniform magnetic field is $\vec{B} = b(0.4\ \mu s)\hat{k}$. The electromagnetic force on a proton with the given velocity is $\vec{F} = e(\vec{E} + \vec{v} \times \vec{B})$.

EVALUATE   Substituting the values given, we find the net force on the proton to be

$$\vec{F} = e(\vec{E} + \vec{v} \times \vec{B}) = (1.6 \times 10^{-19}\ \text{C})\left[\frac{1}{2}(2.1\ \text{T/ms})(5\ \text{cm})(-\hat{j}) + (4.8 \times 10^6\ \text{m/s})\hat{j} \times (2.1\ \text{T/ms})(0.4\mu s)\hat{k} \right]$$

$$= -(8.40 \times 10^{-18}\ \text{N})\hat{j} + (6.45 \times 10^{-16}\ \text{N})\hat{i}$$

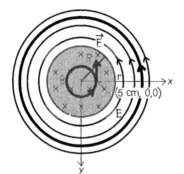

ASSESS   In the region $r < R$, the induced electric field rises linearly with r. Since the proton is moving, the net force on the proton is a vector sum of electric and magnetic force (i.e., the electromagnetic force).

53.   **INTERPRET**   This is a problem about induction in a circular loop, with the flux change caused by a changing magnetic field.

DEVELOP   The flux through a stationary loop perpendicular to a magnetic field is $\Phi_B = BA$, so Faraday's law (Equation 27.2) and Ohm's law (Equation 24.5) relate this to the magnitude of the induced current:

$$I = \frac{|\varepsilon|}{R} = \frac{|d\Phi_B/dt|}{R} = \frac{A|dB/dt|}{R} = \frac{\pi a^2 |dB/dt|}{R}$$

In terms of the charge moving through the circuit during the interval the magnetic field is changing, $I = dq/dt$, or $dq = I dt$.

EVALUATE   Carrying out the integration, we find the total charge moving around the loop to be

$$Q = \int_1^2 dq = \frac{\pi a^2}{R} \int_1^2 dB = \frac{\pi a^2}{R}(B_2 - B_1)$$

ASSESS   In the situation where the magnetic field is constant, $B_1 = B_2$, there would be no induced current, and the charge moved around the loop would be zero.

55.   **INTERPRET**   This problem is about the time constant of a series RL circuit.

DEVELOP   In a series RL circuit, the current as a function of time is given by Equation 27.7:

$$I(t) = I_\infty(1 - e^{-Rt/L})$$

where $I_\infty = \varepsilon_0/R$ is the final current.

EVALUATE   From Equation 27.7, the time constant is

$$\tau_L = \frac{L}{R} = -\frac{t}{\ln(1 - I/I_\infty)} = -\frac{7.6 \text{ s}}{\ln(1 - 1/2)} = \frac{7.6 \text{ s}}{\ln 2} = 11.0 \text{ s}$$

ASSESS   The time constant is inversely proportional to the resistance R. The physical meaning of the time constant $\tau_L = L/R$ is that significant changes in current cannot occur on time scales much shorter than τ_L.

57.   INTERPRET   In this problem we are asked about the inductance and the long-time behavior of a series RL circuit.

DEVELOP   In a series RL circuit, the rising current as a function of time is given by Equation 27.7:

$$I(t) = I_\infty(1 - e^{-Rt/L})$$

where $I_\infty = \varepsilon_0/R$ is the final current.

EVALUATE   (a) The current has risen to half its final value in 30 μs. Thus, the above equation gives

$$\frac{I(t)}{I_\infty} = 1 - e^{-Rt/L} \quad \rightarrow \quad \frac{1}{2} = 1 - e^{-Rt/L}$$

which can be solved to yield $L = \frac{Rt}{\ln 2} = \frac{(2.5 \text{ k}\Omega)(30 \text{ }\mu\text{s})}{\ln 2} = 108$ mH.

(b) After a long time ($t \rightarrow \infty$), the exponential term in Equation 27.7 is negligible, and the current is

$$I_\infty = \frac{\varepsilon_0}{R} = \frac{50 \text{ V}}{2.5 \text{ k}\Omega} = 20 \text{ mA}$$

ASSESS   After many time constants, there is no induced emf in the inductor and we can simply think of the inductor is a conducting wire connecting the different parts of the circuit. Note that $I_\infty = \varepsilon_0/R$ is what you would get by neglecting the inductance and using Ohm's law.

59.   INTERPRET   This problem is about the rate of change of current in a series RL circuit at different instants.

DEVELOP   In a series RL circuit, the rising current as a function of time is given by Equation 27.7:

$$I(t) = I_\infty(1 - e^{-Rt/L})$$

where $I_\infty = \varepsilon_0/R$ is the final current. The rate of change of current is $\frac{dI}{dt} = \frac{\varepsilon_0}{L}e^{-Rt/L}$.

EVALUATE   (a) For $t = 0$, we have

$$\frac{dI}{dt} = \frac{\varepsilon_0}{L} = \frac{60 \text{ V}}{1.5 \text{ H}} = 40 \text{ A/s}$$

(b) Similarly, for $t = 0.1$ s, the rate is

$$\frac{dI}{dt} = \frac{\varepsilon_0}{L}e^{-Rt/L} = \left(\frac{dI}{dt}\right)_0 e^{-Rt/L} = (40 \text{ A/s})e^{-(22 \text{ }\Omega)(0.1 \text{ s})/(1.5 \text{ H})} = 9.23 \text{ A/s}$$

ASSESS   The rate of change of current decreases exponentially with time. After many time constants, the current is approximately equal to $I_\infty = \varepsilon_0/R$, and the rate of change becomes negligibly small.

61.   INTERPRET   This problem is about the buildup and decay of current in a series RL circuit.

DEVELOP   In a series RL circuit, the rising current as a function of time is given by Equation 27.7:

$$I(t) = I_\infty(1 - e^{-Rt/L})$$

where $I_\infty = \varepsilon_0/R$ is the final current. On the other hand, as the switch is thrown back to position B, the battery is removed and the current decays according to Equation 27.8:

$$I(t) = I_0 e^{-Rt/L}$$

EVALUATE   (a) When the current is building up from zero, Equation 27.7 gives

$$I(5 \text{ s}) = \frac{\varepsilon_0}{R}(1 - e^{-Rt/L}) = \frac{12 \text{ V}}{2.7 \text{ }\Omega}\left(1 - e^{-(2.7 \text{ }\Omega)(5 \text{ s})/(20 \text{ H})}\right) = 2.18 \text{ A}$$

(b) At $t = 10$ s, the current has built up to

$$I(10 \text{ s}) = \frac{12 \text{ V}}{2.7 \text{ }\Omega}\left(1 - e^{-(2.7 \text{ }\Omega)(10 \text{ s})/(20 \text{ H})}\right) = 3.29 \text{ A}$$

This current decays when the switch is thrown back to B. Equation 27.8 (where t is the time since 10 s) gives

$$I = (3.29 \text{ A})\, e^{-(2.7\ \Omega)(15\ \text{s}-10\ \text{s})/(20\ \text{H})} = 1.68 \text{ A}$$

ASSESS   The time constant of the circuit is $\tau_L = L/R = (20 \text{ H})/(2.7\ \Omega) = 7.4$ s, so $t = 10$ s corresponds to only $1.35\ \tau_L$. At this instant, the current is only 74% of its long-time value $I_\infty = \varepsilon_0/R = (12 \text{ V})/(2.7\ \Omega) = 4.44$ A. One needs to wait many time constants for the current to approach I_∞.

63.   **INTERPRET**   This problem is about the magnetic energy stored in a series RL circuit as a function of time.

DEVELOP   The magnetic energy stored in an inductor is given by Equation 27.9, $U(t) = LI(t)^2/2$, where $I(t) = I_\infty(1 - e^{-Rt/L})$. Combining the two equations, the stored magnetic energy as a function of time is

$$U(t) = \frac{1}{2}LI_\infty^2(1 - e^{-Rt/L})^2 = U_\infty(1 - e^{-Rt/L})^2$$

where $U_\infty = LI_\infty^2/2$ is the steady-state value of the magnetic energy. When the stored energy is half its steady-state value, $U(t_u)/U_\infty = 1/2$, we have

$$\frac{1}{2} = \frac{U(t_u)}{U_\infty} = (1 - e^{-Rt_u/L})^2 \;\rightarrow\; \frac{1}{\sqrt{2}} = 1 - e^{-Rt_u/L}$$

EVALUATE   Solving the above equation for t_u yields

$$t_u = \frac{L}{R}\ln\left(\frac{\sqrt{2}}{\sqrt{2}-1}\right) = 1.28\frac{L}{R}$$

On the other hand, the current is half its steady-state value when $t_1 = (L/R)\ln 2 = 1$ ms . Dividing these results, we find

$$t_u = \frac{1.28}{\ln 2}\, t_1 = \frac{1.28}{\ln 2}(1 \text{ ms}) = 1.77 \text{ ms}$$

ASSESS   Since $t_u > t_1$, the magnetic energy reaches half its steady-state value after the current has already surpassed half its steady-state value. In fact, the current at $t = t_u$ is

$$I(t_u) = I_\infty(1 - e^{-Rt_u/L}) = \frac{I_\infty}{\sqrt{2}} = 0.707 I_\infty$$

65.   **INTERPRET**   This problem is about the magnetic energy stored and dissipated in a superconducting solenoid.

DEVELOP   The magnetic energy stored in the superconducting solenoid can be found by using Equation 27.9, $U(t) = LI(t)^2/2$. To answer parts **(b)** and **(c)**, we note that the current in an inductor cannot change instantaneously, and the power dissipated in the copper is $P = I^2R$. The power drops to one half its maximum original value, when the current drops to $1/\sqrt{2}$ times its initial value.

EVALUATE   **(a)** In its superconducting state, the solenoid's stored energy is

$$U = \frac{1}{2}LI^2 = \tfrac{1}{2}(3.5 \text{ H})(1.8 \text{ kA})^2 = 5.67 \text{ MJ}$$

(b) If the power dissipated in the copper just after a sudden loss of superconductivity must be less than 100 kW, then

$$R \le \frac{P}{I^2} = \frac{100 \text{ kW}}{(1.8 \text{ kA})^2} = 30.9 \text{ m}\Omega$$

The maximum resistance is therefore equal to 30.9 mΩ.

(c) The decay current as a function of time is given by Equation 27.8, $I(t) = I_0 e^{-Rt/L}$. Thus, the decay time for $I(t)/I_0 = 1/\sqrt{2}$ is

$$t = \frac{L}{R}\ln\left(\frac{I_0}{I}\right) = \frac{3.5 \text{ H}}{30.9 \text{ m}\Omega}\ln\sqrt{2} = 39.3 \text{ s}$$

ASSESS   Since the resistance of the superconducting solenoid is very small, its time constant is large ($\tau_L = L/R = (3.5 \text{ H})/(30.9 \text{ m}\Omega) = 113$ s). This means that it takes much longer for the current to decay, compared to that in an ordinary conductor.

67.   **INTERPRET**   In this problem we are asked to compare the magnetic energy density between a current-carrying loop and a solenoid.

DEVELOP   The magnetic field at the center of the loop is $B_{\text{loop}} = \mu_0 I/2R$ (see Equation 26.9). On the other hand, the magnetic field inside a solenoid is $B_{\text{solenoid}} = \mu_0 n I$. The energy density can be found by using Equation 27.10: $u_B = B^2/2\mu_0$.

EVALUATE   Using the equations above, we find the energy density at the center of the loop to be

$$u_B^{(\text{loop})} = \frac{B_{\text{loop}}^2}{2\mu_0} = \frac{1}{2\mu_0}\left(\frac{\mu_0 I}{2R}\right)^2 = \frac{\mu_0 I^2}{8R^2}$$

In a long thin solenoid of the same radius,

$$u_B^{(\text{solenoid})} = \frac{B_{\text{solenoid}}^2}{2\mu_0} = \frac{1}{2\mu_0}(\mu_0 n I)^2 = \frac{\mu_0 n^2 I^2}{2}$$

so the ratio is

$$\frac{u_B^{(\text{loop})}}{u_B^{(\text{solenoid})}} = \frac{\mu_0 I^2/8R^2}{\mu_0 n^2 I^2/2} = \frac{1}{4n^2 R^2}$$

ASSESS   As expected, the ratio is dimensionless (recall that n represents the number of turns per unit length of a solenoid). The result shows that the energy density inside a solenoid is greater than that at the center of a loop by a factor of $4n^2 R^2$.

69.   **INTERPRET**   In this problem we want to verify that the time integral of the power dissipated through the resistor in a series RL circuit is equal to the energy initially stored in the inductor.

DEVELOP   Equation 27.8, $I(t) = I_0 e^{-Rt/L}$, gives the current decaying through a resistor connected to an inductor carrying an initial current I_0. The instantaneous power dissipated in the resistor is $P_R = I^2 R$.

EVALUATE   (a) Using the two equations above, the power dissipated in the resistor as a function of time is

$$P_R = I^2 R = I_0^2 R\, e^{-2Rt/L}$$

(b) In a time interval dt, the energy dissipated is $dU = P_R dt$, so the total energy dissipated is

$$U = \int_0^{\infty} I_0^2 R e^{-2Rt/L}\, dt = I_0^2 R\, \frac{e^{-2Rt/L}}{(-2R/L)}\bigg|_0^{\infty} = \frac{I_0^2 RL}{2R} = \frac{1}{2}LI_0^2$$

ASSESS   This is precisely the energy initially stored in the inductor. The result must hold true by energy conservation.

71.   **INTERPRET**   This is an induction problem with changing magnetic flux due to a change in area.

DEVELOP   The magnetic flux through the loop is proportional to the vertical distance, y, it falls into the field region, $\Phi_B = BA = Bwy$. The rate of change of the flux is

$$\frac{d\Phi_B}{dt} = \frac{d}{dt}(Bwy) = Bwv$$

where $v = dy/dt$.

EVALUATE   (a) As long as the flux through the loop is changing, there is an upward magnetic force on the induced current in the bottom wire, which may cancel the downward force of gravity on the loop.

(b) Faraday's and Ohm's laws give a magnetic force proportional to the vertical speed,

$$F = IwB = \frac{|\varepsilon_i|}{R}wB = \frac{|-d\Phi_B/dt|\, wB}{R} = \frac{vw^2 B^2}{R}$$

When this equals mg, the terminal speed is

$$v_t = \frac{mgR}{w^2 B^2}$$

(c) The flux of $\vec{B}$ is positive (into the plane of Fig. 27.42) for clockwise circulation around the loop, so the induced current must be negative, or counterclockwise. The motional emf, $dq\vec{v} \times \vec{B}$, on positive charge carriers in the bottom wire, is to the right, and the magnetic force on them, $Id\vec{L} \times \vec{B}$, is upward.)

ASSESS The terminal speed is proportional to the resistance R. This is not surprising because if R is very small, then the induced current would be very large, giving rise to a large upward magnetic force to counter the gravitational force mg, and hence the terminal speed would also be small.

73. **INTERPRET** We qualitatively sketch current and power in a loop as a magnet passes through the loop.
DEVELOP As the magnet is approaching the loop, the flux through the loop is increasing to the right, which creates a current opposing this change in flux. As the magnet leaves the loop, the flux is in the same direction but *decreasing*, so the current is in the opposite direction.
Power depends on current squared, so it is positive in both cases.
EVALUATE See figure below.

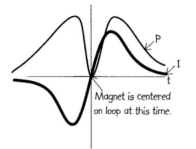

ASSESS The graph is symmetric because the magnet is moving at a steady rate.

75. **INTERPRET** We find the total flux through a toroidal coil with circular cross-section by finding the flux through the coil and using the equation for self-inductance. To find the flux through the coil, we use the magnetic field in a toroidal coil from Chapter 26, and integrate.
DEVELOP Since the magnetic field varies across the interior of the coil, we'll have to integrate. A diagram will be helpful to show how we're doing the integration, as shown below. We will use x for our integration variable, integrating from $x = -R$ to $x = R$. The field inside the coil is given by Equation 26.21 as $B = \dfrac{\mu_0 NI}{2\pi r}$, where in this case $r = A + x$. We find the total flux through one loop by integrating the flux $d\Phi$ through each strip of width dx shown in the figure, $d\Phi = B(r)dA = \dfrac{\mu_0 NI}{2\pi r}(2ydr)$, where $y = \sqrt{R^2 - x^2} = \sqrt{R^2 - (r - A)^2}$. Finally, we multiply this flux by N loops in the toroid and find the self-inductance by using Equation 27.3, $L = \frac{\Phi}{I}$.
EVALUATE

$$\Phi = N\int d\Phi = \int_{A-R}^{A+R} \frac{\mu_0 N^2 I}{2\pi r} 2ydr = \int_{A-R}^{A+R} \frac{\mu_0 N^2 I \sqrt{R^2 - (r - A)^2}}{\pi r} dr$$

$$\rightarrow \Phi = \frac{\mu_0 N^2 I}{\pi}[\pi(A - \sqrt{A^2 - R^2})]$$

$$\rightarrow L = \frac{\Phi}{I} = \mu_0 N^2 (A - \sqrt{A^2 - R^2})$$

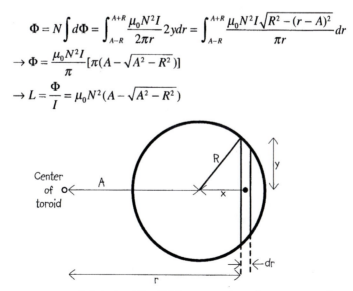

ASSESS The more common approach is to treat the field as approximately constant and equal to the value at $r \rightarrow A$. For further details, see Problem 27.76.

77. **INTERPRET** We use Kirchhoff's laws to find the current in a circuit element. The circuit includes an inductor, so the equations we obtain using Kirchhoff's laws will be differential equations.

DEVELOP We start by drawing our loops and nodes as shown in the figure below. Node A and node B give the same information, so we will use only loop 1, loop 2, and node A. From node A we get $I_1 - I_2 - I_3 = 0$. From loop 1, we obtain $\varepsilon - I_1 R_1 - I_2 R_2 = 0$, and from loop 2 we have $I_2 R_2 - L\frac{dI_3}{dt} = 0$. We want to find the current I_2 as a function of time.

EVALUATE From node A: $I_1 = I_2 + I_3$.

Substitute this into loop 1:

$$\varepsilon - (I_2 + I_3)R_1 - I_2 R_2 = 0 \rightarrow I_2 = \frac{\varepsilon - I_3 R_1}{R_1 + R_2}$$

Substitute into loop 2:

$$\left(\frac{\varepsilon - I_3 R_1}{R_1 + R_2}\right)R_2 - L\frac{dI_3}{dt} = 0 \rightarrow \frac{dI_3}{dt} = \left(\frac{\varepsilon - I_3 R_1}{R_1 + R_2}\right)\frac{R_2}{L} = \frac{\varepsilon R_2}{L(R_1 + R_2)} - \left(\frac{R_1 R_2}{L(R_1 + R_2)}\right)I_3$$

We guess at a solution of the form $I_3(t) = A + Be^{Ct}$, with initial condition $I_3(0) = 0$ since the inductor acts as an open circuit at first.

$$\frac{dI_3}{dt} = BCe^{Ct} = \frac{\varepsilon R_2}{L(R_1 + R_2)} - \left(\frac{R_1 R_2}{L(R_1 + R_2)}\right)(A + Be^{Ct})$$

$$\rightarrow C = -\frac{R_1 R_2}{L(R_1 + R_2)}$$

$$\rightarrow A\left(\frac{R_1 R_2}{L(R_1 + R_2)}\right) = \frac{\varepsilon R_2}{L(R_1 + R_2)} \rightarrow A = \frac{\varepsilon}{R_1}$$

$$I_3(0) = 0 \rightarrow \left(\frac{\varepsilon}{R_1} + B\right) = 0$$

$$\rightarrow B = -\frac{\varepsilon}{R_1}$$

So

$$I_3(t) = \frac{\varepsilon}{R_1}\left(1 - e^{-\left(\frac{R_1 R_2}{L(R_1 + R_2)}\right)t}\right)$$

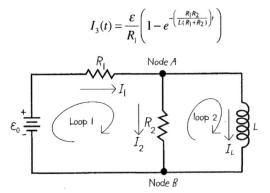

ASSESS The current gradually increases, with a time constant $R_1 R_2/L(R_1 + R_2)$.

79. **INTERPRET** We find the energy stored in a toroidal coil with square cross-section, given the geometry of the toroid and the current it carries. We will use Ampère's law to find the magnetic field inside the coil, integrate to find the total flux, and then find the inductance and thus the energy stored.

DEVELOP The energy stored in an inductor is $U = \frac{1}{2}LI^2$. We will check the manufacturer's claims that the energy stored with a current $I = 63$ A is $U = 100$ J. First, we integrate the flux through strips as shown in the figure below. The field at a distance r from the center is $B = \frac{\mu_0 NI}{2\pi r}$, and $d\Phi = Bl\,dr$. The dimensions of the coil are $R = 1.5$ m, $l = 0.228$ m, $N = 2500$, and $I = 63$ A.

EVALUATE

$$\Phi = N\int d\Phi = \frac{\mu_o N^2 I}{2\pi}l\int_R^{R+l}\frac{dr}{r} = \frac{\mu_0 N^2 Il}{2\pi}\ln\left(\frac{R+l}{R}\right)$$

$$\rightarrow L = \frac{\mu_0 N^2 l}{2\pi}\ln\left(1+\frac{l}{R}\right)$$

$$\rightarrow U = \tfrac{1}{2}LI^2 = \frac{\mu_o N^2 lI^2}{4\pi}\ln\left(1+\frac{l}{R}\right) = 80\text{ J}$$

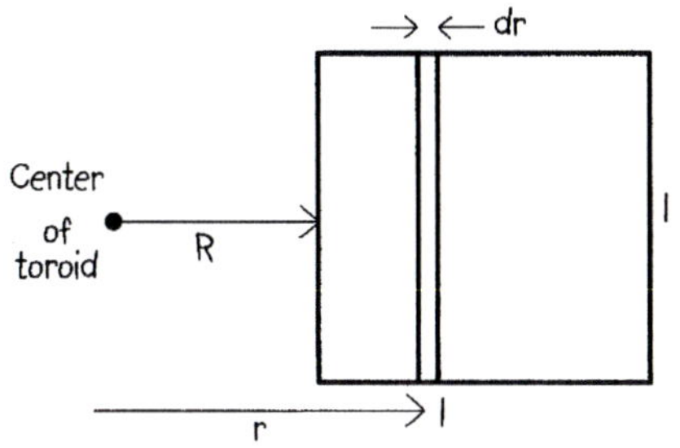

ASSESS The inductor will not perform as specified.

81. **INTERPRET** Here we investigate the feasibility of using the Earth's magnetic field to induce an electric field along the blades of a windmill. The blades sweep out area at a certain rate, and the resulting change in flux creates a voltage by Faraday's law.

DEVELOP Each blade sweeps out a circle once per second, so the rate of change in area is $\frac{dA}{dt} = \frac{\pi R^2}{1\text{ s}}$ and the voltage induced between center and edge is $\varepsilon = -\frac{\pi R^2 B}{1\text{ s}}$. The size of the windmill is $R = 50$ m, and the horizontal component of the Earth's magnetic field is $B = 1\times 10^{-5}$ T.

EVALUATE

$$\varepsilon = -\frac{\pi R^2 B}{1\text{ s}} = -79\text{ mV}$$

ASSESS Go ahead and use a generator: this voltage is not worth the effort.

EXERCISES

Section 28.1 Alternating Current

15. **INTERPRET** We're given the rms voltage and asked to find the peak voltage.

    DEVELOP The rms voltage V_{rms} and the peak voltage V_p are related by Equation 28.1, $V_{rms} = V_p/\sqrt{2}$.

    EVALUATE Equation 28.1 gives $V_p = \sqrt{2}V_{rms} = \sqrt{2}(6.3 \text{ V}) = 8.91 \text{ V}$.

    ASSESS The rms (root-mean-square) voltage is obtained by squaring the voltage and taking its time average, and then taking the square root. Therefore, it is smaller than the peak voltage by a factor of $\sqrt{2}$.

17. **INTERPRET** We're given the AC current in terms of sinusoidal function, and asked to deduce the rms current and the frequency of the current.

    DEVELOP As shown in Equation 28.3, the AC current can be written as

    $$I = I_p \sin(\omega t + \phi_I)$$

    where I_p is the peak current amplitude, ω is the angular frequency, and ϕ_I is the phase constant. Comparison of the current with Equation 28.3 shows that its amplitude and angular frequency are $I_p = 495$ mA and $\omega = 9.43$ (ms)$^{-1}$.

    EVALUATE (a) Applying Equation 28.1 gives $I_{rms} = I_p/\sqrt{2} = 495 \text{ mA}/\sqrt{2} = 350 \text{ mA}$.

    (b) Similarly, using Equation 28.2 we have $f = \omega/2\pi = 9.43 \text{ ms}^{-1}/2\pi = 1.50$ kHz.

    ASSESS The phase ϕ_I is zero in this problem. Note that since the rms (root-mean-square) current is obtained by squaring the current and taking its time average, and then taking the square root, it is smaller than the peak current by a factor of $\sqrt{2}$.

Section 28.2 Circuit Elements in AC Circuits

19. **INTERPRET** In this problem we want to find the rms current in a capacitor connected to an AC power source.

    DEVELOP The amplitude of the current in a capacitor is given by Equation 28.5:

    $$I_p = \frac{V_p}{X_C} = \frac{V_p}{1/\omega C} = \omega C V_p$$

    Using Equation 28.1, the corresponding rms current is $I_{rms} = \omega C V_{rms}$.

    EVALUATE Substituting the values given in the problem statement, we find the rms current to be

    $$I_{rms} = \omega C V_{rms} = (2\pi \times 60 \text{ Hz})(1 \text{ }\mu\text{F})(120 \text{ V}) = 45.2 \text{ mA}$$

    ASSESS The capacitive reactance is $X_C = 1/\omega C = 2.65 \times 10^3 \text{ }\Omega$. In this circuit, the current in the capacitor leads the voltage across the capacitor by $90°$.

21. **INTERPRET** This problem is about capacitive reactance at various angular frequencies.

    DEVELOP From Equation 28.5, we see that the capacitive reactance is

    $$X_C = \frac{1}{\omega C} = \frac{1}{2\pi f C}$$

EVALUATE   (a) For $f = 60$ Hz, the capacitive reactance is

$$X_C = \frac{1}{2\pi f C} = \frac{1}{2\pi(60 \text{ Hz})(3.3 \times 10^{-6} \text{F})} = 804 \ \Omega$$

(b) For $f = 1.0$ kHz, the capacitive reactance is

$$X_C = \frac{1}{2\pi f C} = \frac{1}{2\pi(1.0 \text{ kHz})(3.3 \times 10^{-6} \text{F})} = 48.2 \ \Omega$$

(c) Similarly, for $f = 20$ kHz, the capacitive reactance is

$$X_C = \frac{1}{2\pi f C} = \frac{1}{2\pi(20 \text{ kHz})(3.3 \times 10^{-6} \text{F})} = 2.41 \ \Omega$$

ASSESS   One can see that a capacitor has the greatest effect (largest reactance) at low frequency.

23.   **INTERPRET**   This problem is about the capacitance of a capacitor that's connected across an AC power source.
DEVELOP   The fact that the capacitor and the resistor both pass the same current implies that

$$I_p = \frac{V_p}{R} = \frac{V_p}{X_C} \quad \rightarrow \quad R = X_C = \frac{1}{\omega C}$$

Therefore, the capacitance is $C = 1/\omega R$.
EVALUATE   Substituting the values given, we obtain

$$C = \frac{1}{\omega R} = \frac{1}{2\pi f R} = \frac{1}{2\pi(60 \text{ Hz})(1.8 \text{ k}\Omega)} = 1.47 \ \mu\text{F}$$

ASSESS   Since $R = X_C$, the greater the value of resistance R, the greater the capacitive reactance, and hence the smaller the capacitance.

Section 28.3 *LC* Circuits

25.   **INTERPRET**   This problem is about the resonant frequency of an *LC* circuit.
DEVELOP   Using Equations 28.2 and 28.10, the resonant frequency can be written as

$$f = \frac{\omega}{2\pi} = \frac{1}{2\pi\sqrt{LC}}$$

EVALUATE   Substituting the values given, we find the resonant frequency to be

$$f = \frac{1}{2\pi\sqrt{LC}} = \frac{1}{2\pi\sqrt{(1.7 \text{ mH})(0.22 \ \mu\text{F})}} = 8.23 \text{ kHz}$$

ASSESS   The mechanical analog of the *LC* circuit is the mass-spring system whose angular frequency is $\omega = \sqrt{k/m}$. Thus, the correspondence between the two systems is: $L \leftrightarrow m$ and $C \rightarrow 1/k$.

27.   **INTERPRET**   In this problem we want to know the capacitance range of a variable capacitor, given the frequency range covered.
DEVELOP   Using Equations 28.2 and 28.10, the resonant frequency can be written as

$$f = \frac{\omega}{2\pi} = \frac{1}{2\pi\sqrt{LC}}$$

which can be solved to give $C = \frac{1}{\omega^2 L} = \frac{1}{4\pi^2 f^2 L}$.
EVALUATE   The frequencies $f_1 = 550$ kHz and $f_2 = 1600$ kHz correspond to

$$C_1 = \frac{1}{4\pi^2 f_1^2 L} = \frac{1}{4\pi^2(550 \text{ kHz})^2(2 \text{ mH})} = 41.9 \text{ pF}$$

$$C_2 = \frac{1}{4\pi^2 f_2^2 L} = \frac{1}{4\pi^2(1600 \text{ kHz})^2(2 \text{ mH})} = 4.95 \text{ pF}$$

Thus, the range of the capacitance is from 41.9 pF down to 4.95 pF.
ASSESS   We find the capacitance to be inversely proportional to f^2. Therefore, lower capacitance covers the higher end of the frequency band.

29.   **INTERPRET**   We find the inductance needed to make an *LC* circuit to have a desired resonance frequency with a given value of capacitance.

DEVELOP   The resonance frequency of an *LC* circuit is $\omega_0 = 1/\sqrt{LC}$, so $f_0 = 1/2\pi\sqrt{LC}$. The frequency desired is $f_o = 89.5 \times 10^6$ Hz, and the capacitance is $C = 47 \times 10^{-12}$ F. We solve for *L*.

EVALUATE $f_0 = 1/2\pi\sqrt{LC} \rightarrow L = 1/4\pi^2 f_0^2 C = 67.3$ nH.

ASSESS   This is a very small inductor. At high frequencies, small inductances such as this can have a large effect: so much so that it becomes important to design circuits so as to minimize the inductance of individual lead wires!

Section 28.4 Driven *RLC* Circuits and Resonance

31.   **INTERPRET**   Here we find the impedance of an *LRC* circuit at a given frequency.

DEVELOP $Z = \sqrt{R^2 + (X_L - X_C)^2}$, where $X_C = \frac{1}{\omega C}$ and $X_L = \omega L$. The frequency given is $f = 10 \times 10^3$ Hz $\rightarrow$ $\omega = 20\pi \times 10^3$ s^{-1}. The values of the circuit elements are $R = 1.5$ kΩ, $C = 5.0 \times 10^{-6}$ F, and $L = 50 \times 10^{-3}$ H.

EVALUATE $Z = \sqrt{R^2 + (X_L - X_C)^2} = 3.48$ kΩ.

ASSESS   Note that at this frequency the capacitor has almost no effect compared to the other two circuit elements.

33.   **INTERPRET**   We find the peak current through an *RLC* circuit at resonance, for three values of *R* in the circuit. We use the fact that at resonance, $Z = R$.

DEVELOP   The peak voltage is $V_{max} = 100$ V. The value of *R* is $R = 10$ kΩ. Since $Z = R$ at resonance, the peak current will be $I_{max} = \frac{V_{max}}{R}$.

EVALUATE   For $\frac{1}{2}R$, $I_{max} = \frac{V_{max}}{R} = 20$ mA. Similarly, for R $I_{max} = 10$ mA and for $2R$ $I_{max} = 5$ mA.

ASSESS   At frequencies off the resonance peak, these calculations become somewhat more complicated. But at resonance, $Z = R$ and everything becomes easy.

Sections 28.5 Power in AC Circuits and 28.6 Transformers and Power Supplies

35.   **INTERPRET**   We use the average power of a lamp, as well as the RMS voltage and the power factor, to calculate the RMS current.

DEVELOP $\langle P \rangle = I_{RMS} V_{RMS} \cos\phi$, where $\cos\phi = 0.85$ is the given power factor. $\langle P \rangle = 40$ W and $V_{RMS} = 120$ V, so we simply solve for I_{RMS}.

EVALUATE $I_{RMS} = \frac{\langle P \rangle}{V_{RMS} \cos\phi} = 390$ mA.

ASSESS   The power factor actually matters with fluorescent lamps. With incandescent lamps, the impedance is almost entirely resistive, so the power factor is almost exactly one.

37.   **INTERPRET**   We use the transformer equation, and conservation of energy, to design a transformer that converts 230 VAC to 120 VAC.

DEVELOP   The transformer equation is $V_2 = \frac{N_2}{N_1} V_1$. The voltages are $V_1 = 230$ V and $V_2 = 120$ V, and $N_1 = 1000$. We solve for N_2. For the second part, we use the idea of conservation of energy and let $P_1 = P_2$, where $P = I_{RMS} V_{RMS}$ and $I_1 = 7.3$ A.

EVALUATE

(a) $N_1 = N_1 \frac{V_2}{V_1} = 522$ turns.

(b) $I_1 V_1 = I_2 V_2 \rightarrow I_2 = I_1 \frac{V_1}{V_2} = 14$ V.

ASSESS   Conservation of energy still holds!

PROBLEMS

39.   **INTERPRET**   This problem is about capacitive and inductive reactances, and how varying frequency affects them.

DEVELOP   From Equations 28.5 and 28.7, the capacitive and inductive reactances can be obtained as

$$X_C = \frac{1}{\omega C} = \frac{1}{2\pi f C}, \quad X_L = \omega L = 2\pi f L$$

EVALUATE   (a) From the above equation, the frequency of the applied voltage is

$$f = \frac{1}{2\pi X_C C} = \frac{1}{2\pi (1 \text{ k}\Omega)(2 \text{ }\mu\text{F})} = 79.6 \text{ Hz}$$

(b) The equality $X_L = X_C$ implies

$$L = \frac{X_C}{\omega} = \frac{X_C}{2\pi f} = \frac{1.0 \text{ k}\Omega}{2\pi(79.6 \text{ Hz})} = 2 \text{ H}$$

(c) Doubling ω doubles X_L and halves X_C so X_L would be four times X_C at $f = 159$ Hz.

ASSESS   Capacitive reactance X_C is inversely proportional to ω, while inductive reactance X_L is proportional to ω. A larger capacitor has lower reactance and a larger inductor has higher reactance.

41.   INTERPRET   In this problem a capacitor is connected across an AC generator. We are given the AC voltage and asked about the current across the capacitor.

DEVELOP   The current across the capacitor is given by Equation 28.4:

$$I(t) = \omega C V_p \cos\omega t = \omega C V_p \sin(\omega t + \pi/2)$$

EVALUATE   (a) The above equation shows that the peak current is

$$I_p = V_p \omega C = (22 \text{ V})(377 \text{ s}^{-1})(1.2 \text{ }\mu\text{F}) = 9.95 \text{ mA}$$

(b) The voltage at $t = 6.5$ ms is (remember that ωt is in radians)

$$V(6.5 \text{ ms}) = V_p \sin\omega t = (22 \text{ V})\sin[2\pi(60 \text{ s}^{-1})(6.5 \text{ ms})] = 14.0 \text{ V}$$

(c) Similarly, the current is

$$I(6.5 \text{ ms}) = I_p \cos\omega t = (9.95 \text{ mA})\cos[2\pi(60 \text{ s}^{-1})(6.5 \text{ ms})] = -7.67 \text{ mA}$$

The magnitude of the current is 7.67 mA.

ASSESS   In a capacitor, the current leads the voltage by 90°.

43.   INTERPRET   In this problem an inductor and a lamp are connected across an AC source. We are given the AC voltage and asked about the rms current across the lamp.

DEVELOP   In a series circuit, the same current flows through the inductor and lamp. Since the ratio of the rms quantities for a given circuit element equals that of the peak values, Equation 28.7 gives

$$I_{\text{rms}} = \frac{V_{\text{rms}}}{X_L} = \frac{V_{\text{rms}}}{\omega L}$$

EVALUATE   Substituting the values given, we find the rms current to be

$$I_{\text{rms}} = \frac{V_{\text{rms}}}{\omega L} = \frac{90 \text{ V}}{(2\pi \times 60 \text{ Hz})(0.75 \text{ H})} = 318 \text{ mA}$$

ASSESS   The current in the inductor and the lamp lags the voltage by 90°.

45.   INTERPRET   This problem asks for the inductance that satisfies the resonance condition for a given range of capacitances and frequencies.

DEVELOP   Using Equations 28.2 and 28.10, the resonant frequency can be written as

$$f = \frac{\omega}{2\pi} = \frac{1}{2\pi\sqrt{LC}}$$

which can be solved to give $L = \frac{1}{\omega^2 C} = \frac{1}{4\pi^2 f^2 C}$.

EVALUATE   Using either condition, $f_1 = 88$ MHz with $C_1 = 16.4$ pF, or $f_2 = 108$ MHz with $C_2 = 10.9$ pF, we find the inductance to be

$$L = \frac{1}{4\pi^2 f_1^2 C_1} = \frac{1}{4\pi^2 (88 \text{ MHz})^2 (16.4 \text{ pF})} = 0.199\mu\text{H}$$

$$L = \frac{1}{4\pi^2 f_2^2 C_2} = \frac{1}{4\pi^2 (108 \text{ MHz})^2 (10.9 \text{ pF})} = 0.199\mu\text{H}$$

ASSESS For a given inductance L, the capacitance is inversely proportional to f^2. Thus, lower capacitance covers the higher end of the frequency band.

47. **INTERPRET** This problem involves an oscillation in which energy is transferred back and forth between electric and magnetic fields.

    DEVELOP The electric energy stored in the capacitor is given by $U_E(t) = q^2(t)/2C$, where $q(t) = q_p \cos \omega t$ (see Equation 28.9). Similarly, the magnetic energy stored in the inductor is $U_B(t) = LI^2(t)/2$, where

    $$I(t) = dq/dt = -q_p \omega \sin \omega t = I_p \cos(\omega t + \pi/2)$$

    The quantities are evaluated at $\omega t = \omega(T/8) = \frac{2\pi}{8} = \frac{\pi}{4} = 45°$ (i.e., $\frac{1}{8}$ cycle). Note that phase constant zero corresponds to a fully charged capacitor at $t = 0$.

    EVALUATE (a) From Equation 28.9, we obtain $\frac{q}{q_p} = \cos 45° = \frac{1}{\sqrt{2}}$.

    (b) From the equation for electric energy, the ratio is

    $$\frac{U_E(t)}{U_{E,p}} = \frac{q^2(t)/2C}{q_p^2/2C} = \frac{q^2(t)}{q_p^2} = \cos^2 \omega t = \cos^2 45° = \frac{1}{2}$$

    (c) The ratio of the current is $I(t)/I_p = \cos(\omega t + \pi/2) = \cos 135° = -\frac{1}{\sqrt{2}}$. The direction of the current is away from the positive capacitor plate at $t = 0$.

    (d) From the equation for magnetic energy, $U_B(t)/U_{B,p} = I^2(t)/I_p^2 = \cos^2 135° = \frac{1}{2}$.

    ASSESS At one-eighth of a cycle, half of the total energy is magnetic and half is electric. This is illustrated in Figure 28.11.

49. **INTERPRET** This problem is about an LC circuit with damping due to the resistance. We are interested in the time it takes for the peak voltage to be halved.

    DEVELOP Equation 28.11 gives the charge as a function of time. Since $V(t) = q(t)/C$, the voltage as a function of time can be written as

    $$V(t) = V_p e^{-Rt/2L} \cos \omega t$$

    The peak voltage decays with time constant $2L/R$. Half the initial peak value is reached after a time $t = (2L/R)\ln 2$ (when $e^{-Rt/2L} = \frac{1}{2}$).

    EVALUATE Since the period of oscillation is $T = \frac{2\pi}{\omega} = 2\pi\sqrt{LC}$, the number of cycles represented by t is

    $$\frac{t}{T} = \frac{(2L/R)\ln 2}{2\pi\sqrt{LC}} = \frac{\ln 2}{\pi R}\sqrt{\frac{L}{C}} = \frac{\ln 2}{\pi(1.6\,\Omega)}\sqrt{\frac{20\text{ mH}}{0.15\,\mu\text{F}}} = 50.4$$

    ASSESS We have an oscillation which is underdamped. The larger the resistance, the more rapidly the oscillation will decay.

51. **INTERPRET** This problem is about a series RLC circuit at resonance. We want to find the smallest resistance that still keeps the capacitor voltage under its rated value.

    DEVELOP In a series RLC circuit at resonance, the peak capacitor voltage is

    $$V_{C,p} = I_p X_C = \frac{V_p/R}{\omega_0 C} = \frac{V_p}{R}\sqrt{\frac{L}{C}}$$

    where $\omega_0 = 1/\sqrt{LC}$ is the resonant angular frequency.

    EVALUATE The condition that $V_{C,p} \leq 400$ V implies

    $$R \geq \frac{V_p}{V_{C,p}}\sqrt{\frac{L}{C}} = \left(\frac{32\text{ V}}{400\text{ V}}\right)\sqrt{\frac{1.5\text{ H}}{250\,\mu\text{F}}} = 6.20\,\Omega$$

    ASSESS Our results shows that $V_{C,p}$ is inversely proportional to R. This means that a larger resistor would be required if the capacitor has a lower voltage rating.

53.   INTERPRET   This problem involves analyzing a phasor diagram for a driven *RLC* circuit.

DEVELOP   Our diagram has three phasors, V_{Rp}, V_{Lp}, and V_{Cp}, representing the voltages across the resistor, the inductor, and the capacitor, respectively. Since the resistor voltage is in phase with the current, I_p is in the same direction as V_{Rp}. The resonant frequency is $\omega_0 = 1/\sqrt{LC}$.

EVALUATE   **(a)** From the observation that $V_{Lp} = I_p \omega L > V_{Cp} = I_p/\omega C$, we conclude that $\omega^2 > 1/LC = \omega_0^2$, the frequency is above resonance. **(b)** The applied voltage phasor is the vector sum of the resistor, capacitor, and inductor voltage phasors, as shown below. The current is in phase with the voltage across the resistor, in this case lagging the applied voltage (since $\phi = \tan^{-1}[(V_{Lp} - V_{Cp})/V_{Rp}] > 0$) by approximately $50°$ (as estimated from the figure).

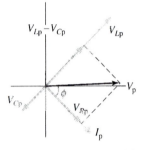

ASSESS   Our circuit is inductive since $V_{Lp} > V_{Cp}$. Note that a positive ϕ means that voltage leads current, and a negative ϕ means voltage lags current; at resonance, $X_L = X_C$ and $\phi = 0$.

55.   INTERPRET   This problem is about power factor in a series *RLC* circuit.

DEVELOP   The power factor of the circuit is

$$\cos\phi \frac{V_{Rp}}{V_p} = \frac{I_p R}{I_p Z} = \frac{R}{Z}$$

The average power in the circuit is given by Equation 28.14:

$$\langle P \rangle = I_{rms} V_{rms} \cos\phi = I_{rms}(I_{rms} Z)(R/Z) = I_{rms}^2 R$$

EVALUATE   **(a)** Substituting the values given, we find the power factor to be

$$\cos\phi = \frac{R}{Z} = \frac{100\ \Omega}{300\ \Omega} = 0.333$$

(b) The above equation gives $\langle P \rangle = I_{rms}^2 R = (200\ \text{mA})^2(100\ \Omega) = 4\ \text{W}$. Note that the average AC power is given by the same expression as the DC power if the rms current is used.

ASSESS   The power factor must be between zero and 1. A purely resistive circuit has a power factor of 1, while a circuit with only capacitance or inductance has a power factor of zero.

57.   INTERPRET   This problem is about the relationship between power lost in transmission and power factor.

DEVELOP   We assume the average power supplied by the city to be $\langle P \rangle = I_{rms} V_{rms} \cos\phi$, where $\cos\phi$ is the power factor. The average power lost in the transmission line is $\Delta P = I_{rms}^2 R$.

EVALUATE   **(a)** With $\cos\phi = 1$, the average power supplied to the city is

$$\langle P \rangle = I_{rms} V_{rms} \cos\phi = (200\ \text{A})(365\ \text{kV})(1) = 73.0\ \text{MW}$$

and that the average power lost in the transmission line is

$$\Delta P = I_{rms}^2 R = (200\ \text{A})^2 (100\ \Omega) = 4\ \text{MW}$$

Thus, the percent lost is $\Delta P/\langle P \rangle = (4/73) \times 100\% \approx 5.5\%$.

(b) If the power factor in part **(a)** were 0.6 instead of 1.0, the percent lost would be $\Delta P/\langle P \rangle = (5.5\%)/0.6 = 9.1\%$.

(c) Since ΔP is a constant, the larger $\langle P \rangle$ (which is proportional to $\cos\phi$), the smaller the fraction of power lost, $\Delta P/\langle P \rangle$. A large power factor is therefore better for the power plant owners.

ASSESS   The power factor $\cos\phi$ is between zero and 1. Our problem demonstrates that low power factors result in wasted energy.

59. **INTERPRET** This problem deals with DC power supplies. If the time constant RC is long enough, the capacitor voltage will only decrease slightly before the AC voltage from the transformer rises again to fully charge the capacitor.

DEVELOP The scenario is depicted in Figure 28.24. From the given DC output, we find the load resistance to be $R = (22 \text{ V})/(150 \text{ mA}) = 147 \text{ }\Omega$. In one period of the input AC ($T = 1/f$), the capacitor voltage must decay by less than 3%, or $e^{-T/RC} \geq 0.97$.

EVALUATE The above condition implies that

$$C \geq -\frac{T}{R \ln(0.97)} = -\frac{1}{(60 \text{ Hz})(147 \text{ }\Omega) \ln(0.97)} = 3.73 \text{ mF}$$

ASSESS If the capacitance is large enough, the load current and voltage can be made arbitrarily smooth with negligible decay.

61. **INTERPRET** We have an AC generator connected to a series RLC circuit, and we want to know its maximum peak voltage when the circuit is at resonance.

DEVELOP The peak capacitor voltage is $V_{Cp} = I_p X_C$. At resonance, the impedance is $Z = R$ and $I_p = V_p/R$. The capacitive reactance is $X_C = 1/\omega_0 C = \sqrt{L/C}$.

EVALUATE The condition that $V_{Cp} = (V_p/R)\sqrt{L/C} \leq 600 \text{ V}$ implies

$$V_p \leq (600 \text{ V})(1.3 \text{ }\Omega)\sqrt{\frac{0.33 \text{ }\mu\text{F}}{27 \text{ mH}}} = 2.73 \text{ V}$$

ASSESS The inductor voltage at resonance is

$$V_{Lp} = I_p X_L = \frac{V_p}{R}\omega_0 L = \frac{V_p}{R}\frac{L}{\sqrt{LC}} = \frac{V_p}{R}\sqrt{\frac{L}{C}}$$

which is the same as V_{Cp}. The two voltages cancel exactly at resonance. Note that V_{Cp} and V_{Lp} are higher than V_p.

63. **INTERPRET** In Example 28.4, we found a frequency at which the current in an RLC circuit is half the maximum. Here we find a second frequency at which the current will be half the maximum. We use the equation for Z.

DEVELOP From the example, we have $C = 11.5 \text{ }\mu\text{F}$, $R = 8.0 \text{ }\Omega$, and $L = 2.2 \text{ mH}$. We also know that $Z = \sqrt{R^2 + (X_L - X_C)^2}$, where $X_C = \frac{1}{\omega C}$ and $X_L = \omega L$. The current is given by Ohm's law, $I = \frac{V}{Z}$, and we are looking for a value of ω such that $Z = 2R$.

EVALUATE

$$Z = \sqrt{R^2 + \left(\omega L - \frac{1}{\omega C}\right)^2} = 2R$$

$$\rightarrow 4R^2 = R^2 + \left(\omega L - \frac{1}{\omega C}\right)^2 \rightarrow 3R^2 = \omega^2 L^2 - 2\frac{L}{C} + \frac{1}{\omega^2 C^2}$$

$$\rightarrow 3R^2\omega^2 = \omega^4 L^2 - 2\frac{L\omega^2}{C} + \frac{1}{C^2} \rightarrow \omega^4 L^2 + \left(-2\frac{L}{C} - 3R^2\right)\omega^2 + \frac{1}{C^2} = 0$$

$$\rightarrow \omega = \{\pm 3882, \pm 10181\}$$

The sign of ω does not matter. We need to convert to frequency using $f = \frac{\omega}{2\pi}$, so $f = \{618 \text{ Hz}, 1620 \text{ Hz}\}$.

ASSESS The 618-Hz answer was given in the example, so the solution we need is $f = 1620 \text{ Hz}$.

65. **INTERPRET** We have two capacitors connected first in series and then in parallel with an AC generator, and we want to know their capacitances.

DEVELOP Equation 28.5 gives the rms current when capacitors are connected to an AC generator, $I_{rms} = V_{rms}/X_C = \omega C V_{rms}$. In the parallel connection, $C = C_1 + C_2$, and we have

$$30 \text{ mA} = (2\pi \times 10^5 \text{ V/s})(C_1 + C_2)$$

while in the series connection, $C = C_1 C_2/(C_1 + C_2)$, and Equation 28.5 gives

$$5.5 \text{ mA} = (2\pi \times 10^5 \text{ V/s})\frac{C_1 C_2}{C_1 + C_2}$$

The two equations can be used to solve for C_1 and C_2.

EVALUATE   Simplifying the above two equations leads to $C_1 + C_2 = 47.7$ nF and $C_1 C_2 = (20.4 \text{ nF})^2$. Eliminating C_1 from the second equation and substituting into the first equation, we obtain the following quadratic equation:

$$C_2^2 - (47.7 \text{ nF})C_2 + (20.4 \text{ nF})^2 = 0$$

Since the initial two equations are symmetric in C_1 and C_2, eliminating C_2 gives the same equation as the above, but with C_2 replaced by C_1. So the solutions for C_1 and C_2 are

$$\frac{1}{2}[(47.7 \text{ nF}) \pm \sqrt{(47.7 \text{ nF})^2 - 4(20.4 \text{ nF})^2}] = 11.5 \text{ nF and } 36.2 \text{ nF}$$

ASSESS   The parallel connection yields a greater capacitance, and hence a larger current compared to the series combination.

67.   **INTERPRET**   This problem involves designing a circuit to be used in our "black box."

DEVELOP   We want the output voltage to lead the input voltage (by 45°). By inspecting Figure 28.16, we readily notice that the voltage across a resistor leads the voltage across a capacitor in series with it. We also see that the voltage across an inductor leads the voltage across a resistor in series with it. Both circuits can be adapted to the criteria of the "black box" in this problem.

EVALUATE   Case (1): RC circuit with AC input. In this circuit, it can be seen that $V = V_R + V_C$ and $I = I_C = I_R$. In the corresponding phasor diagram shown below, V_C lags I by 90°, V_R and I are in phase, and V is the vector sum of these (see Table 28.1). We drew I horizontally for convenience. The impedance is

$$Z = \frac{V_p}{I_p} = \frac{\sqrt{V_{Rp}^2 + V_{Cp}^2}}{I_p} = \sqrt{R^2 + X_C^2} = \sqrt{R^2 + \frac{1}{\omega^2 C^2}}$$

and the phase angle is

$$\tan\phi = -\frac{V_{Cp}}{V_{Rp}} = -\frac{X_C}{R} = -\frac{1}{\omega RC}$$

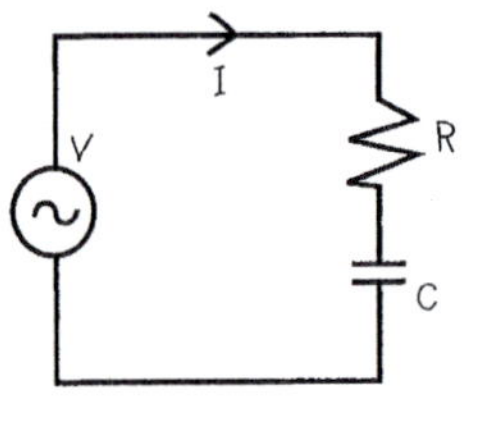

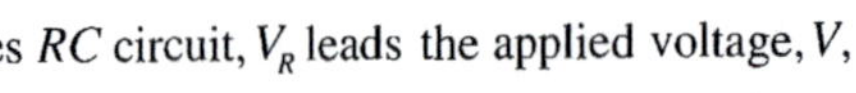

The current I always leads V, because $\phi < 0$. (Recall that ϕ is defined by $I = I_p \sin\omega t$ when $V = V_p \sin(\omega t + \phi)$.)

The result implies that in a series RC circuit, V_R leads the applied voltage, V, by an angle $\tan^{-1}(1/\omega RC)$, which may be adjusted to 45° if $\omega RC = 1$. The peak voltage across the entire resistance is $V_{Rp} = V_p \cos 45° = V_p/\sqrt{2}$, so if we divide the resistance into two parts, $R_1 + R_2 = R$, with $R_2/R = 1/\sqrt{2}$, then the peak voltage across R_2 will be $(1/\sqrt{2})V_{Rp} = \frac{1}{2}V_p$, as desired (rms voltages have the same ratio as peak voltages).

Case (2): RL circuit with AC input. When a capacitor is replaced with an inductor, the phasors for V_R and I are still parallel, but V_L leads I by 90°. The voltage V is the vector sum of V_R and V_L, so the impedance is

$$Z = \frac{V_p}{V_p} = \frac{\sqrt{V_{Rp}^2 + V_{Lp}^2}}{V_p} = \sqrt{R^2 + X_L^2} = \sqrt{R^2 + \omega^2 L^2}$$

and the phase angle is

$$\tan\phi = \frac{V_{Lp}}{V_{Rp}} = \frac{X_L}{R} = \frac{\omega L}{R}$$

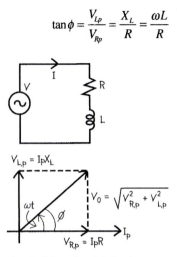

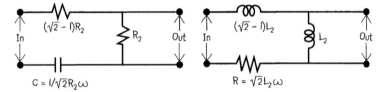

In this case, I always lags V, because $\phi > 0$. (Negative ϕ is in the same sense as ωt, measured from V.)

In a series RL circuit, V_L leads V by $90° - \tan^{-1}(\omega L/R)$, which equals $45°$ if $\omega L = R$. Again, $V_{Lp} = V_p/\sqrt{2}$, so if we divide L into $L_1 + L_2$, with $L_2 = L/\sqrt{2}$, the peak voltage across L_2 is $V_p/2$. Both circuits are sketched below.

ASSESS   We have shown how the circuit can be designed in two different ways to adapt to the criteria of the "black box." Our circuit conditions can be verified explicitly. Note that V_{out} is the open-circuit output voltage. If a load is connected across the output terminals, the magnitude and the phase of the voltage will be changed accordingly.

69.   **INTERPRET**   This problem involves a series RLC circuit at resonance. We are asked to find the resistance, the inductance, and the capacitance.

DEVELOP   At resonance, the impedance is $Z = R$, and the current is $I_p = V_p/Z = V_p/R$ and $X_L - X_C = 0$. Away from resonance, $Z = V_p/I_p$, and $|X_L - X_C| = \sqrt{Z^2 - R^2}$.

EVALUATE   The resonance condition gives

$$R = \frac{V_p}{I_p} = \frac{20\ \text{V}}{50\ \text{mA}} = 400\ \Omega$$

On the other hand, at half the resonant frequency, $1\ \text{kHz} = \frac{1}{2}(2\ \text{kHz})$, the impedance is

$$Z = \frac{V_p}{I_p} = \frac{20\text{V}}{15\ \text{mA}} = 1.33\ \text{k}\Omega = (10/3)R$$

which gives

$$|X_L - X_C| = \sqrt{Z^2 - R^2} = \sqrt{(10/3)^2 - 1}\ R = \frac{\sqrt{91}}{3}R$$

With $X_L = \omega L$ and $X_C = 1/\omega C$, we obtain the following conditions:

$$\frac{1}{\omega_0 C} - \omega_0 L = 0, \qquad \frac{1}{\frac{1}{2}\omega_0 C} - \frac{1}{2}\omega_0 L = \frac{\sqrt{91}}{3}R$$

These equations can be solved for C and L, with the following result:

$$L = \frac{2\sqrt{91}\,R}{9\omega_0} = \frac{2\sqrt{91}\,(400\ \Omega)}{9(2\pi \times 2\ \text{kHz})} = 67.5\ \text{mH}$$

$$C = \frac{1}{\omega_0^2 L} = \frac{1}{(2\pi \times 2\ \text{kHz})^2 (67.5\ \text{mH})} = 93.8\ \text{nF}$$

ASSESS   Below resonance, capacitive reactance dominates, with $X_C > X_L$.

71.   **INTERPRET**   In this problem we are asked to derive the Q-factor of the RLC circuit.

DEVELOP   To derive the expression for Q, we first need to know the power in the circuit. From Equations 28.12 and 28.14 (with rms values), and

$$\cos\phi = \frac{V_{Rp}}{V_p} = \frac{I_p R}{I_p Z} = \frac{R}{Z}$$

the average power in a series RLC circuit can be written as

$$\langle P \rangle = I_{\text{rms}} V_{\text{rms}} \cos\phi = (V_{\text{rms}}/Z)\,V_{\text{rms}}\,(R/Z) = \frac{V_{\text{rms}}^2 R}{Z^2}$$

The above expression show the power falls to half its resonance value (V_{rms}^2/R) when $Z = \sqrt{2}R$, or when $|X_L - X_C| = R$. In terms of the resonant frequency, $\omega_0 = 1/\sqrt{LC}$, this condition becomes

$$\left| \omega L - \frac{1}{\omega C} \right| = L\left| \omega - \frac{\omega_0^2}{\omega} \right| = R \;\; \rightarrow \;\; \omega^2 - \omega_0^2 = \pm\frac{R}{L}\omega$$

The solutions of these quadratics, with $\omega > 0$, are

$$\omega_\pm = \frac{1}{2}\left[\pm\frac{R}{L} + \sqrt{\frac{R^2}{L^2} + 4\omega_0^2} \right]$$

The Q-factor is then equal to $\omega_0/\Delta\omega$, where $\Delta\omega = \omega_+ - \omega_-$.

EVALUATE   If $\frac{R}{L} \ll \omega_0$ (equivalent to $R \ll \sqrt{L/C}$), we can neglect the first term under the square root sign compared to the second, obtaining $\omega \approx \omega_0 \pm R/2L$. The difference between these two values of ω is $\Delta\omega = R/L$, from which we obtain $Q = \omega_0/\Delta\omega = \omega_0 L/R$.

ASSESS   The Q-factor measures the "quality" of oscillation. The smaller the resistance, the higher the Q-factor. In the absence of resistance ($R \rightarrow 0$), the LC circuit can oscillate indefinitely.

73.   **INTERPRET**   We find the RMS voltage of a sawtooth wave by finding the average value of the square of the voltage.

DEVELOP   The voltage goes linearly from $-V_p$ to V_p in period T, so the equation for $V(t)$ (for one period) is $V(t) = V_p(\frac{2t}{T} - 1)$. We integrate V^2 over one period, then divide by the period to find the average value of V^2, then take the square root to find V_{RMS}.

EVALUATE

$$\langle V^2 \rangle = \frac{1}{T}\int_0^T V^2\,dt = \frac{1}{T}\int_0^T V_p^2 \left(\frac{2t}{T} - 1 \right)^2 dt = \frac{V_p^2}{T}\int_0^T \left(\frac{4t^2}{T^2} - \frac{4t}{T} + 1 \right) dt$$

$$\rightarrow \langle V^2 \rangle = \frac{V_p^2}{T}\left[\frac{4}{3}T - 2T + T \right] = \frac{1}{3}V_p^2$$

$$V_{RMS} = \sqrt{\langle V^2 \rangle} = \frac{V_p}{\sqrt{3}}$$

ASSESS   This seems reasonable, since for a sine wave $V_{RMS} = \frac{V_p}{\sqrt{2}}$.

75. **INTERPRET**   We find the voltage across each element in an *RLC* circuit, using the total impedance, the impedance of each element, and the phase angles.

DEVELOP   The voltage generated is $V = V_p \sin \omega t$. The RMS voltage generated is $V_{RMS} = 120 \text{ V} = \frac{V_p}{\sqrt{2}}$. The frequency of the AC signal is $\omega = 377 \text{ s}^{-1}$, impedance is $Z = \sqrt{R^2 + (X_L - X_C)^2}$, and the current in the circuit is $I = \frac{V}{Z}$. Voltage in a resistor is in phase with the current, voltage in a capacitor is 90° behind the current, and voltage in an inductor is 90° ahead of the current.

We can use $I_p = \frac{V_p}{Z}$ to find the current in the circuit, then use $V_p = I_p X$ to find the peak voltage in each element with reactance X. Once we have that peak value, we can use the phase of the generator's voltage to find the phase of the voltage in each element and thus the actual voltage in each element.

The component values in the circuit are $R = 50 \ \Omega$, $L = 0.5 \text{ H}$, and $C = 20 \times 10^{-6}$ F.

EVALUATE   We need the reactance of each element, and the impedance of the entire circuit. $X_L = \omega L = 188.5 \ \Omega$, $X_C = \frac{1}{\omega C} = 132.6 \ \Omega$, and $Z = 75.0 \ \Omega$. The peak current is then $I_p = \frac{V_p}{Z} = \frac{V\sqrt{2}}{Z} = 2.26$ A. The phase difference between the peak voltage V_p and peak current I_p is $\phi = \tan^{-1}\left(\frac{X_L - X_C}{R}\right) = 48.2°$.

Now we find a time that fits the description for the voltage from the generator: $V(t) = V_p \sin \omega t = 45 \text{ V} \rightarrow$ $t = 0.00762$ s. (Note that we had to take the second solution to t in this case, to meet the stipulation that the voltage is decreasing.) The phase angle of the voltage at this time is $\theta = \omega t = 2.873 \text{ rad} = 164°$.

For the resistor, voltage is in phase with current, which is 48.2° degrees behind the generator voltage. So $V_R = I_p R \sin \theta_R = I_p R \sin(164.6° - 48.2°) = 101.2$ V.

For the capacitor, voltage is 90° behind the current, so $V_C = I_p X_C \sin \theta_C = I_p \frac{1}{\omega C} \sin(164.6° - 48.2° - 90°) = 133.3$ V.

For the inductor, voltage is 90° ahead of the current, so $V_L = I_p X_L \sin \theta_C = I_p \omega L \sin(164.6° - 48.2° + 90°) = -189.4$ V.

ASSESS   Note that the sum of the voltages on each element is 45 V, which is just the voltage provided by the generator at that instant.

77. **INTERPRET**   We plot two sine waves with given frequencies, amplitudes, and relative phases.

DEVELOP $V(t) = V_{RMS} \sqrt{2} \sin(\omega t + \phi)$. The frequency of each wave needs to be 1.0 kHz, so $\omega = 2000\pi \text{ s}^{-1}$ for each. The RMS amplitude of the first one is 10 V, and the second, 7.1 V. The phase of the second one is $\phi = -90°$ relative to the first.

EVALUATE   See figure below.

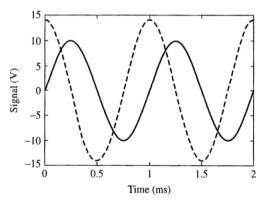

ASSESS   The actual phases are not stated: you may use any you like as long as the difference between the two is 90°.

79. **INTERPRET**   We find the maximum current in an *RLC* circuit at resonance.

DEVELOP   At resonance, the impedance Z is just the resistance R, and the current is the same in all series-circuit elements, so the maximum current in the inductor is just $I_{max} = \frac{V_{max}}{R}$. The max voltage is $V_{max} = 8.0$ V. The resistance is $R = 5.5 \ \Omega$, and we really don't care what the inductor and capacitor values are.

EVALUATE $I_{max} = \frac{8.0 \text{ V}}{5.5 \ \Omega} = 1.45$ A.

ASSESS   This current is within the safe limit.

MAXWELL'S EQUATIONS AND ELECTROMAGNETIC WAVES

EXERCISES

Section 29.2 Ambiguity in Ampère's Law

13. **INTERPRET** In this problem we are asked to find the displacement current through a surface.

    DEVELOP As shown in Equation 29.1, Maxwell's displacement current is

    $$I_d = \varepsilon_0 \frac{d\Phi_E}{dt} = \varepsilon_0 \frac{d(EA)}{dt} = \varepsilon_0 A \frac{dE}{dt}$$

    EVALUATE The above equation gives

    $$I_d = \varepsilon_0 A \frac{dE}{dt} = (8.85 \times 10^{-12}\,\text{C}^2/\text{N} \cdot \text{m}^2)(1\,\text{cm}^2)(1.5\,\text{V/m} \cdot \mu\text{s}) = 1.33\,\text{nA}$$

    ASSESS Displacement current arises from changing electric flux and has units of amperes (A), just like ordinary current. ·

Section 29.4 Electromagnetic Waves

15. **INTERPRET** We are given the electric and magnetic fields of an electromagnetic wave and asked to find the direction of propagation.

    DEVELOP The direction of propagation of the electromagnetic wave is the same as the direction of the cross product $\vec{E} \times \vec{B}$.

    EVALUATE When $\vec{E}$ is parallel to $\hat{j}$ and $\vec{B}$ is parallel to $\hat{i}$, the direction of propagation is parallel to $\vec{E} \times \vec{B}$, or $\hat{j} \times \hat{i} = -\hat{k}$.

    ASSESS For electromagnetic waves in vacuum, the directions of the electric and magnetic fields, and of wave propagation, form a right-handed coordinate system.

Section 29.5 Properties of Electromagnetic Waves

17. **INTERPRET** This problem is about measuring the distance between the Sun and the Earth using light-minutes.

    DEVELOP A light-minute (abbreviated as c-min) is approximately equal to

    $$1\,\text{c-min} = (3 \times 10^8\,\text{m/s})(60\,\text{s}) = 1.8 \times 10^{10}\,\text{m}$$

    On the other hand, the mean distance of the Earth from the Sun (an Astronomical Unit) is about $R_{SE} = 1.5 \times 10^{11}\,\text{m}$.

    EVALUATE In units of c-min, R_{SE} can be rewritten as

    $$R_{SE} = (1.5 \times 10^{11}\,\text{m}) \frac{1\,\text{c-min}}{1.8 \times 10^{10}\,\text{m}} = 8.33\,\text{c-min}$$

    ASSESS The result implies that it takes about 8.33 minutes for the sunlight to reach the Earth.

19. **INTERPRET** In this problem we want to deduce the airplane's altitude by measuring the travel time of a radio wave signal it sends out.

    DEVELOP The speed of light is $c = 3 \times 10^8$ m/s and the total distance traveled is $\Delta r = 2h$.

    EVALUATE Since $\Delta r = 2h = c\Delta t$ (for waves traveling with speed c), the altitude is

    $$h = \frac{c\Delta t}{2} = \frac{(3 \times 10^8\,\text{m/s})(50\,\mu\text{s})}{2} = 7.5\,\text{km}$$

ASSESS   The airplane is flying lower than the typical cruising altitude of 12,000 m (35,000 ft) for commercial jet airplanes.

21. **INTERPRET**   This problem is about the round-trip time delay for radio signals traveling between the Earth and the Moon.

DEVELOP   The time it takes to get a reply is twice the distance (out and back) divided by the speed of light.

EVALUATE   The distance between the Earth and the Moon is $R_{ME} = 3.85 \times 10^8$ m, so the time required is

$$\Delta t = \frac{2R_{ME}}{c} = \frac{2(3.85 \times 10^8 \text{ m})}{3 \times 10^8 \text{ m/s}} = 2.57 \text{ s}$$

ASSESS   The signal has to travel a very long distance, so a time delay of 2.57 seconds is not surprising. The time delay via geostationary satellite communication is typically between 240 ms and 280 ms.

23. **INTERPRET**   In this problem we are asked to find the wavelength of the electromagnetic radiation, given its frequency.

DEVELOP   The wavelength of the electromagnetic wave can be calculated using Equation 29.16c: $f\lambda = c$.

EVALUATE   The wavelength in a vacuum (or air) is

$$\lambda = \frac{c}{f} = \frac{3 \times 10^8 \text{ m/s}}{60 \text{ Hz}} = 5 \times 10^6 \text{ m}$$

ASSESS   The wavelength is almost as large as the radius of the Earth.

25. **INTERPRET**   This problem is about the direction of polarization of an electromagnetic wave.

DEVELOP   The direction of propagation of the electromagnetic wave is the same as the direction of the cross product $\vec{E} \times \vec{B}$. In our case, we have $\hat{E} \times \hat{B} = \hat{k}$, where $\hat{k}$ is the unit vector in the $+ z$ direction.

EVALUATE   Since the magnetic field points in the $+y$ direction, $\vec{B} = B\hat{j}$, we must have $\vec{E} = E\hat{i}$, so that $\hat{i} \times \hat{j} = \hat{k}$. The wave is linearly polarized.

ASSESS   For electromagnetic waves in vacuum, the directions of the electric and magnetic fields, and of wave propagation, form a right-handed coordinate system. One may write $\hat{E} \times \hat{B} = \hat{n}$, where $\hat{n}$ is the unit vector in the direction of propagation.

27. **INTERPRET**   This problem is about the intensity of the light beam after emerging from a polarizer.

DEVELOP   The intensity of the light after emerging from a polarizer is given by the Law of Malus (Equation 29.18), $S = S_0 \cos^2 \theta$, where θ is the angle between the field and the preferred direction.

EVALUATE   From Equation 29.18, we have $S/S_0 = \cos^2 70° = 11.7\%$.

ASSESS   The intensity depends on $\cos^2 \theta$. The limit $\theta = 0$ corresponds to the situation where the direction of polarization of the incident light is the same as the preferred direction specified by the polarizer, and $S = S_0$. On the other hand, when $\theta = 90°$, then no light passes through the polarizer.

Section 29.8 Energy and Momentum in Electromagnetic Waves

29. **INTERPRET**   This problem is about the average intensity of a laser beam associated with a given electric field strength.

DEVELOP   The average intensity of an electromagnetic wave is given by Equation 29.20:

$$\bar{S} = \frac{E_p B_p}{2\mu_0} = \frac{cB_p^2}{2\mu_0} = \frac{E_p^2}{2\mu_0 c}$$

EVALUATE   With $E_p = 3.0 \times 10^6$ V/m, the average intensity is

$$\bar{S} = \frac{E_p^2}{2\mu_0 c} = \frac{(3 \times 10^6 \text{ V/m})^2}{2(4\pi \times 10^{-7} \text{ N/A}^2)(3 \times 10^8 \text{ m/s})} = 1.19 \times 10^{10} \text{ W/m}^2$$

ASSESS   We need a very powerful laser to produce the breakdown field strength. The laser intensity can be compared to the average solar intensity which is about 1370 W/m^2.

31. **INTERPRET** This problem is about the average intensity of a radio signal associated with a given electric field strength.

 DEVELOP The average intensity of an electromagnetic wave is given by Equation 29.20:

$$\bar{S} = \frac{E_p B_p}{2\mu_0} = \frac{cB_p^2}{2\mu_0} = \frac{E_p^2}{2\mu_0 c}$$

 EVALUATE With $E_p = 450\ \mu\text{V/m}$, the average intensity is

$$\bar{S} = \frac{E_p^2}{2\mu_0 c} = \frac{(450\ \mu\text{V/m})^2}{2(4\pi \times 10^{-7}\ \text{N/A}^2)(3 \times 10^8\ \text{m/s})} = 2.69 \times 10^{-10}\ \text{W/m}^2$$

 ASSESS The intensity is very low since the signal is so weak.

33. **INTERPRET** In this problem we are asked to find the intensity of a signal at a distance from the source.

 DEVELOP We can use Equation 29.21, $S = P/4\pi r^2$, for the intensity at a distance r from an isotropic point source, provided we neglect the possible effects of reflections from the ground, etc.

 EVALUATE With $P = 1.0$ kW, the intensity at $r = 5.0$ km is

$$S = \frac{P}{4\pi r^2} = \frac{10^3\ \text{W}}{4\pi(5\ \text{km})^2} = 3.18\ \mu\text{W/m}^2$$

 ASSESS The intensity at this distance is rather weak. Note that S decreases like $1/r^2$.

PROBLEMS

35. **INTERPRET** This problem is about the rate of change of electric field that results in a magnetic field.

 DEVELOP The electric and magnetic fields are related by Equation 29.1. If we evaluate the integrals around the circular field line of radius r shown in Figure 29.15, and the plane area it bounds, we obtain:

$$\oint \vec{B} \cdot d\vec{r} = 2\pi r B = \mu_0 \varepsilon_0 \frac{d}{dt} \int \vec{E} \cdot d\vec{A} = \frac{1}{c^2} \pi r^2 \frac{dE}{dt}$$

 where we have used $c = 1/\sqrt{\varepsilon_0 \mu_0}$.

 EVALUATE **(a)** Thus, the rate of change of electric field is

$$\frac{dE}{dt} = \frac{2c^2 B}{r} = \frac{2(3 \times 10^8\ \text{m/s})^2 (2\mu\text{T})}{0.5\ \text{m}} = 7.20 \times 10^{11}\ \text{V/m} \cdot \text{s}$$

 (b) A circulation of $\vec{B}$ clockwise around the circle gives a positive displacement current into the page, so $\vec{E}$ is increasing in this direction.

 ASSESS Any change in electric flux results in a displacement current which produces a magnetic field. The displacement current encircled by the loop of radius $r = 0.5$ m is

$$I_d = \varepsilon_0 \frac{d\Phi_E}{dt} \int \vec{E} \cdot d\vec{A} = \varepsilon_0 \pi r^2 \frac{dE}{dt} = 5\ \text{A}$$

 This is precisely the current a long wire must carry in order to produce the same magnetic field strength at a distance $r = 0.5$ m from its center.

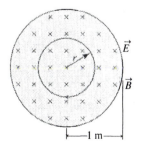

37. **INTERPRET** Given the magnetic field strength, we are asked to find the corresponding electric field strength in an electromagnetic wave.

 DEVELOP For an EM wave in free space, Equation 29.17 gives $E = cB$.

EVALUATE   Given that $B = 50\mu T$, the electric field strength is

$$E = cB = (3 \times 10^8 \text{ m/s})(0.5 \times 10^{-4} \text{ T}) = 15 \text{ kV/m}$$

ASSESS   In an EM wave, both the field strengths E and B are not independent; once one quantity is determined, the other can be found via the relation $E = cB$.

39.   **INTERPRET**   Given the electric field strength, we are asked to find the corresponding magnetic field strength in an electromagnetic wave.
   DEVELOP   For an EM wave in free space, Equation 29.17 gives $E = cB$.
   EVALUATE   Given that $E = 320\mu V/m$, the corresponding magnetic field strength is

$$B = \frac{E}{c} = \frac{320 \ \mu V/m}{3 \times 10^8 \text{ m/s}} = 1.07 \text{ pT}$$

   ASSESS   This is a very small magnetic field. Note that in an EM wave, both the field strengths E and B are not independent; once one quantity is determined, the other can be found via the relation $E = cB$.

41.   **INTERPRET**   This problem is about the fraction of light transmitted as the direction of polarization changes.
   DEVELOP   The fraction of light transmitted can be found by the Law of Malus (Equation 29.18), $S = S_0 \cos^2 \theta$, where θ is the angle between the field and the preferred direction.
   EVALUATE   Since 72° is 18° short of 90° (the alignment for 100% transmission), from Equation 29.18, we find that only $S/S_0 = \cos^2 18° = 90.5\%$ is transmitted.
   ASSESS   The intensity of the laser beam depends on $\cos^2 \theta$. The limit $\theta = 0$ corresponds to the situation where the direction of polarization of the laser beam is the same as the preferred direction specified by the polarizer, and $S = S_0$.

43.   **INTERPRET**   This problem is about the intensity of the light beam after passing through two polarizers that are oriented at different angles to the polarization direction of the light.
   DEVELOP   The intensity of the light after emerging from a polarizer is given by the Law of Malus (Equation 29.18), $S = S_0 \cos^2 \theta$, where θ is the angle between the field and the preferred direction. With two polarizers, we apply the equation twice.
   EVALUATE   Two successive applications of Equation 29.18 yield

$$\frac{S}{S_0} = \cos^2 60° \cos^2 (90° - 60°) = \left(\frac{1}{4}\right)\left(\frac{3}{4}\right) = 0.1875 \approx 18.8\%$$

   ASSESS   To see that the result makes sense, let's solve the problem in two steps. The intensity of the beam after passing the first polarizer with $\theta_1 = 60°$ is $S_1 = S_0 \cos^2 \theta_1$. Since the angle between the first and the second polarizers is $\theta_2 = 90° - 60° = 30°$, upon emerging from the second polarizer, the intensity becomes

$$S_2 = S_1 \cos^2 \theta_2 = (S_0 \cos^2 \theta_1)\cos^2 \theta_2$$

   which is the same as above.

45.   **INTERPRET**   We want to show that the forms of electric and magnetic fields given in the problem satisfy the wave equations.
   DEVELOP   The two relations the fields satisfy are

$$\frac{\partial E}{\partial x} = -\frac{\partial B}{\partial t} \qquad \frac{\partial B}{\partial x} = -\varepsilon_0 \mu_0 \frac{\partial E}{\partial t}$$

To verify the relations, let f and g be arbitrary functions of the variable $u = kx \pm \omega t$. Then

$$\frac{\partial f}{\partial x} = \frac{df}{du} \cdot \frac{\partial u}{\partial x} = f'k \qquad \frac{\partial f}{\partial t} = \frac{df}{du} \cdot \frac{\partial u}{\partial t} = f'(\pm\omega)$$

Similar equations hold for g. We complete the proof by setting $\vec{E} = f(u)\hat{j}$ and $\vec{B} = g(u)\hat{k}$ (a generalization of Equations 29.10 and 29.11).

EVALUATE Equations 29.12 and 29.13 imply $kf' = \pm \omega g'$ and $kg' = \pm \mu_0 \varepsilon_0 \omega f'$. Both conditions are satisfied, provided $\omega/k = 1/\sqrt{\mu_0 \varepsilon_0} = c$, and $f' = g'/\sqrt{\mu_0 \varepsilon_0} = cg'$ (i.e., take f proportional to g). In other words, $\vec{E} = f(kx \pm \omega t)\hat{j}$ and $\vec{B} = \sqrt{\mu_0 \varepsilon_0}\, f(kx \pm \omega t)\hat{k}$ satisfy the wave equation and Maxwell's equations, where $k = \omega\sqrt{\mu_0 \varepsilon_0}$, and f is an arbitrary function.

ASSESS If we differentiate Equation 29.12 with respect to x and Equation 29.13 with respect to t, we get

$$\frac{\partial^2 E}{\partial x^2} = -\frac{\partial^2 B}{\partial t \partial x} \qquad \frac{\partial^2 B}{\partial x \partial t} = -\varepsilon_0 \mu_0 \frac{\partial^2 E}{\partial t^2}$$

Since $\partial^2 B/\partial x \partial t = \partial^2 B/\partial t \partial x$, the two equations can be combined to give

$$\frac{\partial^2 E}{\partial x^2} - \varepsilon_0 \mu_0 \frac{\partial^2 E}{\partial t^2} = \left(\frac{\partial^2}{\partial x^2} - \frac{1}{c^2}\frac{\partial^2}{\partial t^2}\right)E = 0$$

Thus, we see that $\vec{E}$ satisfies the wave equation. One may also verify that $\vec{B}$ satisfies the same equation as well.

47. **INTERPRET** This problem is about the leakage of microwave power within the US safety standard.

DEVELOP The power corresponding to the safety standard of intensity, uniformly distributed over the window area, is $P_{leak} = (10 \text{ mW/cm}^2)(40 \times 17 \text{ cm}^2) = 6.8$ W.

EVALUATE We see that $P_{leak} = 6.8$ W is 1.09% of the microwave's 625 W power output.

ASSESS This is a very small fraction of the overall microwave power. Since the intensity decreases like $1/r^2$, if you stand far away from the microwave, then very little radiation would reach you.

49. **INTERPRET** The problem asks for a comparison of power output between a star and a quasar.

DEVELOP The average intensity of radiation received determines the apparent brightness, so $S_{quasar} = S_{star}$.

EVALUATE From Equation 29.21, $S = P/4\pi r^2$, we see that the above condition implies that $(P/r^2)_{quasar} = (P/r^2)_{star}$, if both behave like isotropic sources. Thus,

$$\frac{P_{quasar}}{P_{star}} = \left(\frac{10^{10} \text{ ly}}{5 \times 10^4 \text{ ly}}\right)^2 = 4 \times 10^{10}$$

ASSESS The luminosity of a quasar is comparable to a galaxy of stars.

51. **INTERPRET** This problem is about the peak electric and magnetic field strengths, given the average power of the light source.

DEVELOP For an isotropic source of electromagnetic waves (in a medium with vacuum permittivity and permeability), Equations 20.20 and 20.21 give

$$\bar{S} = \frac{P}{4\pi r^2} = \frac{E_p B_p}{2\mu_0} = \frac{c B_p^2}{2\mu_0} = \frac{E_p^2}{2\mu_0 c}$$

EVALUATE The peak electric field strength is

$$E_p = \sqrt{\frac{2\mu_0 cP}{4\pi r^2}} = \sqrt{\frac{2(4\pi \times 10^{-7} \text{ N/A}^2)(3 \times 10^8 \text{ m/s})(60 \text{ W})}{4\pi (1.5 \text{ m})^2}} = 40 \text{ V/m}$$

Using Equation 29.17, we find the peak magnetic field strength to be $B_p = E_p/c = 133$ nT.

ASSESS The field strengths due to the light bulb are rather small.

53. **INTERPRET** This problem asks for the energy and momentum carried by the light from a camera flash.

DEVELOP The energy U is the average power times the duration of the flash, and the momentum is simply given by $p = U/c$.

EVALUATE **(a)** The total energy the flash carries is $U = Pt = (2.5 \text{ kW})(1 \text{ ms}) = 2.5$ J.

(b) Similarly, the momentum is

$$p = \frac{U}{c} = \frac{2.5 \text{ J}}{3 \times 10^8 \text{ m/s}} = 8.33 \times 10^{-9} \text{ kg} \cdot \text{m/s}$$

ASSESS A camera flash works by storing energy in a capacitor; releasing the energy then causes a quick bright flash of light.

55.   **INTERPRET**   Given the intensity of the laser beam, we are asked to find the corresponding radiation pressure it exerts on a light-absorbing surface.

DEVELOP   The radiation pressure generated by a totally absorbed electromagnetic wave of given average intensity can be calculated using Equation 29.22: $P_{rad} = \bar{S}/c$.

EVALUATE   Substituting the values given, the radiation pressure is

$$P_{rad} = \frac{\bar{S}}{c} = \frac{180 \text{ W/cm}^2}{3 \times 10^8 \text{ m/s}} = 6 \text{ mPa}$$

ASSESS   The pressure is a lot smaller compared to the normal atmospheric pressure of $P_{atm} = 1.013 \times 10^5$ Pa.

57.   **INTERPRET**   This problem is about solar sailing based on the radiation pressure from sunlight.

DEVELOP   The force on the sail is the radiation pressure times the area, $F = (2S/c)A = ma$ since the sail is perfectly reflecting.

EVALUATE   Thus, the sail area necessary for propulsion of the specified spacecraft is

$$A = \frac{mac}{2S} = \frac{(10^3 \text{ kg})(1 \text{ m/s}^2)(3 \times 10^8 \text{ m/s})}{2(1368 \text{ W/m}^2)} = 1.10 \times 10^8 \text{ m}^2$$

where $S = 1368$ W/m^2 is the sunlight intensity at Earth's orbit. Note that the acceleration of gravity at the Earth's orbit, due to the Sun, is

$$\frac{GM_\odot}{r^2} = g_\odot \left(\frac{R_\odot}{r}\right)^2 = 5.90 \times 10^{-3} \text{ m/s}^2$$

This is small compared to 1 m/s^2, and can be neglected in estimating the sail area.

ASSESS   The area is about 27,000 acres (1 acre = 4047 m^2), making the feasibility of the proposal doubtful.

59.   **INTERPRET**   This problem is about the power of the light source, given the thrust it yields.

DEVELOP   The thrust of a (photon) rocket is the rate that momentum is carried away by its (beam) exhaust:

$$F_{th} = \frac{\Delta p}{\Delta t} = \frac{\Delta U/c}{\Delta t} = \frac{\Delta U}{c\Delta t}$$

EVALUATE   To yield a thrust of $F_{th} = 3.5 \times 10^7$ N, the beam power must be

$$\frac{\Delta U}{\Delta t} = cF_{th} = (3 \times 10^8 \text{ m/s})(3.5 \times 10^7 \text{ N}) = 1.05 \times 10^{16} \text{ W}$$

This is 10^4 times the world's electric power generating capacity.

ASSESS   This is not a practical means of launching payloads off the Earth. The momentum per unit energy of a light beam ($pc/E = 1$) is greater than that of any particle beam ($pc/E = v/c < 1$), so a photon rocket could be more energy-efficient in long-distance, low-thrust space travel.

61.   **INTERPRET**   This problem is about radiation pressure on an absorbing object at the Sun's surface.

DEVELOP   For an isotropic source, the radiation pressure on an opaque (perfectly absorbing) object is

$$P_{rad} = \frac{\bar{S}}{c} = \frac{P}{4\pi r^2 c}$$

EVALUATE   The luminosity of the Sun (power radiated) is $P = 3.85 \times 10^{26}$ W, and the radius of the Sun is $R_S = 6.96 \times 10^8$ m. Thus, the radiation pressure is

$$P_{rad} = \frac{\bar{S}}{c} = \frac{P}{4\pi R_S^2 c} = \frac{3.85 \times 10^{26} \text{ W}}{4\pi(6.96 \times 10^8 \text{ m})^2(3 \times 10^8 \text{ m/s})} = 211 \text{ mPa}$$

ASSESS   This is much greater compared to 4.6 μPa on the surface of the Earth.

63.   **INTERPRET**   Based on the peak electric field strength given at different locations from the light source, we want to deduce the shape of the source.

DEVELOP   As shown in Equation 29.20, $\bar{S} = E_p^2/2\mu_0 c$, the intensity is proportional to the square of the peak electric field. The given data gives

$$\frac{\bar{S}_1}{\bar{S}_2} = \frac{E_{1p}^2}{E_{2p}^2} = \left(\frac{150}{122}\right)^2 = 1.51 \approx \frac{12}{8} = \frac{r_2}{r_1}$$

implying that the ratio of intensities is proportional to that of the inverse distances.

EVALUATE   The above analysis shows that $\bar{S}r$ is roughly constant. The intensity near a long, cylindrically symmetric source, where r is the axial distance, has this space dependence (see Problem 52).

ASSESS   The result can be contrasted with a point source, in which the intensity varies as $1/r^2$.

65. **INTERPRET**   We show directly that a wave moves at speed $\frac{\omega}{k}$.

   DEVELOP   The equation for the electric field in this case is $\vec{E}(x,t) = E_p \sin(kx - \omega t)\hat{j}$, where the phase is $\phi(x,t) = xk - \omega t = \frac{\pi}{2}$. If the phase remains constant, then its derivative is zero: $d\phi/dt = \partial\phi/\partial x\, dx/dt + d\phi/dt = 0$. We will solve this for $x(t)$, which will tell us the velocity.

   EVALUATE

$$\frac{d\phi}{dt} = \frac{\partial\phi}{\partial x}\frac{dx}{dt} + \frac{d\phi}{dt} = k\frac{dx}{dt} - w = 0 \rightarrow v_x = \frac{\omega}{k}$$

   ASSESS   We have shown what was required.

67. **INTERPRET**   From the transmission percentage of a stack of polarizers, we determine how many polarizers the stack contains. We use Malus' law.

   DEVELOP   The Law of Malus is $S = S_0 \cos^2\theta$, where in this case $\theta = 14°$. The first polarizer eliminates 50% of the initially unpolarized light, and each subsequent polarizer is equivalent to multiplying the amount of light remaining by $\cos^2\theta$, so the total percentage of the light that comes through the stack of n polarizers is $S = S_0 \frac{1}{2}(\cos^2\theta)^{n-1}$. $S = 0.37 S_0$, so solve for n.

   EVALUATE

$$0.37\,\cancel{S_0} = \cancel{S_0}\frac{1}{2}(\cos^2\theta)^{n-1} \rightarrow \ln(0.74) = (n-1)\ln(\cos^2\theta) \rightarrow \frac{\ln(0.74) + \ln(\cos^2\theta)}{\ln(\cos^2\theta)} = n = 6$$

   ASSESS   The stack has six sheets. If you got 17.5 sheets, you probably forgot that the first sheet eliminates *half* the unpolarized incident light.

69. **INTERPRET**   We want to make a quarter-wave antenna optimized for a given frequency.

   DEVELOP   We use the relationship between frequency and wavelength, $c = f\lambda$, to find the wavelength of the signal, then divide by 4. $f = 27.3 \times 10^6$ Hz.

   EVALUATE $L = \frac{1}{4}\lambda = \frac{c}{4f} = 2.75$ m.

   ASSESS   This seems reasonable—it's about the length of the long antennas typically seen on pickup trucks.

71. **INTERPRET**   We will use the inverse-square law to find how long it would take to collect 100 joules of energy from the Voyager 2 spacecraft, with a 70-m radio antenna.

   DEVELOP   We will assume that the transmitter on the Voyager 2 spacecraft radiates its $P_t = 21$ W of power uniformly in all directions, so that the intensity back here at Earth is $S = \frac{P_t}{4\pi r^2}$, where $r = 4.5 \times 10^9$ m. The antenna collecting this signal is a "mere" 70 meters in diameter, with area $A = \pi r'^2 (r' = 35$ m). We multiply the signal intensity by the antenna area to find the reception power, then find the time necessary to collect $E = 100$ J of energy.

   EVALUATE

$$P_r t = E = \frac{P_t}{4\cancel{\pi} r^2}(\cancel{\pi} r'^2)t \rightarrow t = \frac{4E}{P_t}\left(\frac{r}{r'}\right)^2 = 3.15 \times 10^{17}\ \text{s}$$

   ASSESS   Just for comparison, the age of the universe is about 4.4×10^{17} s. Another thing to consider is that the transmission antenna does *not* radiate uniformly in all directions, so the actual time may be somewhat shorter; but even with that correction, there's not much to recommend this as an energy source.

REFLECTION AND REFRACTION

EXERCISES

Section 30.1 Reflection

13. **INTERPRET** This problem involves sketching the path of the light ray reflected from the surfaces of two mirrors.

    DEVELOP The path of the reflected ray can be constructed using the law of reflection which states that the angle of incidence equals the angle of reflection: $\theta_1' = \theta_1$.

    EVALUATE The first reflected ray leaves the upper mirror at a grazing angle of 30°, and therefore strikes the lower mirror normally. It is then reflected twice more in retracing its path in the opposite direction.

    ASSESS Our double-mirror arrangement is a retroreflector that sends light rays back to their point of origin. Retroreflection has many practical applications (e.g., taillights, stop signs, etc.).

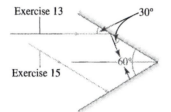

15. **INTERPRET** This problem is about the direction of the light ray after undergoing reflection from a two-mirror arrangement.

    DEVELOP The path of the reflected ray can be constructed using the law of reflection: the angle of incidence equals the angle of reflection: $\theta_1' = \theta_1$.

    EVALUATE Entering parallel to the top mirror, a ray makes an angle of incidence of 30° with the bottom mirror. It then strikes the top mirror also at 30° incidence, and is reflected out of the system parallel to the bottom mirror (see the figure). The total deflection is 240° CCW (or 120° CW) from the incident direction.

    ASSESS The reflected path follows from the law of reflection.

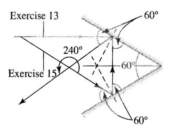

Section 30.2 Refraction

17. **INTERPRET** This problem is about the pit depth of a CD, given the wavelength of the laser light used and the index of refraction of the CD.

    DEVELOP Using Equation 30.2, $n = c/v$, and the reasoning in Example 30.3, the wavelength in the plastic can be written as $\lambda = \lambda_{air}/n$. The pit depth is $\lambda/4$.

EVALUATE Substituting the values given, we find the wavelength of the laser light to be

$$\lambda = \frac{\lambda_{air}}{n} = \frac{780 \text{ nm}}{1.55} = 503 \text{ nm}$$

The pit depth is one quarter of this, or 126 nm.

ASSESS A typical pit on a CD is about 100 nm deep and 500 nm wide. Our result in within this range.

19. **INTERPRET** This problem is about refraction at an interface. We apply Snell's law to find the index of refraction of the material.

    DEVELOP Snell's law (Equation 30.3) states that $n_1 \sin\theta_1 = n_2 \sin\theta_2$, where n_1 and n_2 are the refractive indices of the two media, and θ_1 and θ_2 are the angles the light ray makes with the normal of the surface.

    EVALUATE With air as medium 1, we get

    $$n_2 = \frac{n_1 \sin\theta_1}{\sin\theta_2} = \frac{1(\sin 24°)}{\sin 15°} = 1.57$$

    ASSESS Since $\theta_2 < \theta_1$, the light ray bends toward the normal and we expect $n_2 > n_1$.

21. **INTERPRET** This problem asks for the polarizing angle of the air-diamond interface.

    DEVELOP Using Equation 30.4, the polarizing angle for light in air reflected from diamond is $\theta_p = \tan^{-1}(n_{diamond}/n_{air})$.

    EVALUATE With $n_{diamond} = 2.419$, we find the polarizing angle to be

    $$\theta_p = \tan^{-1}(n_{diamond}/n_{air}) = \tan^{-1}(2.419/1) = 67.5°$$

    ASSESS At this angle, the reflected light ray is perpendicular to the transmitted light ray, as illustrated in Figure 30.9.

Section 30.3 Total Internal Reflection

23. **INTERPRET** We want to find in this problem the critical angle for total internal reflection in various media.

    DEVELOP For $n_{air} \approx 1$, using Equation 30.5, the critical angle for total internal reflection in a medium of refractive index n is (air is medium-2) $\theta_c = \sin^{-1}(n_2/n_1)$.

    EVALUATE (a) From Table 30.1, we find $n = 1.309$ for ice, so $\theta_c = \sin^{-1}(1/1.309) = 49.8°$.

    (b) With $n = 1.49$ for polystyrene, $\theta_c = \sin^{-1}(1/1.49) = 42.2°$.

    (c) Similarly, for rutile, $n = 2.62$ and $\theta_c = \sin^{-1}(1/2.62) = 22.4°$.

    ASSESS The larger n, the smaller critical angle. Light incident at $\theta \geq \theta_c$ cannot escape from the medium.

25. **INTERPRET** In this problem we want to find the critical angle for total internal reflection in various media.

    DEVELOP The critical angle in medium-1, at an interface with medium-2, is given by Equation 30.5: $\theta_c = \sin^{-1}(n_2/n_1)$, where $n_1 > n_2$.

    EVALUATE (a) For glass ($n_1 = 1.52$) immersed in water ($n_2 = 1.333$), the critical angle is $\theta_c = \sin^{-1}(n_2/n_1) = \sin^{-1}(1.333/1.52) = 61.3°$.

    (b) The same glass immersed in benzene has $\theta_c = \sin^{-1}(1.501/1.52) = 80.9°$.

    (c) Since the index of refraction of diiodomethane ($n_2 = 1.738$) is not smaller than that for this glass, there is no total internal reflection for light propagating in the glass. However, for light originating in the liquid, the critical angle at the glass interface is $\theta_c' = \sin^{-1}(1.52/1.738) = 61.0°$.

    ASSESS This problem shows that for total internal reflection to take place as light propagates from medium 1 to medium 2, we must have $n_1 > n_2$.

Section 30.4 Dispersion

27. **INTERPRET** This problem is about the angle between the red and the blue light in glass due to dispersion.

    DEVELOP Using Snell's law the angle of refraction for either wavelength is $\theta_2 = \sin^{-1}(\sin 50°/n)$. The angle between the two laser beams is $\theta_{2,red} - \theta_{2,blue}$.

    EVALUATE Substituting the values given in the problem statement, we get

    $$\theta_{2,red} - \theta_{2,blue} = \sin^{-1}(\sin 50°/1.621) - \sin^{-1}(\sin 50°/1.680) = 28.20° - 27.13° = 1.07°$$

    ASSESS The refractive index of a material is higher for blue light than the red light. So, from Equation 30.2, we expect red light to travel at a greater speed than the blue light.

PROBLEMS

29. **INTERPRET** This problem involves sketching the path of the light ray reflected from the surfaces of two mirrors.
DEVELOP The path of the reflected ray can be constructed using the law of reflection: the angle of incidence equals the angle of reflection: $\theta'_1 = \theta_1$.
EVALUATE Now, after the first reflection, the ray leaves the top mirror at a grazing angle of $37\frac{1}{2}°$, and so makes a grazing angle of $180° - 75° - 37\frac{1}{2}° = 67\frac{1}{2}°$ with the bottom mirror. It is therefore deflected through an angle of $2(37\frac{1}{2}°) + 2(67\frac{1}{2})° = 210°$ CW as it exits the system, after being reflected once from each mirror.
ASSESS The reflected path follows from the law of reflection.

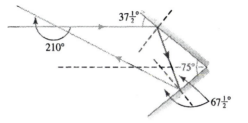

31. **INTERPRET** This problem is about the light ray reflected from an arrangement of two mirrors at an angle.
DEVELOP Consider the figure shown. A ray incident on the first mirror at a grazing angle α is deflected through an angle 2α (this follows from the law of reflection). It strikes the second mirror at a grazing angle β, and is deflected by an additional angle 2β.
EVALUATE The total deflection is

$$\Delta\theta = 2\alpha + 2\beta = 2(180° - \phi)$$

ASSESS When $\phi = 90°$, the total deflection is $\Delta\theta = 180°$. This means that the final direction of the light ray is opposite of its original direction.

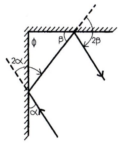

33. **INTERPRET** This problem is about refraction at the air-water interface. The mark you see is the point at the bottom of the tank along the refracted path.
DEVELOP As shown in the diagram, the mark seen on the meter stick is at position

$$x = x_1 + x_2 = (L - h) + h\tan\theta_2$$

where $\theta_2 = \sin^{-1}(\sin 45°/1.333)$. Thus, $x = 40$ cm $- 0.374h$.
EVALUATE **(a)** For $h = 0$ (empty), $x = 40$ cm.
(b) For $h = 20$ cm (half full), $x = 40$ cm $- 0.374(20$ cm$) = 32.5$ cm.
(c) Similarly, for $h = 40$ cm (full), $x = 40$ cm $- 0.374(40$ cm$) = 25.04$ cm.
ASSESS The more water in the tank, the more "bending" the path, and hence the smaller mark on the meter stick you see.

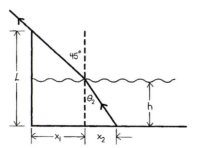

35.   **INTERPRET**   This problem is about the refraction of light rays in water.

DEVELOP   If the beam enters at the rim of the tank, the maximum angle of refraction it can have and still reach the bottom is

$$\theta_2^{max} = \tan^{-1}\left(\frac{11}{10}\right) = 47.7°$$

From Snell's law, the maximum angle of incidence is given by $n_1 \sin\theta_1^{max} = n_2 \sin\theta_2^{max}$.

EVALUATE   Using the above equation, we see that the angle of incidence in air must be less than

$$\theta_1^{max} = \sin^{-1}(n_2 \sin\theta_2^{max}/1) = \sin^{-1}(1.33\sin 47.7°) = 80.5°$$

In other words, the grazing angle α (with the horizontal water surface) must be greater than

$\alpha^{min} = 90° - \theta_1^{max} = 9.48°$.

ASSESS   Below $\alpha^{min} = 9.48°, \theta_2 > \theta_2^{max} = 47.7°$ and the refracted path will not reach the bottom of the tank.

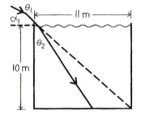

37.   **INTERPRET**   This problem is about the direction of refracted light rays in water.

DEVELOP   From Figure 30.22, we have

$$x_1 = h_1 \tan\theta_1 = (0.5 \text{ m})\tan 40° = 0.420 \text{ m}$$
$$x_2 = h_2 \tan\theta_2 = (1.6 \text{ m})\tan\theta_2$$

From Snell's law, $\sin\theta_2 = \sin\theta_1/n_2$, or $\theta_2 = \sin^{-1}(\sin 40°/1.333) = 28.8°$ ($n_1 = 1$ for air). This gives $x_2 = 0.881$ m.

EVALUATE   Thus, the total horizontal distance from the dock is $x_1 + x_2 = 1.30$ m.

ASSESS   The horizontal distance increases with θ_1. So, the smaller the angle θ_1, the closer the keys are to the dock.

39.   **INTERPRET**   This problem is about light incident on a prism. We want to find the angle through which the light beam is deflected.

DEVELOP   From Snell's law and plane geometry, we have $\theta_2 = \sin^{-1}(\sin\theta_1/n)$ (Snell's law for the first refraction, with $n_1 = 1$ and $n_2 = n$), $\phi_2 = \alpha - \theta_2$, where α is the exterior angle to the triangle formed by the ray segment in the prism and the normals to the surfaces, $\phi_1 = \sin^{-1}(n\sin\phi_2)$ (Snell's law for the second refraction).

The total deflection is the sum of the deflections at each refraction, taking clockwise deflection to be positive in Fig. 30.23. Substituting the expressions obtained above, one gets (see Problem 62)

$$\delta = \theta_1 - \theta_2 + \phi_1 - \phi_2 = \theta_1 + \phi_1 - \alpha$$
$$= \theta_1 + \sin^{-1}\left\{n\sin\left[\alpha - \sin^{-1}\left(\frac{\sin\theta_1}{n}\right)\right]\right\} - \alpha$$

EVALUATE   For the data in this problem, the other angles and the deflection are:

$$\theta_2 = \sin^{-1}(\sin 37°/1.52) = 23.3°$$
$$\phi_2 = 60° - \theta_2 = 36.7°$$
$$\phi_1 = \sin^{-1}(1.52\sin 36.7°) = 65.2°$$

Therefore, the total deflection is

$$\delta = \theta_1 + \phi_1 - \alpha = 37° + 65.2° - 60° = 42.2°$$

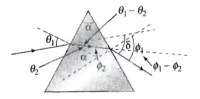

ASSESS   The total deflection δ is a complicated nonlinear function of θ_1, as shown in the figure.

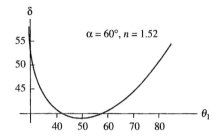

41.   **INTERPRET**   We find the minimum index of refraction for a prism (shown in Figure 30.11) so that total internal reflection occurs. The prism is surrounded by air and we will use the equation for critical angle.
      DEVELOP   The equation for critical angle is $\sin\theta_c = n_2/n_1$, where $n_2 \approx 1$ and by inspection of the figure, $\theta_c = 45°$. We solve for n_1.
      EVALUATE

$$n_1 = \frac{n_2}{\sin\theta_c} = 1.41$$

ASSESS   Most types of glass and clear plastic have an index of refraction greater than 1.5, so this type of prism is not unusual. Take apart a pair of binoculars some time, and you'll see several of these.

43.   **INTERPRET**   This problem is about the speed of light in a medium, given its critical angle at an interface with air.
      DEVELOP   From Equations 30.2 and 30.5, the relationship between the critical angle and the speed of light in a material can be written as

$$\sin\theta_c = \frac{n_{air}}{n} \approx \frac{2}{n} = \frac{v}{c}$$

EVALUATE   With $\theta_c = 61°$, the speed of light in the medium is
$$v = c\sin\theta_c = (3\times 10^8 \text{ m/s})\sin 61° = 2.62\times 10^8 \text{ m/s}$$

ASSESS   The critical angle and the speed of light in a material are both related to the index of refraction. When $n = n_{air}$, $\sin\theta_c = 1$ and $v = c$, as expected.

45.   **INTERPRET**   In this problem we look for the connection between the critical angle and the polarizing angle.
      DEVELOP   For an interface with air, the critical angle is usually specified for incident light in the material ($n_1 = n$ and $n_2 = 1$ in Equation 30.5), so $\sin\theta_c = 1/n$. However, the polarizing angle for the same interface is usually specified for incident light in air ($n_1 = 1$ and $n_2 = n$ in Equation 30.4), so $\tan\theta_p = n$.
      EVALUATE   Combining the two equations, we obtain

$$\frac{1}{n} = \sin\theta_c = \frac{1}{\tan\theta_p} = \cot\theta_p$$

ASSESS   This is not a fundamental relation; it merely reflects the fact that both angles depend on the relative index of refraction.

47.   **INTERPRET**   This problem is about the critical angle for total internal reflection at the glass-air interface.
      DEVELOP   The critical angle for the blue light and the red light are
$$\theta_{c,\text{blue}} = \sin^{-1}(1/1.680) = 36.5°$$
$$\theta_{c,\text{red}} = \sin^{-1}(1/1.621) = 38.1°$$

EVALUATE   For incidence angles between these values ($\theta_{c,\text{blue}} < \theta < \theta_{c,\text{red}}$), blue light will be totally reflected, while some red light is refracted at the glass-air interface.
ASSESS   Since the refractive index for the blue light is greater than that of the red light ($n_{\text{blue}} > n_{\text{red}}$), the critical angle for blue light is less than for red light.

49.   **INTERPRET**   This problem is about the connection between the speed of light in a medium, and its critical angle at an interface with air.

DEVELOP   The speed of light in a medium with refractive index n is (Equation 30.2) $v = c/n$. On the other hand, from Equation 30.5, the critical angle of the medium is $\sin\theta_c = 1/n$. Thus, the relationship between the critical angle and the speed of light in a material can be written as

$$\sin\theta_c = \frac{n_{air}}{n} \approx \frac{2}{n} = \frac{v}{c}$$

EVALUATE   From the above equation, we find the speed of light in the medium is

$$v = c\sin\theta_c$$

ASSESS   The critical angle and the speed of light in a material are both related to the index of refraction. When $n = n_{air}$, $\sin\theta_c = 1$ and $v = c$, as expected.

51.   **INTERPRET**   We are asked to compute the speed of light in the crystal, given the incidence angle and the angle of refraction.

DEVELOP   The speed of light in a medium with refractive index n is (Equation 30.2) $v = c/n$. On the other hand, Snell's law in Equation 30.3 gives $\sin\theta_1 = n\sin\theta_2$, where medium 1 is air having $n_1 = 1$.

EVALUATE   Combining Equations 30.2 and 30.3, we find

$$v = \frac{c}{n} = \frac{c\sin\theta_2}{\sin\theta_1} = \frac{(3\times10^8 \text{ m/s})\sin 22°}{\sin 35°} = 1.96\times10^8 \text{ m/s}$$

ASSESS   The speed of light in the crystal is lower than in vacuum by $n = \sin\theta_1/\sin\theta_2 = 1.53$.

53.   **INTERPRET**   This problem is about refraction of sunlight. We want to know the diameter of the tank such that sunlight can reach part of the tank bottom whenever the Sun is above the horizon.

DEVELOP   The rays of sunlight which first hit the bottom of the tank just skim the opposite edge of the rim. The diameter and depth of the tank (d and h) are related to the angle of refraction by $\tan\theta_2 = d/h$. Combining this with Snell's law (Equation 30.3), we find ($n_1 = 1$ for air and $n_2 = 1.333$ for water)

$$d = h\tan\theta_2 = h\tan[\sin^{-1}(n_1\sin\theta_1/n_2)]$$

EVALUATE   If we let θ_1 approach 90° (Sun angle approaches 0°), then the tank diameter becomes

$$d \rightarrow (2.4 \text{ m})\tan[\sin^{-1}(n_1/n_2)] = \frac{(2.4 \text{ m})}{\sqrt{(n_2/n_1)^2 - 1}} = 2.72 \text{ m}$$

where we have used $\tan\theta = \sin\theta/\sqrt{1 - \sin^2\theta}$.

ASSESS   If the diameter is smaller than 2.72 m, then in order for the sunlight to reach the bottom of the tank, a smaller value of θ_1 would be required. The diameter of the tank as a function of θ_1 is depicted in the figure.

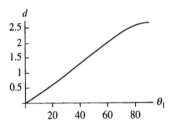

55.   **INTERPRET**   This problem involves refraction in three media, each having a different index of refraction. We want to show that the index of the intermediate material does not matter.

DEVELOP   Since medium-2 has parallel interfaces with media-1 and 3, the angle of refraction at the 1-2 interface equals the angle of incidence at the 2-3 interface, as shown. The normals to the interfaces are also parallel, so the alternate angles, marked θ_2, are equal.

EVALUATE Thus Snell's law implies

$$n_1 \sin\theta_1 = n_2 \sin\theta_2 = n_3 \sin\theta_3$$

Thus, the angles in media-1 and 3 are related ($n_1 \sin\theta_1 = n_3 \sin\theta_3$) as if media-2 were not present.

ASSESS A parallel-sided slab does not deflect rays of light; the rays are displaced somewhat, but remain parallel to their incident direction.

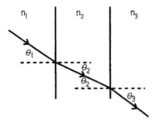

57. **INTERPRET** This problem is about finding the index of refraction of the medium, with the path of the light ray specified.

DEVELOP In the figure below, the incident ray appears to hit the cube at the center of a side. Since the thickness of each material is the same, we have $x_1 = L \tan\theta_1$, $x_2 = L \tan\theta_2$, and $x_1 + x_2 = L$. Thus $\tan\theta_1 + \tan\theta_2 = 1$, where θ_1 and θ_2 are the angles of refraction in the two materials, as shown. From Snell's law,

$$\sin 35° = n_1 \sin\theta_1 = n_2 \sin\theta_2$$

so θ_1 and θ_2 can be eliminated in terms of the indices of refraction of the two materials, and since n_1 is given, n_2 can be easily determined.

EVALUATE Substituting the values given, we have

$$\theta_1 = \sin^{-1}(\sin 35°/1.43) = 23.6°$$
$$\theta_2 = \tan^{-1}(1 - \tan\theta_1) = 29.3°$$

Using Snell's law, the refractive index of the right-hand slab is $n_2 = \sin 35°/\sin\theta_2 = 1.17$.

ASSESS In order to write the solution compactly for general values of n_1 and n_2, first notice that

$$\tan\theta = \frac{\sin\theta}{\cos\theta} = \frac{\sin\theta}{\sqrt{1 - \sin^2\theta}} = \frac{1}{\sqrt{\csc^2\theta - 1}}$$

and that $n_1 \csc 35° = \csc\theta_1$ and $n_2 \csc 35° = \csc\theta_2$ (recall that the cosecant is the reciprocal of the sine). Then

$$\tan\theta_1 + \tan\theta_2 = \frac{1}{\sqrt{n_1^2 \csc^2 35° - 1}} + \frac{1}{\sqrt{n_2^2 \csc^2 35° - 1}} = 1$$

Since $n_1 = 1.43$ one finds $\sqrt{n_2^2 \csc^2 35° - 1} = 1.78$, and $n_2 = 1.17$, which is the same as before.

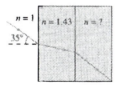

59. **INTERPRET** This problem involves two refractions and a total internal reflection as the light ray passes through a spherical raindrop.

DEVELOP The angle ϕ can be found by totaling the deflections each time the ray in Figure 30.25 is refracted or reflected. The deflection at A is $\theta - \theta'$, at B is $180° - 2\theta'$, and at C is $\theta - \theta'$. The sum is

$$\delta = 2(\theta - \theta') + 180° - 2\theta' = 180° + 2\theta - 4\theta'$$

and is related to ϕ by $180° = \phi + \delta$, so $\phi = 4\theta' - 2\theta$.

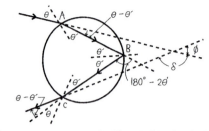

EVALUATE   By eliminating θ' using Snell's law, $\sin\theta' = \sin\theta/n$, the desired expression

$$\phi = 4\sin^{-1}(\sin\theta/n) - 2\theta$$

is obtained. Note that light incident at the boundaries of the drop, at A, B, and C, is partially reflected and partially refracted; we show only the rays relevant to the formation of a rainbow.

ASSESS   The angle ϕ is a complicated nonlinear function of θ, as shown on the right (with $n = 1.333$). The maximum value of ϕ is approximately equal to $42.1°$. This is the average angle, above the anti-solar direction, that an observer sees a rainbow, because n is the average index of refraction for visible wavelengths.

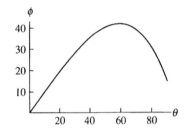

61.   **INTERPRET**   This problem involves two refractions and two total internal reflections as the light ray passes through a spherical raindrop.

DEVELOP   The analysis of the secondary rainbow is similar to that of the primary rainbow (see Problems 58 and 59). The angles for an incident ray, which suffers two internal reflections in a spherical drop of water, are shown for the emergent ray traveling downward to an observer on the ground. The total deflection for two refractions and two internal reflections,

$$\delta = 2(\theta - \theta') + 2(180° - 2\theta')$$

is related to the observation angle from the anti-solar direction, ϕ, by $\delta = 180° + \phi$.

EVALUATE   Combining the two equations, we obtain

$$\phi = 2\theta - 6\theta' + 180° = 2\theta - 6\sin^{-1}(\sin\theta/n) + 180°$$

If we differentiate Snell's law with respect to θ, and substitute for θ' in terms of ϕ and θ, we get

$$\frac{d}{d\theta}(\sin\theta) = \cos\theta = \frac{d}{d\theta}(n\sin\theta') = n\cos\theta'\frac{d\theta'}{d\theta} = n\sqrt{1-\sin^2\theta'}\,\frac{d}{d\theta}\left[\frac{1}{6}(2\theta + 180° - \phi)\right]$$

$$= \sqrt{n^2 - \sin^2\theta}\left(\frac{1}{3} - \frac{1}{6}\frac{d\phi}{d\theta}\right)$$

A concentrated beam is formed for the incident angle which makes $d\phi/d\theta = 0$. Thus,

$$\cos\theta_m = \frac{1}{3}\sqrt{n^2 - \sin^2\theta_m}$$

which implies $\cos^2\theta_m = (n^2 - 1)/8$, and $\sin^2\theta_m = (9 - n^2)/8 = n^2\sin^2\theta_m'$. Finally, the maximum value of ϕ is

$$\phi_{max} = 2\theta_m - 6\theta_m' + 180° = 2\sin^{-1}\left(\sqrt{\frac{9-n^2}{8}}\right) - 6\sin^{-1}\left(\sqrt{\frac{9-n^2}{8n^2}}\right) + 180°$$

For $n = 1.333$, the average angle is $50.9°$. However, substituting $n_{red} = 1.330$ and $n_{violet} = 1.342$, we obtain $\phi_{max,red} = 50.10°$ and $\phi_{max,violet} = 53.22°$ for the secondary rainbow.

ASSESS   Since $\phi_{max,red} < \phi_{max,violet}$, the colors appear in the reverse order from that in the primary rainbow. Although the deflection for violet rays is always larger than that for red rays (no matter how many internal reflections are considered), the relation between ϕ and δ depends on the quadrant of δ, and is different for the primary and secondary rainbows.

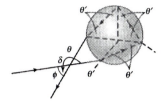

63.   **INTERPRET**   This problem is about the reversal of the direction of an incident light ray by a corner reflector.

DEVELOP   A single plane mirror reverses the direction of just the normal component of a ray striking its surface. For example, a ray incident in the direction

$$\hat{q} = \cos\alpha_x \hat{i} + \cos\alpha_y \hat{j} + \cos\alpha_z \hat{k}$$

on a mirror normal to the x axis, is reflected into the direction

$$-\cos\alpha_x \hat{i} + \cos\alpha_y \hat{j} + \cos\alpha_z \hat{k}$$

In our notation, $\hat{q}$ is a unit vector, and $\cos^2\alpha_x + \cos^2\alpha_y + \cos^2\alpha_z = 1$.

EVALUATE   If the ray also strikes mirrors which are normal to the y and z axes, as in a corner reflector, it emerges in the direction $\hat{q}' = -\hat{q}$, or opposite to the initial direction.

ASSESS   In order to strike all three mirrors, the direction cosines of the incident ray must have magnitudes greater than some minimum non-zero value, depending on the size of the reflector (i.e., $|\cos\alpha_i| > 0$ for $i = x$, y, and z).

65.   **INTERPRET**   This problem is to demonstrate that Snell's law for refraction follows Fermat's principle.

DEVELOP   Take coordinate system of the problem is shown on the right. Since the velocity of light for the two rays AP and PB, which cross the interface, is different, Fermat's principle requires that

$$t = \frac{AP}{v_1} + \frac{PB}{v_2} = \frac{\sqrt{x^2 + y_A^2 + z^2}}{v_1} + \frac{\sqrt{(x_B - x)^2 + y_B^2 + z^2}}{v_2}$$

be an extremum.

EVALUATE   The condition $\partial t/\partial z = 0$ implies both rays lie in the plane of incidence (i.e., $z = 0$). The condition

$$0 = \frac{\partial t}{\partial x} = \frac{x}{v_1\sqrt{x^2 + y_A^2}} - \frac{x_B - x}{v_2\sqrt{(x_B - x)^2 + y_B^2}}$$

then gives Snell's law, when the angles of incidence ($\sin\theta_1 = x/\sqrt{x^2 + y_A^2}$) and refraction ($\sin\theta_2 = (x_B - x)/\sqrt{(x_B - x)^2 + y_B^2}$) are substituted.

ASSESS   Snell's law (law of refraction) as well as the law of reflection follow directly from Fermat's principle which states that light takes the path that requires the shortest time.

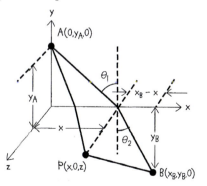

67.   **INTERPRET**   We find the wavelength of light in a material. We use the angle of incidence, the angle of refraction, and Snell's law to find the index of refraction, then use $\lambda' = \lambda/n$.

DEVELOP   Snell's law states that $n_1 \sin\theta_1 = n_2 \sin\theta_2$. The angle of incidence is $\theta_1 = 50°$, the angle of refraction is $\theta_2 = 27°$, and $n_1 \approx 1$, so we can find n_2. We will then use $\lambda_2 = \lambda_1/n_2$ to find the wavelength in the material, which must be between 350 and 400 nm in order for the device to function properly. The initial wavelength of the light is $\lambda = 633$ nm.

EVALUATE

$$n_2 = n_1 \frac{\sin\theta_1}{\sin\theta_2} = 1.687 \rightarrow \lambda_2 = \frac{633 \text{ nm}}{1.687} = 375 \text{ nm}$$

ASSESS   The wavelength meets the requirements.

69.   **INTERPRET**   We find the critical angle for an interface between glass and titanium dioxide, given the index of refraction of both materials.

DEVELOP   The critical angle is given by $\sin\theta_c = n_2/n_1$, where in this case $n_2 = 1.60$, and $n_1 = 2.62$. In order for the device to work, the critical angle must be larger than 60°.

EVALUATE

$$\theta_c = \sin^{-1}\frac{n_2}{n_1} = 37.6°$$

ASSESS   This critical angle is less than 60°, so it will not meet the specifications given in the problem.

IMAGES AND OPTICAL INSTRUMENTS

EXERCISES

Section 31.1 Images with Mirrors

17.   **INTERPRET** This problem is about image formation in a plane mirror.

   DEVELOP A small mirror (M) on the floor intercepts rays coming from a customer's shoes (O), which are traveling nearly parallel to the floor. The angle to the customer's eye (E) from the mirror is twice the angle of reflection, so

$$\tan 2\alpha = \frac{h}{d}$$

   EVALUATE Solving for the angle, we find

$$\alpha = \frac{1}{2}\tan^{-1}\left(\frac{h}{d}\right) = \frac{1}{2}\tan^{-1}\left(\frac{140 \text{ cm}}{50 \text{ cm}}\right) = 35.2°$$

for the given distances. Therefore, the plane of the mirror should be tilted by $35.2°$ from the vertical to provide the customer with a floor-level view of her shoes.

   ASSESS The angle tilted decreases if d is increased, and vice versa. This is what we expect from our experience at shoe stores.

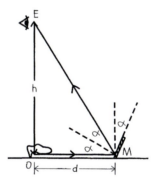

19.   **INTERPRET** We have an image formation problem involving a concave mirror. We want to know the orientation of the image as well as its height compared to the object.

   DEVELOP The magnification M, the ratio of the image height h' to object height h, is given by Equation 31.1:

$$M = \frac{h'}{h} = -\frac{s'}{s}$$

where s and s' are object and image distances to the mirror. The two quantities s and s' are related by the mirror equation (Equation 31.2):

$$\frac{1}{s} + \frac{1}{s'} = \frac{1}{f}.$$

where f is the focal length of the mirror.

   EVALUATE **(a)** One can solve Equation 31.2 for s', and substitute into Equation 31.1, to yield

$$M = \frac{h'}{h} = -\frac{s'}{s} = -\frac{sf/(s-f)}{s} = -\frac{f}{s-f} = -\frac{f}{5f-f} = -\frac{1}{4}$$

   (b) A negative magnification applies to a real, inverted image.

ASSESS   The situation corresponds to the first case shown in Table 31.1. From the ray diagram, we see that the image is real, inverted, and reduced in size. Since $s' > 0$, the image is in front of the mirror.

21.   **INTERPRET**   This problem is about image formation in a concave mirror. We want to know the distance the object should be placed in order to produce a full-size image.

DEVELOP   The magnification M, the ratio of the image height h' to object height h, is given by Equation 31.1:

$$M = \frac{h'}{h} = -\frac{s'}{s}$$

where s and s' are object and image distances to the mirror. A full-size image means that $|M| = 1$.

EVALUATE   (a) For a full-sized image, we require

$$M = \frac{h'}{h} = -\frac{s'}{s} = -\frac{sf/(s-f)}{s} = -\frac{f}{s-f} = -1$$

or $s = 2f = R$. That is, put the object at the center of curvature of the mirror (or at twice its focal length).

(b) From Equation 31.2, we have

$$s' = \frac{sf}{s-f} = \frac{(2f)f}{2f-f} = 2f > 0$$

Thus, the image is real.

ASSESS   For a spherical mirror, $|M| = 1$ applies only to a real image, $h' = -h$, unless one allows the plane mirror, $f = \infty$, as a special case.

Section 31.2 Images with Lenses

23.   **INTERPRET**   We have an image formation problem involving a converging lens. We want to know magnification as well as the orientation of the image.

DEVELOP   The magnification of a thin lens, for paraxial rays, is given by Equation 31.4,

$$M = \frac{h'}{h} = -\frac{s'}{s} = -\frac{sf/(s-f)}{s} = -\frac{f}{s-f}$$

where we used the lens equation (Equation 31.5) to eliminate s'.

EVALUATE   (a) If $s = 1.5f$, then $M = -\frac{f}{s-f} = -\frac{f}{1.5f-f} = -2$. The minus sign means that the image is real and inverted.

ASSESS   The lens in this problem corresponds to the second case shown in Table 31.2. With $2f > s\ (1.5f) > f$, we get a real, inverted, and enlarged image.

25.   **INTERPRET**   In this problem we are asked about the focal length of a magnifying glass which is a converging lens, given the object and image distances.

DEVELOP   The focal length f is related to the object and image distances through the lens equation (Equation 31.5):

$$\frac{1}{f} = \frac{1}{s} + \frac{1}{s'} \quad \rightarrow \quad f = \frac{ss'}{s+s'}$$

EVALUATE   Since the distances are given from the lamp (object), $s = 25$ cm and $s' = 1.6$ m $- 25$ cm $= 135$ cm. Substituting these into the lens equation, we find

$$f = \frac{ss'}{s+s'} = \frac{(25 \text{ cm})(135 \text{ cm})}{25 \text{ cm} + 135 \text{ cm}} = 21.1 \text{ cm}$$

ASSESS   Since $f < s < 2f$, we expect the image to be real, inverted, and enlarged.

27.   **INTERPRET**   We find the focal length of a magnifying glass from the magnification at a certain distance. We will use the lens equation and the definition of magnification.

DEVELOP   For a single lens such as this, magnification is defined as $M = -s'/s$. We are told that the object distance $s = 9.0$ cm and the magnification is $M = 1.5$, so we can find the image distance s'. The lens equation relates the image and object distances to the focal length by $1/s + 1/s' = 1/f$, so knowing the distances we can find the focal length.

EVALUATE

$$s' = -Ms \rightarrow \frac{1}{s} - \frac{1}{Ms} = \frac{1}{f} \rightarrow \frac{1}{s}\left(1 - \frac{1}{M}\right) = \frac{1}{f} \rightarrow f = \frac{s}{1 - \frac{1}{M}} = 27 \text{ cm}$$

ASSESS Be careful about the signs in problems such as these. We must remember that magnification is the *negative* of the ratio of distances: if we'd missed that detail we'd have gotten an answer of $f = 5.4$ cm.

Section 31.3 Refraction in Lenses: The Details

29. **INTERPRET** This problem involves image formation by a single refracting interface between two media.

    DEVELOP The image formed by a refracting interface is described by Equation 31.6,

    $$\frac{n_1}{s} + \frac{n_2}{s'} = \frac{n_2 - n_1}{R}$$

    For the flat surface of the wading pool, the radius of curvature is $R = \infty$. The index of refraction is $n_1 = 1.333$ for water, s is the depth of the pool (your feet, the object, are on the bottom), $n_2 = 1$ for air, and $s' = -30$ cm (for a virtual image at the apparent depth).

    EVALUATE Thus, the above equation gives

    $$\frac{1.333}{s} + \frac{1}{-30 \text{ cm}} = \frac{n_2 - n_1}{\infty} = 0$$

    or $s = 40$ cm.

    ASSESS This problem could also be solved directly from Snell's law, without the paraxial ray approximation. Note that the image formed by a flat refracting surface is always on the same side of the surface as the object.

31. **INTERPRET** This problem involves image formation by a curved refracting interface between two media.

    DEVELOP The image formed by a refracting interface is described by Equation 31.6,

    $$\frac{n_1}{s} + \frac{n_2}{s'} = \frac{n_2 - n_1}{R}$$

    The equation can be used to solve for the apparent distance s'.

    EVALUATE With $R = -2$ mm (concave surface facing object), $n_1 = 1.333$ for dew, $s = 2$ mm $- 1$ mm $= 1$ mm (distance of object from surface), and $n_2 = 1$ for air, gives an image distance of

    $$s' = \frac{n_2 Rs}{(n_2 - n_1)s - n_1 R} = \frac{(1)(-2 \text{ mm})(1 \text{ mm})}{(1 - 1.333)(1 \text{ mm}) - (1.333)(-2 \text{ mm})} = -0.857 \text{ mm}$$

    ASSESS A negative s' means that the image is virtual.

Section 31.4 Optical Instruments

33. **INTERPRET** This problem is about the power of the lens required to correct the vision. The uncorrected eye has a near point of 55 cm, whereas the corrected eye has a near point of 25 cm.

    DEVELOP The lens equation (Equation 31.5) relates the focal length to the object and image distances:

    $$\frac{1}{s} + \frac{1}{s'} = \frac{1}{f}$$

    With the focal length f in meters, $1/f$ is the power, in diopters.

    EVALUATE The above equation gives

    $$\frac{1}{s'_{\text{retina}}} = \frac{1}{f_{\text{eye}}} - \frac{1}{55 \text{ cm}} = \frac{1}{f_{\text{eye}}} - \frac{1}{f_{\text{cor}}} - \frac{1}{25 \text{ cm}}$$

    The image of the near point is on the retina. (The focal length of the eye plus the corrective lens, two closely spaced lenses, essentially in contact, is given in Problem 73.)

    EVALUATE The power of the lens is

    $$P = \frac{1}{f_{\text{cor}}} = \frac{1}{0.25 \text{ m}} - \frac{1}{0.55 \text{ m}} = 2.18 \text{ diopters}$$

ASSESS   Alternatively, Example 31.6 argues that the corrective lens should produce a virtual image of an object at 25 cm (the standard near point) at a distance of 55 cm (the uncorrected near point), so

$$\frac{1}{f_{cor}} = \frac{1}{0.25 \text{ m}} - \frac{1}{0.55 \text{ m}} = 2.18 \text{ diopters}$$

which is the same as above.

35.   INTERPRET   We find the power of a lens that would make distant objects appear to be at a closer distance, using the lens equation.

DEVELOP   The nearsighted patient described in the problem can see clearly at distances up to 80 cm. We want to have a lens that produces an image of distant ($s \approx \infty$) objects at this distance $s' = -80$ cm. We use the lens equation

$$\frac{1}{s} + \frac{1}{s'} = \frac{1}{f}$$

and report our answer in diopters, which is the inverse focal length in meters.

EVALUATE

$$\frac{1}{s} + \frac{1}{s'} = \frac{1}{f} \rightarrow f = s' = -0.80 \text{ m}$$
$$\rightarrow P = \frac{1}{f} = -1.25 \text{ diopters}$$

ASSESS   Note the sign of s'. It's negative, since the image is on the same side of the lens as the object.

37.   INTERPRET   This problem is about the magnification of a compound microscope.

DEVELOP   The magnification of a compound telescope is given by Equation 31.9:

$$M = -\frac{L}{f_0}\left(\frac{25 \text{ cm}}{f_e}\right)$$

EVALUATE   Substituting the values given, we find the overall magnification of the microscope to be

$$M = -\frac{L}{f_0}\left(\frac{25 \text{ cm}}{f_e}\right) = -\frac{83 \text{ mm}}{6.1 \text{ mm}}\left(\frac{25 \text{ cm}}{1.7 \text{ cm}}\right) = -200$$

ASSESS   The image is magnified 200 times. The minus sign means that the image is inverted.

PROBLEMS

39.   INTERPRET   We have an image formation problem involving a concave mirror. We want to know the position of the image, its height, and its orientation.

DEVELOP   The magnification M, the ratio of the image height h' to object height h, is given by Equation 31.1:

$$M = \frac{h'}{h} = -\frac{s'}{s}$$

where s and s' are object and image distances to the mirror. The two quantities s and s' are related by the mirror equation (Equation 31.2):

$$\frac{1}{s} + \frac{1}{s'} = \frac{1}{f}$$

where f is the focal length of the mirror.

EVALUATE   (a) The position of the image is at

$$s' = \frac{fs}{s-f} = \frac{(17 \text{ cm})(10 \text{ cm})}{10 \text{ cm} - 17 \text{ cm}} = -24.3 \text{ cm}$$

The negative sign means that the image is behind the mirror.

(b) The magnification of the image is

$$M = \frac{h'}{h} = -\frac{s'}{s} = -\frac{sf/(s-f)}{s} = -\frac{f}{s-f} = -\frac{17 \text{ cm}}{10 \text{ cm} - 17 \text{ cm}} = 2.43$$

Therefore, the height of the image is $h' = Mh = (2.43)12 \text{ mm} = 29.1 \text{ mm}$.

(c) The image is virtual ($s' < 0$), upright ($M > 0$), and enlarged ($|M| > 1$).

ASSESS   The situation corresponds to the third case depicted in Table 31.1. The mirror is concave with $s < f$.

41.   **INTERPRET**   This problem is about image formation in a concave mirror. We want to know the position of the object, given its magnification and the focal length of the mirror.

DEVELOP   Using Equations 31.1 and 31.2, the magnification of the image can be written as

$$M = \frac{h'}{h} = -\frac{s'}{s} = -\frac{sf/(s-f)}{s} = -\frac{f}{s-f}$$

The fact that the image is upright tells us that $M > 0$. This equation allows us to solve for s, the position of the object.

EVALUATE   With $M = +3$ and $f = 27$ cm, we get

$$s = f\left(1 - \frac{1}{M}\right) = (27 \text{ cm})\left(1 - \frac{1}{3}\right) = 18 \text{ cm}$$

The object is in front of the mirror.

ASSESS   The image distance is $s' = -Ms = -3(18 \text{ cm}) = -54$ cm. The situation corresponds to the third case depicted in Table 31.1. The mirror is concave with $s < f$, and the image is virtual, upright, and enlarged.

43.   **INTERPRET**   This problem is about the image formed by a telescope. The main mirror of a telescope is concave, since only such a mirror collects light from a distant object into a real image.

DEVELOP   Inspection of Figure 31.6a (partially redrawn below) shows that the angular size of the object and image are equal

$$\frac{h}{s} = \frac{h'}{s'} \approx \theta = \frac{\text{size}}{\text{distance}}$$

Since the object distance is astronomical,

$$0 \approx \frac{1}{s} = \frac{1}{f} - \frac{1}{s'} \quad \rightarrow \quad s' \approx f$$

That is, the image distance is approximately equal to the focal length. It follows from the law of reflection that the angular size of the object and image are the same, as seen from the mirror.

EVALUATE   Combining the two equations above, the image size is

$$h' = \theta s' \approx \theta f = (0.5°)\left(\frac{\pi \text{ rad}}{180°}\right)(8.5 \text{ m}) = 7.42 \text{ cm}$$

ASSESS   The situation corresponds to the first case shown in Table 31.1. From the ray diagram, we see that the image is real, inverted, and reduced in size. Since $s' > 0$, the image is in front of the mirror.

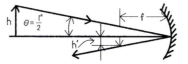

45.   **INTERPRET**   This problem is about image formation in a concave mirror. We are given the focal length of the mirror and the magnification and asked about the positions of the object and its image.

DEVELOP   The magnification M, the ratio of the image height h' to object height h, is given by Equation 31.1:

$$M = \frac{h'}{h} = -\frac{s'}{s}$$

where s and s' are object and image distances to the mirror. The two quantities s and s' are related by the mirror equation (Equation 31.2):

$$\frac{1}{s} + \frac{1}{s'} = \frac{1}{f}$$

where f is the focal length of the mirror. The two equations can be combined to give

$$M = \frac{h'}{h} = -\frac{s'}{s} = -\frac{sf/(s-f)}{s} = -\frac{f}{s-f}$$

The fact that the image is inverted tells us that $M < 0$. This equation allows us to solve for s and s'.

EVALUATE   (a) With $M = -1.5$ and $f = 1.2$ m, we get

$$s = f\left(1 - \frac{1}{M}\right) = (1.2 \text{ m})\left(1 - \frac{1}{-1.5}\right) = 2.0 \text{ m}$$

Since $s > 0$, the object is in front of the mirror.

(b) From Equation 31.1, the position of the image is

$$s' = -Ms = -(-1.5)(2.0 \text{ m}) = 3.0 \text{ m}$$

A real image is, of course, in front of the mirror.

ASSESS   The situation corresponds to the second case shown in Table 31.1. From the ray diagram, we see that the image is real $(s' > 0)$, inverted $(M < 0)$, and enlarged $(|M| > 1)$ in size.

47.   **INTERPRET**   This problem is about the image formed by a converging lens.

DEVELOP   The magnification of a thin lens, for paraxial rays, is given by Equation 31.4,

$$M = \frac{h'}{h} = -\frac{s'}{s} = -\frac{sf/(s-f)}{s} = -\frac{f}{s-f}$$

where we have used the lens equation (Equation 31.5) to eliminate s'. The equation is what we shall use to solve for s.

EVALUATE   When a virtual, upright image is formed by a converging lens, the magnification is positive, $M = 1.6$. Therefore,

$$s = f\left(1 - \frac{1}{M}\right) = (32 \text{ cm})\left(1 - \frac{1}{1.6}\right) = 12 \text{ cm}$$

ASSESS   The image formation in this problem corresponds to the third case shown in Table 31.2. With $s < f$, we get a virtual, upright, and enlarged image. Note that a diverging lens always produces a reduced image.

49.   **INTERPRET**   This problem is about the image formed by a converging lens. We are given the focal length of the lens and the object height, and asked about the image height and type.

DEVELOP   Using the lens equation (Equation 31.5) and Equation 31.4 for the magnification for a thin converging (positive f) lens, we obtain

$$M = \frac{h'}{h} = -\frac{s'}{s} = -\frac{sf/(s-f)}{s} = -\frac{f}{s-f}$$

so $h' = Mh = -fh/(s-f)$.

EVALUATE   (a) If $f = 35$ cm and $s = f + 10$ cm $= 45$ cm, then

$$h' = Mh = -\frac{fh}{s-f} = -\frac{(35 \text{ cm})(2.2 \text{ cm})}{45 \text{ cm} - 35 \text{ cm}} = -7.7 \text{ cm}$$

A negative image height signifies a real, inverted image.

(b) If $s = f - 10$ cm, then

$$h' = Mh = -\frac{fh}{s-f} = -\frac{(35 \text{ cm})(2.2 \text{ cm})}{25 \text{ cm} - 35 \text{ cm}} = +7.7 \text{ cm}$$

which represents a virtual, erect image of the same size.

ASSESS   The image formed in (a) corresponds to thimagee second case shown in Table 31.2. With $2f > s > f$, we get a real, inverted, and enlarged image. The situation in (b) corresponds to the third case shown in Table 31.2. With $s < f$, we get a virtual, upright, and enlarged image.

51.   **INTERPRET**   This problem is about the image formed by a convex lens. We are given the focal length of the distance between the object and the image, and asked to locate the lens.

DEVELOP   Since s and s' are both positive for a real image, and the distance $a \equiv s + s' = 70$ cm is fixed, the lens equation (Equation 31.5) can be rewritten as

$$\frac{1}{f} = \frac{1}{s} + \frac{1}{s'} = \frac{s+s'}{ss'} = \frac{a}{ss'} \rightarrow af = (s+s')f = ss' = s(a-s) = s'(a-s')$$

Solving the quadratic equation yields the two positions we look for.

EVALUATE   The solutions for s (or s'), are

$$s = \frac{1}{2}a[1 \pm \sqrt{1 - 4f/a}] = \frac{1}{2}(70 \text{ cm})[1 \pm \sqrt{1 - 4(17 \text{ cm})/(70 \text{ cm})}] = 29.1 \text{ cm or } 40.9 \text{ cm},$$

which are the desired lens locations. Note that this situation has a real solution only if $0 < 4f \leq a$.

ASSESS   Both s and s' are positive, as required for a real image.

53.   **INTERPRET**   This problem is about the image formed by a converging lens.

We are given the focal length of the lens and the magnification, and asked about the position of the object.

DEVELOP   Using the lens equation (Equation 31.5) and Equation 31.4 for the magnification for a thin converging (positive f) lens, we obtain

$$M = \frac{h'}{h} = -\frac{s'}{s} = -\frac{sf/(s-f)}{s} = -\frac{f}{s-f}$$

so $s = f(1 - 1/M)$.

EVALUATE   From the above equation, we get

$$s = f\left(1 - \frac{1}{M}\right) = (25 \text{ cm})\left(1 - \frac{1}{1.8}\right) = 11.1 \text{ cm}$$

ASSESS   Since $s < f$, the situation corresponds to the third case shown in Table 31.2. The image is virtual, upright, and enlarged.

55.   **INTERPRET**   This problem involves image formation by a refracting plano-convex lens.

DEVELOP   The focal length of the lens is given by the lens maker's formula in Equation 31.7,

$$\frac{1}{f} = (n-1)\left(\frac{1}{R_1} - \frac{1}{R_2}\right)$$

With $R_1 = 26$ cm and $R_2 = \infty$ (or $R_1 = \infty$ and $R_2 = -26$ cm), the focal length is

$$\frac{1}{f} = (n-1)\left(\frac{1}{R_1} - \frac{1}{R_2}\right) = \frac{n-1}{R_1} = \frac{1.62 - 1}{26 \text{ cm}} = \frac{1}{41.9 \text{ cm}} \rightarrow f = 41.9 \text{ cm}$$

EVALUATE   An object at $s = 68$ cm has its image at

$$\frac{1}{s'} = \frac{1}{f} - \frac{1}{s} = \frac{1}{41.9 \text{ cm}} - \frac{1}{68 \text{ cm}} = \frac{1}{109 \text{ cm}} \rightarrow s' = 109 \text{ cm}$$

This is a real, inverted image, on the opposite side of the lens from the object.

ASSESS   The image formed corresponds to the second case shown in Table 31.2. With $2f > s > f$ and $s' > 2f$, we get a real, inverted, and enlarged image.

57.   **INTERPRET**   As seen in Example 31.4, this problem involves image formation by the aquarium wall, which is a two-dimensional concave refracting spherical surface.

DEVELOP   The image formed by a refracting interface is described by Equation 31.6,

$$\frac{n_1}{s} + \frac{n_2}{s'} = \frac{n_2 - n_1}{R}$$

The equation can be used to solve for the apparent distance s'.

EVALUATE   With $R = -35$ cm (concave surface facing object), $n_1 = 1.333$ for water, $s = 70$ cm $- 15$ cm $= 55$ cm (distance from the near wall), and $n_2 = 1$ for air, gives an image distance of

$$s' = \frac{n_2 R s}{(n_2 - n_1)s - n_1 R} = \frac{(1)(-35 \text{ cm})(55 \text{ cm})}{(1 - 1.333)(55 \text{ cm}) - (1.333)(-35 \text{ cm})} = -67.9 \text{ cm}$$

A negative s' means that the image is virtual.

ASSESS   In this case, the object is closer to the refracting surface than its image (see sketch and compare with Figure 31.23*b*).

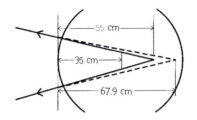

59.   **INTERPRET**   This problem involves image formation by a refracting crystal ball. We are interested in the index of refraction of the ball.

DEVELOP   The outer speck appears $\frac{1}{3}$ the distance to the center of the ball, $s' = -|R|/3$, since the speck at the center appears at the center (see Problem 56). The actual distance of the outer speck is given as $s = |R|/2$. We can then solve for n_1 using Equation 31.6,

$$\frac{n_1}{s} + \frac{n_2}{s'} = \frac{n_2 - n_1}{R}$$

EVALUATE   Equation 31.6 (with n_1 for the ball's material, $n_2 = 1$ for air, and $R = -|R|$ for a concave surface toward the object) gives

$$\frac{n_1}{|R|/2} + \frac{1}{-|R|/3} = \frac{1 - n_1}{-|R|}$$

This simplifies to $2n_1 - 3 = n_1 - 1$, or $n_1 = 2$.

ASSESS   The index of refraction of crystal indeed is about 2.0.

61.   **INTERPRET**   We have a plano-convex lens, and we want to know the relationship between its index of refraction and radius of curvature of the curved surface.

DEVELOP   The focal length of the lens is given by the lens maker's formula in Equation 31.7,

$$\frac{1}{f} = (n-1)\left(\frac{1}{R_1} - \frac{1}{R_2}\right)$$

The plano-convex lens has $R_1 = R$ and $R_2 = \infty$ (or $R_1 = \infty$ and $R_2 = -R$). Thus, it has a focal length of

$$\frac{1}{f} = (n-1)\left(\frac{1}{R} - \frac{1}{\infty}\right) = \frac{n-1}{R} \quad \rightarrow \quad f = \frac{R}{n-1}$$

EVALUATE   If $f = R$, then the index of refraction is $n = 2$.

ASSESS   The smaller R, the more curved the lens and the more it bends light. This implies a shorter focal length. The higher refraction index n, the greater the refraction, and the shorter focal length.

63.   **INTERPRET**   This problem is about the image formed by a double-convex lens. Given the radii of curvature and the object and image distances, we are asked to find the index of refraction of the lens.

DEVELOP   Using Equations 31.5 and 31.7, the focal length of a lens can be written as

$$\frac{1}{f} = \frac{1}{s} + \frac{1}{s'} = (n-1)\left(\frac{1}{R_1} - \frac{1}{R_2}\right)$$

With $s, s', R_1,$ and R_2 given, the above equation can be used to calculate the refractive index, n.

EVALUATE   Substituting the values given in the problem statement, we find

$$\frac{1}{s} + \frac{1}{s'} = (n-1)\left(\frac{1}{R_1} - \frac{1}{R_2}\right) \quad \rightarrow \quad \frac{1}{30 \text{ cm}} + \frac{1}{128 \text{ cm}} = (n-1)\left(\frac{1}{35 \text{ cm}} - \frac{1}{-35 \text{ cm}}\right) = \frac{1}{24.3 \text{ cm}}$$

Therefore, $n = 1 + \frac{35 \text{ cm}}{2(24.3 \text{ cm})} = 1.72$.

ASSESS   If the focal length f is fixed and $R_1 = -R_2 = |R|$, the refractive index can be written as

$$n = 1 + \frac{|R|}{2f}$$

Thus, we see that increasing the curvature radii increases the index of refraction.

65.   **INTERPRET**   Diamond and glass have different index of refraction. So this problem is about the effect on the image when the index of refraction of the lens is changed.

DEVELOP   Using the lens equation (Equation 31.5) and Equation 31.4 for the magnification for a thin converging (positive f) lens, we obtain

$$M = \frac{h'}{h} = -\frac{s'}{s} = -\frac{sf/(s-f)}{s} = -\frac{f}{s-f}$$

Magnification is positive for a virtual image. Thus, for a crown glass lens we have

$$M_g = 2 = -\frac{f_g}{s - f_g}$$

which can be solved to give $f_g = 2s = 30$ cm. The focal length of a diamond lens with the same radii of curvature is (using Equation 31.7 and Table 30.1)

$$f_d(n_d - 1) = f_g(n_g - 1) \;\rightarrow\; f_d = \left(\frac{n_g - 1}{n_d - 1}\right) f_g = \left(\frac{1.520 - 1}{2.419 - 1}\right) 30 \text{ cm} = 11.0 \text{ cm}$$

EVALUATE   An object 15 cm from the diamond lens produces a real, inverted image (negative M) magnified by

$$M_d = -\frac{f_d}{s - f_d} = -\frac{11.0 \text{ cm}}{15 \text{ cm} - 11.0 \text{ cm}} = -2.75$$

ASSESS   In the case of a crown glass, $f_g > s$ and the image is virtual. For the diamond lens, since $s > f_d$, the image is real with $M_d < 0$.

67.   **INTERPRET**   This problem is about the type of lens that must be used in order to improve the closeup capability of a camera.

DEVELOP   For an object at 20 cm, the auxiliary lens should produce a virtual image at 60 cm from either lens (the distance between the lenses is negligible).

EVALUATE   Thus, the required power is

$$P_{aux} = \frac{1}{f_{aux}} = \frac{1}{0.20 \text{ m}} + \frac{1}{-0.60 \text{ m}} = \frac{1}{0.3 \text{ m}} = 3.33 \text{ diopters}$$

Since the focal length is positive, the lens is converging.

ASSESS   The camera without the auxiliary lens can be compared to the eye, in Example 31.6, with a receding near point.

69.   **INTERPRET**   This problem is about the angular magnification of an astronomical telescope.

DEVELOP   For a refracting telescope, the angular magnification is given by Equation 31.10:

$$m = \frac{\beta}{\alpha} = \frac{f_o}{f_e}$$

where α and β are the angles subtended by the actual object and the final image, respectively, while f_o and f_e are the focal lengths of the objective lens and the eye piece, respectively. This is the equation we shall use to solve for β.

EVALUATE   Substituting the values given, the apparent angular size is

$$\beta = \alpha \left(\frac{f_o}{f_e}\right) = 50'' \left(\frac{1 \text{ m}}{40 \text{ mm}}\right) = 1250'' = 20.8' \approx \frac{1}{3}{}^\circ$$

ASSESS   The magnification in this case is $m = 25$. Note that a two-lens refracting telescope gives an inverted (real) image.

71.   **INTERPRET**   This problem is about the image formed by a reflecting ball which is a convex mirror.

DEVELOP   Using Equations 31.1 and 31.2, the magnification of the image can be written as

$$M = \frac{h'}{h} = -\frac{s'}{s} = -\frac{sf/(s-f)}{s} = -\frac{f}{s-f}$$

which yields $f = sM/(M-1)$. The diameter of the ball is $D = 2R = 4|f|$, where f is the focal length (negative in this case).

EVALUATE Using the two equations above, the diameter of the ball is

$$D = 2R = 4|f| = 4s\left|\frac{1}{1 - 1/M}\right| = 4(6\text{ cm})\left|\frac{1}{1 - 4/3}\right| = 72\text{ cm}$$

ASSESS The situation corresponds to case 4 illustrated in Table 31.1. With a convex mirror ($f < 0$), the image is virtual, upright, and reduced.

73. **INTERPRET** This problem is about the corrective power of lenses. We want to show that for closely spaced lenses, the powers are additive.

DEVELOP For two closely-spaced thin lenses, distances measured from either lens are the same. We can consider that the image which would be produced by the first lens alone acts as an object for the second lens, where $s_2 = -s_1'$, because this image, if real, is on the other side of the second lens, or if virtual (i.e., $s_1' < 0$), is on the same side. The lens equations for each lens are

$$\frac{1}{f_1} = \frac{1}{s_1} + \frac{1}{s_1'} \quad \text{and} \quad \frac{1}{f_2} = \frac{1}{s_2} + \frac{1}{s_2'} = \frac{1}{-s_1'} + \frac{1}{s_2'}$$

EVALUATE Adding the two equations yields

$$\frac{1}{f} = \frac{1}{f_1} + \frac{1}{f_2} = \frac{1}{s_1} + \frac{1}{s_2'}$$

where f is the focal length of the lens combination. Since the dioptric power of a lens is the reciprocal of its focal length in meters, the additivity of this quantity follows, under the conditions stated.

ASSESS The additive nature of the corrective power allows an optometrist, during an eye exam, to continue to add lenses until the desired power is attained.

75. **INTERPRET** In this problem we are asked to generalize the lens maker's equation to the situation where the lens is immersed in an external medium with refractive index n_{ext}.

DEVELOP Refraction at the two lens surfaces in Figure 31.24, when the surrounding medium has index of refraction n_{ext} (instead of $n_1 = 1$), is described by equations analogous to the two preceding Equation 31.6:

$$\frac{n_{ext}}{s_1} + \frac{n_{lens}}{s_1'} = \frac{n_{lens} - n_{ext}}{R_1} \quad \text{and} \quad \frac{n_{lens}}{t - s_1'} + \frac{n_{ext}}{s_2'} = \frac{n_{ext} - n_{lens}}{R_2}$$

(These are just Equation 31.6 applied to the left- and right-hand surfaces.) We then take the limit $t \to 0$ to get the desired result.

EVALUATE For $t \to 0$, there is no distinction between distances measured from either surface, so adding the equations and dropping the subscripts 1 and 2, we find

$$\frac{n_{ext}}{s} + \frac{n_{ext}}{s'} = (n_{lens} - n_{ext})\left(\frac{1}{R_1} - \frac{1}{R_2}\right)$$

Division by n_{ext} gives the sought-for generalization of Equation 31.7.

ASSESS The ratio n_{lens}/n_{ext} is the relative index of refraction. When $n_{ext} = 1$, we recover the lens maker's equation given in Equation 31.7.

77. **INTERPRET** In this problem we are asked to analyze a lens of finite thickness t.

DEVELOP Our plan is to add the two equations, one for the left-hand surface and one for the right-hand surface, and then make further simplification by eliminating s_1'.

EVALUATE Adding the two equations gives

$$\frac{1}{s_1} + \frac{1}{s_2'} + n\left(\frac{1}{s_1'} + \frac{1}{t - s_1'}\right) = \frac{1}{s_1} + \frac{1}{s_2'} + \frac{nt}{s_1'(t - s_1')} = (n-1)\left(\frac{1}{R_1} - \frac{1}{R_2}\right)$$

Eliminating s_1' from the third term in the middle member of the above equation, by using either of the original two equations, we arrive at the following result:

$$\frac{nt}{s_1'(t - s_1')} = -\frac{[(n-1)s_1 - R_1]^2 t}{s_1 R_1[t(s_1 + R_1) + ns_1(R_1 - t)]} = -\frac{[(n-1)s_2' + R_2]^2 t}{s_2' R_2[t(R_2 - s_2') + ns_2'(R_2 + t)]}$$

Then the desired relation between s_1 and s_2' (with t, n, R_1 and R_2 as parameters) is achieved. (Note that the object distance $s_1 = s$, and the image distance $s_2' = s'$, are measured from different lens surfaces. The subscripts 1 and 2 are retained as a reminder of this.) For example, from the first of the original equations,

$$\frac{n}{s_1'} = \frac{n-1}{R_1} - \frac{1}{s_1} = \frac{(n-1)s_1 - R_1}{s_1 R_1}$$

Then

$$s_1' = \frac{n s_1 R_1}{(n-1)s_1 - R_1} \quad \rightarrow \quad t - s_1' = -\frac{n s_1 (R_1 - t) + t(R_1 + s_1)}{(n-1)s_1 - R_1}$$

Multiplication and division by nt produces the first expression for the third term above. The other expression comes from a similar treatment of the second original equation,

$$\frac{n}{t - s_1'} = -\frac{(n-1)}{R_2} - \frac{1}{s_2'}$$

Then

$$t - s_1' = -\frac{n s_2' R_2}{(n-1)s_2' + R_2} \quad \rightarrow \quad s_1' = \frac{n s_2'(R_2 + t) + t(R_2 - s_2')}{(n-1)s_2' + R_2}$$

and the second expression for $nt/s_1'(t - s_1')$ follows.

ASSESS   In the limit $t \rightarrow 0$, the third term on the left-hand side vanishes and we readily recover the lens maker's equation for a thin lens shown in Equation 31.7.

79.   INTERPRET   This problem is about Galileo's telescope which consists of a double-concave eyepiece and the usual double-convex objective lens.

DEVELOP   The usual configuration for a Galilean telescope has focal points of the objective and eyepiece coincident, producing a telescope tube of manageable length. The intermediate image, P', can be found from rays through the near focal point and the center of the objective. The former ray, which is parallel to the axis between the lenses, diverges away from the near focal point of the eyepiece, after passing through. A third ray, through the center of the eyepiece and the intermediate image, P', locates the final image, P''.

EVALUATE   The ray diagram is depicted below. This is seen to be virtual and upright.

ASSESS   Repeated use of the lens equation confirms this conclusion, and for objects at infinity, gives f_1/f_2 for the angular magnification. Disadvantages of the Galilean telescope are its limited field of view and inability to incorporate cross hairs.

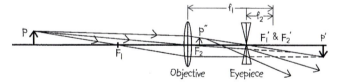

81.   INTERPRET   We show algebraically that the maximum angular magnification of a simple magnifier is $m = 1 + \frac{25\ cm}{f}$. We will use the definition of linear magnification, and the lens equation.

DEVELOP   The angular magnification is defined by the ratio of the apparent size of the image to the apparent size of the object at the near point. $m = (h'/_{25\ cm} / h/_{25\ cm}) = h'/h$. This is just the linear magnification, $M = h'/h = -s'/s$. The image distance is $s' = 25$ cm. We will solve the lens equation for s, and substitute into the equation for m.

EVALUATE

$$\frac{1}{s} + \frac{1}{s'} = \frac{1}{f} \rightarrow s = \frac{f s'}{s' - f}$$

$$m = \frac{h'}{h} = -\frac{s'}{s} = -s'\left(\frac{s' - f}{f s'}\right) = -\frac{s'}{f} + 1 = 1 + \frac{25\ cm}{f}$$

ASSESS   We have shown what was required.

83. **INTERPRET** We apply the result of Problem 31.82 to an actual lens, given the wavelength dependence of the index of refraction for the material.

DEVELOP Our result from Problem 31.82 is $df/f = -dn/n - 1$. The focal length of the lens is $f = 30$ cm at $\lambda = 550$ nm, and the index of refraction is $n = n_0 - b\lambda$ where $n_0 = 1.546$ and $b = 4.47 \times 10^{-5}$ nm^{-1}. We want to find the total variation in focal length df over a 10-nm spread in wavelength centered on 550 nm.

EVALUATE $dn \approx (10 \text{ nm}) \times (-4.47 \times 10^{-5} \text{ nm}^{-1}) = -4.47 \times 10^{-4}$, so $df = -f(dn/n) - 1 = 0.025$ cm.

ASSESS This is a small variation, because of the very small change in wavelength. For full-spectrum visible-light applications, the spread of wavelengths is about 200 nm, and the magnitude of df can be significant.

85. **INTERPRET** We will use the lens equation to find the focal length needed for a simple single-lens camera.

DEVELOP We are told the image and object distances: $s = 80$ cm and $s' = 4.77$ cm. We use $1/s + 1/s' = 1/f$ to calculate the focal length f.

EVALUATE $f = 4.50$ cm.

ASSESS This makes sense, because the object distance is quite large compared to the image distance, and f is approximately the image distance. If the object distance s were infinite, then f would be exactly the image distance.

EXERCISES

Section 32.2 Double-Slit Interference

11. **INTERPRET** This problem is about double-slit interference. We are interested in the spacing between adjacent bright fringes.

    DEVELOP We assume that the geometrical arrangement of the source, slits, and screen is that for which Equations 32.2a and 32.2b apply. The location of bright fringes is given by

    $$y_{\text{bright}} = m\frac{\lambda L}{d}$$

    where m is the order number.

    EVALUATE The spacing of bright fringes is

    $$\Delta y = (m+1)\frac{\lambda L}{d} - m\frac{\lambda L}{d} = \frac{\lambda L}{d} = \frac{(550 \text{ nm})(75 \text{ cm})}{0.025 \text{ mm}} = 1.65 \text{ cm}$$

    ASSESS Since $\lambda \ll d$, the spacing between bright fringes is much smaller than L, as it should.

13. **INTERPRET** This problem is about double-slit interference. We are interested in the wavelength of the light source.

    DEVELOP For small angles, the interference fringes are evenly spaced, with $\Delta\theta = \lambda/d$ (see Equation 31.1a).

    EVALUATE Substituting the values given, we obtain $\lambda = d\Delta\theta = (0.37 \text{ mm})(0.065°)(\pi/180°) = 420 \text{ mm}$.

    ASSESS The wavelength λ is much smaller than the slit spacing d, as expected.

Section 32.3 Multiple-Slit Interference and Diffraction Gratings

15. **INTERPRET** The setup is a multi-slit interference experiment. We want to know the number of minima (destructive interferences) between two adjacent maxima.

    DEVELOP In an N-slit system with slit separation d (illuminated by normally incident plane waves), the main maxima occur for angles (see Equation 32.1a) $\sin\theta = m\lambda/d$, and minima for angles (see Equation 32.4) $\sin\theta = m'\lambda/Nd$ (excluding m' equal to zero or multiples of N).

    EVALUATE Between two adjacent maxima, say $m' = mN$ and $(m+1)N$, there are $N-1$ minima. The number of integers between mN and $(m+1)N$ is

    $$(m+1)N - mN - 1 = N - 1$$

    because the limits are not included. Therefore, For $N = 5$, the number of minima is 4.

    ASSESS The interference pattern resembles that shown in Figure 32.8.

17. **INTERPRET** In this problem we want to locate certain maxima and minima in a multi-slit interference experiment.

    DEVELOP According to Equation 32.1a, primary maxima occur at angles $\theta = \sin^{-1}(m\lambda/d)$. On the other hand, minima occur at angles (see Equation 32.4)

    $$\theta' = \sin^{-1}(m'\lambda/Nd), \quad m' = \pm 1, \ \pm 2, \ \ldots,$$

    where m' is an integer but not an integer multiple of N.

    EVALUATE (a) Using the above equation, the first two (after the central peak, $m = 0$) are for $m = 1$ and 2 at

    $$\theta_1 = \sin^{-1}(633 \text{ nm}/7.5\mu\text{m}) = 4.84°$$
    $$\theta_2 = \sin^{-1}(2 \times 633 \text{ nm}/7.5\mu\text{m}) = 9.72°$$

(b) With $N = 5$ excluded, the third minimum is for $m' = 3$, and the sixth for $m' = 7$ (because $m' = 5$ doesn't count). Then

$$\theta_3' = \sin^{-1}(3\lambda/5d) = 2.90°$$
$$\theta_7' = \sin^{-1}(7\lambda/5d) = 6.79°$$

ASSESS The minima would be difficult to observe because the secondary maxima between them are faint.

19. **INTERPRET** This problem is about diffraction gratings. For a given wavelength, we are interested in the highest visible order.

DEVELOP The grating condition is $\sin\theta = m\lambda/d$, and, of course, for the diffracted light to be visible, $\theta < 90°$, or $m\lambda/d < 1$. Therefore, the highest order visible is the greatest integer m less than d/λ.

EVALUATE **(a)** For this grating, $d = 1\,cm/10^4 = 10^3\,nm$, so for $\lambda = 450$ nm the highest visible order is less than $10^3/450 = 2.22$, or $m_{max} = 2$.

(b) Similarly, for $\lambda = 650$ nm, the highest visible order is less than $10^3/650 = 1.54$, or $m_{max} = 1$.

ASSESS Increasing wavelength lowers m_{max}. This can be seen from Equation 32.1a, $d\sin\theta = m\lambda$.

Section 32.4 Interferometry

21. **INTERPRET** The problem involves interference of a thin film. We want to find the minimum film thickness that results in constructive interference.

DEVELOP The condition for constructive interference from a soap film is Equation 32.7:

$$2nd = \left(m + \frac{1}{2}\right)\lambda$$

The minimum thickness corresponds to the integer $m = 0$.

EVALUATE Substituting the values given, we get

$$2nd_{min} = \frac{1}{2}\lambda \rightarrow d_{min} = \frac{\lambda}{4n} = \frac{550\,nm}{4(1.33)} = 103\,nm$$

Note that Equation 32.7 applies to normal incidence on a thin film in air.

ASSESS The typical thickness of a thin film is on the order of 100 nm. Thin-film interference accounts for the bands of color seen in a soap film or oil slick.

23. **INTERPRET** The enhanced reflection is a consequence of constructive interference, so we look for the range of wavelengths that satisfies such a condition.

DEVELOP Equation 32.7 gives the condition for constructive interference from a given thickness of glass surrounded by air, so

$$\lambda = \frac{4nd}{2m+1} = \frac{4(1.65)(450\,nm)}{2m+1} = \frac{2970\,nm}{2m+1}$$

EVALUATE Integers giving wavelengths in the visible range (400 to 750 nm) are $m = 2$ and 3, corresponding to $\lambda = 594$ nm and 424 nm, respectively.

ASSESS The wavelengths correspond to orange and blue colors.

25. **INTERPRET** We use interference to find the portion of a thin film that appears dark. We interpret this to mean that portion of the film that is so thin that no constructive interference occurs for any visible wavelength.

DEVELOP We start by finding the minimum thickness that will cause constructive interference for the minimum visible wavelength, which is $\lambda = 400$ nm. The portion of the film that is thinner than this minimum will appear dark. We use $2nd = (m + \frac{1}{2})\lambda$, with $m = 1$ and $n = 1.333$, and solve for d.

EVALUATE

$$2nd = \frac{3\lambda}{2} \rightarrow d = \rightarrow \frac{3\lambda}{4n} = 0.225\,\mu m$$

Since the film goes from a thickness of zero at the top to 1.0 μm at the bottom, this means that the top 22.5% of the film is dark.

ASSESS Since the loop in Example 32.4 is 20 cm high, this means that the top 4.5 cm is dark.

Section 32.5 Huygens' Principle and Diffraction

27.   **INTERPRET**   This problem involves a single-slit diffraction of light. We are interested in the angular width of the central peak.

DEVELOP   The condition for destructive interference in a single-slit diffraction is given by Equation 32.8:

$$a\sin\theta = m\lambda, \quad m = \pm1, \pm2, \ldots$$

The first minima ($m = \pm1$) occur at

$$\sin\theta = \pm\frac{\lambda}{a} = \pm\frac{633\text{ nm}}{2.5\ \mu\text{m}} \rightarrow \theta = \pm14.7°$$

EVALUATE   The total angular width of the diffracted beam is $2|\theta| = 29.3°$.

ASSESS   The case $m = 0$ is excluded in Equation 32.8 because it corresponds to the central maximum in which all waves are in phase.

29.   **INTERPRET**   This problem is about intensity of a diffraction maximum relative to the central peak.

DEVELOP   The intensity as a function of angle in single-slit diffraction is given by Equation 32.10:

$$\bar{S} = \bar{S}_0 \left(\frac{\sin(\phi/2)}{\phi/2} \right)^2$$

The second and third minima lie at angles $\sin\theta_2 = 2\lambda/a$ and $\sin\theta_3 = 3\lambda/a$.

EVALUATE   If we take the mid-value to be at $\sin\theta = 5\lambda/2a$, then the intensity at this angle, relative to the central intensity, is

$$\frac{\bar{S}}{\bar{S}_0} = \left(\frac{\sin(5\pi/2)}{5\pi/2} \right)^2 = \frac{4}{25\pi^2} = 1.62\times10^{-2}$$

ASSESS   The intensity at the second secondary maximum is only about 1.62% of the central-peak intensity.

Section 32.6 The Diffraction Limit

31.   **INTERPRET**   We use the Rayleigh criterion to determine how large an aperture we need on a telescope that is to resolve a given angle.

DEVELOP   We use the Rayleigh criterion for circular apertures: $\theta_{\min} = 1.22\lambda/D$. The wavelength is $\lambda = 500$ nm and the angular resolution needed is $\theta_{\min} = 0.35$ arcseconds $= 9.72\times10^{-5}$ degrees $= 1.70\times10^{-6}$ rad. We solve for D.

EVALUATE

$$D = \frac{1.22\lambda}{\theta_{\min}} = 0.360\text{ m} = 36\text{ cm}$$

ASSESS   Make sure that you always use radians for your angle measurements in this type of problem!

33.   **INTERPRET**   This is a problem about the diffraction limit with a circular aperture. We want to know the aperture diameter needed to achieve a certain resolution.

DEVELOP   As shown in Equation 32.11b, the minimum resolvable source separation for a circular aperture is (Rayleigh criterion)

$$\theta_{\min} = \frac{1.22\lambda}{D}$$

where D is the aperture diameter. The angle subtended by a human feature 5 cm across at 100 km is

$$\theta_{\min} = 5\text{ cm}/100\text{ km} = 5\times10^{-7}\text{ (radians)}$$

EVALUATE   The Rayleigh criterion for a diffraction-limited telescope, using light of wavelength $\lambda = 550$ nm, requires an aperture of

$$D = \frac{1.22\lambda}{\theta_{\min}} = \frac{1.22(550\text{ nm})}{5\times10^{-7}} = 1.34\text{ m}$$

ASSESS   Atmospheric turbulence would limit the resolution to no better than $\frac{1}{2}'' = 2.4\times10^{-6}$ radians.

PROBLEMS

35. **INTERPRET** The concept behind this problem is double-slit interference. The object of interest is the phase difference at a given point.

DEVELOP The difference in path for waves arriving from the two slits is

$$\Delta r = d \sin\theta \approx d \tan\theta = d\frac{y}{L}$$

since $\lambda \ll d$ and the small-angle approximation can be used.

EVALUATE The phase difference is

$$\Delta\phi = \left(\frac{2\pi}{\lambda}\right)\Delta r = \frac{2\pi}{\lambda}\frac{yd}{L} = \frac{2\pi(0.56 \text{ cm})(0.035 \text{ mm})}{(500 \text{ nm})(1.5 \text{ m})} = 2\pi(0.261 \text{ cycles}) = 1.64 \text{ rad} = 94.1°$$

ASSESS Constructive interference corresponds to $\Delta\phi = 2\pi m$, or $yd/\lambda L = m$, where m is an integer.

37. **INTERPRET** We have a double-slit experiment here. Given the slit spacing, we are asked to find the highest-order bright fringes.

DEVELOP The maximum diffraction angle for which light hits the screen is

$$\theta_{\text{max}} = \tan^{-1}(0.5 \text{ m}/2.0 \text{ m}) = 14.0°$$

Bright fringes will appear on the screen in orders of interference for which

$$\theta = \sin^{-1}\left(\frac{m\lambda}{d}\right) < 14.0° \quad \rightarrow \quad m < \frac{d\sin 14°}{633 \text{ nm}} = d(3.83\times 10^5 \text{ m}^{-1})$$

EVALUATE **(a)** For $d = 0.10$ mm $= 10^{-4}$ m, $m_{\text{max}} = 38$.

(b) For $d = 10^{-5}$ m, $m_{\text{max}} = 3$.

ASSESS The maximum order m_{max} increases with slit spacing d.

39. **INTERPRET** Our light source for the double-slit experiment has two wavelengths. The angular position where interference is constructive for one wavelength is destructive for the other.

DEVELOP For a bright fringe of order m_1, from wavelength λ_1, to have the same angular position, in a double-slit apparatus of the type described in the text, as a dark fringe of order m_2, from wavelength λ_2, we must have

$$m_1\lambda_1 = (m_2 + 1/2)\lambda_2$$

For $\lambda_1 = 550$ nm and $\lambda_2 = 400$ nm, one finds

$$11m_1 = 8m_2 + 4$$

EVALUATE By inspection, the smallest integer values satisfying this condition are $m_1 = 4$ and $m_2 = 5$; that is, the fourth bright fringe of wavelength 550 nm coincides with the sixth dark fringe of wavelength 400 nm (recall that the first bright fringe has $m = 1$, while the first dark fringe has $m = 0$).

ASSESS This problem demonstrates the role played by the wavelength in determining the nature of the interference at an angular position.

41. **INTERPRET** The concept behind the grating spectrometer is multiple-slit interference. We are interested in the angular separation between two spectral lines produced by different wavelengths.

DEVELOP We use $\theta = \sin^{-1}(3\lambda/d)$ for the angular positions of the third-order grating maximum, with $d^{-1} = 3500/\text{cm} = 3.5\times 10^{-4}/\text{nm}$.

EVALUATE Then for H_α light, $\theta_\alpha = \sin^{-1}(3\times 656\times 3.5\times 10^{-4}) = 43.5°$ and for Na D-light, $\theta_D = \sin^{-1}(3\times 589\times 3.5\times 10^{-4}) = 38.2°$. The difference is $\Delta\theta = 5.33°$.

ASSESS A difference of 5.33° is adequate to distinguish the two wavelengths.

43. **INTERPRET** The concept behind the grating spectrometer is multiple-slit interference, and we explore different slit spacings. The visible spectrum has wavelength between 400 nm and 700 nm.

DEVELOP We use $\theta = \sin^{-1}(m\lambda/d)$ for the angular positions of the mth-order grating maximum. The fourth-order spectrum has angles satisfying

$$\frac{4(400 \text{ nm})}{d} < \sin\theta < \frac{4(700 \text{ nm})}{d}$$

EVALUATE   (a) The spectrum overlaps the fifth-order spectrum down to wavelengths with

$$\frac{4\lambda}{d} = \frac{5(400 \text{ nm})}{d} \rightarrow \lambda = 500 \text{ nm}$$

and up to wavelengths in fifth order with

$$\frac{5\lambda}{d} = \frac{4(700 \text{ nm})}{d} \rightarrow \lambda = 560 \text{ nm}$$

(b) This result is independent of d, as long as the orders exist (i.e., $\sin\theta = m\lambda/d < 1$). Any order present in a 3000-line/cm grating is also present in a grating with fewer lines per cm.

(c) For this grating, $d = 1 \text{ cm}/10^4 = 10^3 \text{ nm}$, so for $\lambda = 400 \text{ nm}$ the highest visible order is less than $10^3/400 = 2.5$, or $m_{max} = 2$. On the other hand, for $\lambda = 700 \text{ nm}$, the highest visible order is less than $10^3/700 = 1.43$, or $m_{max} = 1$. Thus, we conclude that the maximum order present in a 10^4-line/cm grating, for any visible wavelength, is 2 (see also solution to Problem 19).

ASSESS   Decreasing d while keeping λ fixed lowers m_{max}. This can be seen from Equation 32.1a, $d\sin\theta = m\lambda$.

45.   INTERPRET   This is a problem about diffraction grating. We have two wavelengths that are just barely resolved.

DEVELOP   The resolving power of a grating is given by Equation 32.5:

$$\frac{\lambda}{\Delta\lambda} = mN$$

EVALUATE   If the entire width of the grating is illuminated, then the number of slits is $N = (400/\text{mm})(3.5 \text{ cm}) = 14,000$ and wavelength difference is

$$\Delta\lambda = \frac{\lambda}{mN} = \frac{560 \text{ nm}}{4(14,000)} = 0.01 \text{ nm}$$

ASSESS   The larger number of slits N, the higher the resolving power, and the smaller the wavelength difference that we can distinguish.

47.   INTERPRET   This problem is about thin-film interference. Three media are involved: toluene, water, and air.

DEVELOP   Since $n_{toluene} > n_{water} > n_{air}$, there is a 180° phase change for reflection only at the air/toluene interface, and not at the toluene/water interface, of the film. Then Equation 32.7 applies for constructive interference (of normally incident rays):

$$2nd = \left(m + \frac{1}{2}\right)\lambda$$

EVALUATE   Solving for the thickness, we get

$$d = \left(m + \frac{1}{2}\right)\frac{\lambda}{2n} = \left(m + \frac{1}{2}\right)\frac{(460 \text{ nm})}{2(1.49)} = (2m+1)(77.2 \text{ nm})$$

The minimum thickness is $d_{min} = 77.2 \text{ nm}$, although odd multiples of this are also possible.

ASSESS   The typical thickness of a thin film is on the order of 100 nm. Thin-film interference accounts for the bands of color seen in a soap film or oil slick.

49.   INTERPRET   In this problem we are asked to compare the resolving power between echelle spectroscopy and grating.

DEVELOP   If the echelle and grating have the same width, then the number of lines in each is proportional to the given spacings, $N/N' = 80/600$. The resolving power of a grating is given by Equation 32.5:

$$\frac{\lambda}{\Delta\lambda} = mN$$

EVALUATE   The ratio of the resolving powers is then

$$\frac{mN}{m'N'} = \frac{12 \times 80}{1 \times 600} = 1.6$$

ASSESS   The echelle spectroscopy with $m = 12$ has about 60% greater resolving power than the grating with $m' = 1$.

51.   **INTERPRET**   This problem is about X-ray diffraction in a crystal. We are interested in the spacing between the crystal planes.

   DEVELOP   Constructive interference in X-ray diffraction is given by the Bragg condition (Equation 32.6):

$$2d \sin\theta = m\lambda, \qquad m = 1,2,3, \ldots$$

   EVALUATE   From the Bragg condition, one finds

$$d = \frac{m\lambda}{2\sin\theta} = \frac{1(97 \text{ pm})}{2\sin 8.5°} = 328 \text{ pm} = 3.28 \text{ Å}$$

   ASSESS   The spacing between crystal planes is typically a few angstroms.

53.   **INTERPRET**   This problem is about thin-film interference. We want to find the minimum film thickness which yields constructive interference.

   DEVELOP   The condition for constructive interference in thin film (of normally incident rays with 180° phase change at one boundary) is:

$$2nd = \left(m + \frac{1}{2}\right)\lambda$$

   On the other hand, if there's a 180° phase change at both boundaries of the film, then the condition would be

$$2nd = m\lambda$$

   EVALUATE   For the soap film in air (a 180° phase change at one boundary), we find the minimum thickness ($m = 0$) to be

$$d_{\text{soap}} = \frac{\lambda}{4n} = \frac{500 \text{ nm}}{4(1.33)} = 94.0 \text{ nm}$$

   For the oil film on water (a 180° phase change at both boundaries), we have

$$d_{\text{oil}} = \frac{\lambda}{2n} = \frac{500 \text{ nm}}{2(1.25)} = 200 \text{ nm}$$

   ASSESS   The reflected and incident rays are 180° out of phase when they reflect at an interface that goes from a lower to a higher refractive index. There is no phase difference on reflection when the incident medium has a higher refractive index.

55.   **INTERPRET**   We treat the plastic tray as a "thin film" so our analysis on thin-film interference applies. We want to find the minimum film thickness which yields constructive interference.

   DEVELOP   The condition for constructive interference (of normally incident rays with 180° phase change at one boundary) is:

$$2nd = \left(m + \frac{1}{2}\right)\lambda$$

   EVALUATE   Setting $m = 0$ (for the minimum thickness) gives

$$d = \frac{\lambda}{4n} = \frac{c}{4nf} = \frac{3 \times 10^8 \text{ m/s}}{4(1.45)(2.4 \text{ GHz})} = 2.16 \text{ cm}$$

   ASSESS   This is a reasonable thickness of a plastic tray.

57.   **INTERPRET**   This problem is about constructive interference for normally incident light on the thin, wedge-shaped film of some liquid between glass surfaces.

   DEVELOP   With N bright bands visible for an air wedge, the maximum m-value in the condition for constructive interference is $N-1$ (see Problem 56). Therefore, the maximum thickness of the wedge is less than that corresponding to the value $m = N$, or

$$d_{\text{max}} < \left(N + \frac{1}{2}\right)\frac{\lambda}{2}$$

When liquid of refractive index n fills the wedge, there is still a 180° phase change at only one surface (regardless of whether $n_{liq} > n_{glass}$ or not). Constructive interference occurs for

$$\left(m+\frac{1}{2}\right)\frac{\lambda}{2n} \le d_{max} < \left(N+\frac{1}{2}\right)\frac{\lambda}{2}$$

EVALUATE   Thus, m_{max} is the largest integer less than $nN + \frac{1}{2}(n-1)$, and the number of bright bands is $m_{max} + 1$.
ASSESS   We count $m = 0$ as the first bright fringe.

59.   INTERPRET   This problem is about the Michelson interferometer. We want to find the wavelength of the light used, given the shift in bright fringes.
DEVELOP   In the Michelson interferometer, each bright fringe shift corresponds to a path difference of one wavelength.
EVALUATE   Since the path changes by twice the distance moved by the mirror, we have

$$550\lambda = 2 \times 0.15 \text{ mm} \rightarrow \lambda = 545 \text{ nm}$$

ASSESS   The wavelength corresponds to yellow light in the visible spectrum.

61.   INTERPRET   This problem is about interference between incoming waves and reflected waves. We would like to adjust the path difference so that the interference is constructive.
DEVELOP   If the incoming waves are roughly perpendicular to the wall, an additional path difference of $\lambda/2$ between the direct and reflected waves, corresponding to a radio receiver displacement of $\lambda/4$ (the path difference of the waves is approximately twice the distance to the wall), will change an interference minimum into a maximum.
EVALUATE   Therefore, the distance you should move away from the wall is

$$\Delta x = \frac{\lambda}{4} = \frac{c}{4f} = \frac{3 \times 10^8 \text{ m/s}}{4(89.5 \text{ MHz})} = 83.8 \text{ cm}$$

ASSESS   The interference between the incoming and reflected signals is constructive when the path difference is an integer multiple of wavelength, and destructive when it is an odd-integer multiple of half-wavelength.

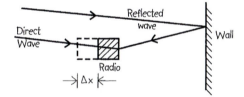

63.   INTERPRET   This is a problem about the diffraction limit with a circular aperture.
DEVELOP   If one of the Keck telescopes were diffraction-limited while observing with 550-nm light, its maximum resolution would be

$$\theta_{min} = \frac{1.22\ \lambda}{D} = \frac{1.22\ (550 \text{ nm})}{10 \text{ m}} = 6.71 \times 10^{-8} \text{ (radians)}$$

At the distance of San Francisco, objects separated by

$$\Delta x = \theta_{min} r = (6.71 \times 10^{-8})(3400 \text{ km}) = 22.8 \text{ cm}$$

could be resolved.
EVALUATE   **(a)** From the analysis above, we see that newspaper fonts could not be resolved.
(b) Billboard-sized letters possibly could.
(c) With an effective aperture five times larger, features one-fifth the size, or 22.8 cm/5 ≈ 4.6 cm, could be resolved. Thus, with the Keck optical interferometer, it might be possible to read very large headlines.
ASSESS   In this problem we have ignored atmospheric turbulence which reduces the effective aperture, and hence the resolving power.

65.   INTERPRET   This problem is about the largest distance within the limitation of human eyes in resolving two headlights separated by a distance.

DEVELOP   If we use the Rayleigh criterion (Equation 32.11b for small angles) to estimate the diffraction-limited angular resolution of the eye, at a pupil diameter of 3.1 mm, in light of wavelength 550 nm, we obtain

$$\theta_{min} = \frac{1.22\lambda}{D} = \frac{1.22(550\ nm)}{3.1\ mm} \approx 2.16 \times 10^{-4} \approx 45''$$

(Actually, the wavelength inside the eye is different, $\lambda' = \lambda/n$, because of the average index of refraction of the eye.)

EVALUATE   This angle corresponds to a linear separation of $y = 1.5$ m at a distance of

$$r = \frac{y}{\theta_{min}} = \frac{1.5\ m}{2.16 \times 10^{-4}} = 6.93\ km \approx 4\ mi$$

ASSESS   Even though other factors determine visual acuity, this is a reasonable ballpark estimate.

67.   INTERPRET   We treat the diamond layer as "thin film" and apply the analysis presented in Section 32.4.

DEVELOP   If the diamond layer is surrounded by air (or material of lesser index of refraction) then Equation 32.7 is the condition for constructive interference for normally incident light. Thus, maximum intensity occurs for wavelengths

$$\lambda = \frac{4nd}{2m+1} = \frac{4(2.42)(250\ nm)}{2m+1} = \frac{2420\ nm}{2m+1}$$

EVALUATE   The only integer which gives a visible wavelength is $m = 2$ (i.e., $2420/700 < 2m + 1 < 2420/400$), for which $\lambda = 2420$ nm/5 = 484 nm.

ASSESS   The wavelength corresponds to green light in the visible spectrum.

69.   INTERPRET   This problem is about the Michelson interferometer. We are told the difference in optical path due to a column of gas compared to a vacuum, and asked to find the refractive index of the gas.

DEVELOP   Since the wavelength of the light is different in a gas (e.g., chlorine or air) and in vacuum ($\lambda_{gas} = \lambda_{vac}/n_{gas}$), there is a difference in the number of wave cycles in the enclosed interferometer arm when the box is evacuated or filled with gas. The light travels the length of the arm twice, out and back, and each cycle of difference results in one fringe shift. Thus, the number of fringes in the shift is

$$m = \frac{2L}{\lambda_{gas}} - \frac{2L}{\lambda_{vac}} = 2L\left(\frac{1}{\lambda_{gas}} - \frac{1}{\lambda_{vac}}\right) = 2L\left(\frac{n_{gas}}{\lambda_{vac}} - \frac{1}{\lambda_{vac}}\right) = \frac{2L}{\lambda_{vac}}(n_{gas} - 1)$$

EVALUATE   From the above equation, we find the refractive index of the gas to be

$$n_{gas} = 1 + \frac{m\lambda_{vac}}{2L}$$

ASSESS   The interferometry allows for the determination of the refractive index of a gas.

71.   INTERPRET   In this problem we want to verify the condition for maximum intensity when light is incident on a diffraction grating at an angle.

DEVELOP   The path difference between two rays (representing a plane wavefront), incident on adjacent slits of a grating (A and B, with spacing $AB = d$), at an angle α with the normal to the grating, is $PA = d\sin\alpha$. The path difference between corresponding outgoing rays, making an angle θ on either side of the normal, is $AQ = BQ' = d\sin\theta$. The total path difference is the sum (or difference) of these, depending on whether θ is on the same (or opposite) side of the normal as α (since we chose both angles to be positive).

EVALUATE   A maximum intensity occurs when the total path difference is an integral number of wavelengths, or $d(\sin\theta \pm \sin\alpha) = m\lambda$.

ASSESS   When $\alpha = 0$, we recover the usual condition given in Equation 32.1a: $d\sin\theta = m\lambda$.

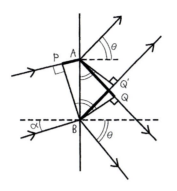

73. **INTERPRET**   We find the first two bright fringes when microwaves are incident on a pair of slits spaced 3λ apart and the resulting pattern is projected on a nearby screen. This could be a very straightforward problem, if we apply Equation 32.2; but we're not given the wavelength so we really don't know if the approximations made to obtain Equation 32.2 apply. The principle of interference *does* apply, of course.

DEVELOP   Our simplest approach would be to just apply $d\sin\theta = m\lambda$, with $d = 3\lambda$, then use the angle θ we find for $m = 1$ and $m = 2$ to find the position of the bright fringes at the screen distance $L = 50$ cm. However, microwaves are typically on the order of 1–10 cm in wavelength, so the approximation $L \gg d$ that was made to obtain Equation 32.2 may not apply.

If we *don't* make the approximation $L \gg d$, though, we'll end up with an equation that is wavelength-dependent, and we don't know the wavelength. So we'll use 32.2, but we'll keep in mind that our answer is at best only approximately correct.

EVALUATE $d = 3\lambda \rightarrow 3\lambda\sin\theta = m\lambda \rightarrow \theta = \sin^{-1}(\frac{m}{3})$. We know that $L = 50$ cm, and $\frac{y}{L} = \tan\theta = \tan(\sin^{-1}(\frac{m}{3}))$. For $m = 1, y = 17.7$ cm and for $m = 2, y = 44.7$ cm.

ASSESS   These values are probably pretty close for short-wavelength microwaves, but at the longer end of the microwave region they will be a bit off.

75. **INTERPRET**   We derive the equation for the intensity of a double-slit interference pattern. The most direct method, integrating over both slits, is suggested in the problem statement. That method will work, but it might be easier to use the result for a single slit (Problem 32.74) and then add the electric field for the second slit.

DEVELOP   The electric field as a function of angle is, according to the solution for Problem 32.74,

$$E = (-E_p\sin\omega t)\frac{\sin\left(\frac{\phi}{2}\right)}{\frac{\phi}{2}} = E_0\frac{\sin\left(\frac{\phi}{2}\right)}{\frac{\phi}{2}}$$

A second slit, located at a distance d from the center of the first slit, will add another component of electric field which is phase shifted by an amount $\delta = d\sin\theta$. For the sake of symmetry (and making the problem easier) we put the origin at the center between the two slits, and shift the upper one back $\frac{\delta}{2}$ and the bottom one forward $\frac{\delta}{2}$. We convert this shift to a phase angle:

$$\frac{\lambda}{\delta/2} = \frac{2\pi}{\varphi} \rightarrow \varphi = \frac{\pi d\sin\theta}{\lambda}$$

so the total electric field due to both slits is

$$E_2 = -E_0\frac{\sin\left(\frac{\phi}{2}\right)}{\left(\frac{\phi}{2}\right)}e^{-i\varphi} - E_0\frac{\sin\left(\frac{\phi}{2}\right)}{\left(\frac{\phi}{2}\right)}e^{+i\varphi}$$

EVALUATE   The total electric field due to both slits is

$$E_2 = -E_0\frac{\sin\left(\frac{\phi}{2}\right)}{\frac{\phi}{2}}e^{-i\varphi} - E_0\frac{\sin\left(\frac{\phi}{2}\right)}{\frac{\phi}{2}}e^{+i\varphi}$$

$$= -E_0\frac{\sin\left(\frac{\phi}{2}\right)}{\frac{\phi}{2}}\cos\varphi$$

Squaring this to obtain the intensity gives us $\overline{S} = \overline{S}_0 \left[\dfrac{\sin\left(\frac{\phi}{2}\right)}{\frac{\phi}{2}} \right]^2 \cos^2 \varphi$.

ASSESS   Adding a second slit gives us the same intensity as before, but multiplied by a second term $\cos^2\left(\frac{\pi d \sin\theta}{\lambda}\right)$.

77.   **INTERPRET**   We are asked to check the resolving power of a grating spectrometer, to see if it meets our experimental needs.

DEVELOP   The resolving power of a grating is given by $\lambda/\Delta\lambda = mN$. The grating has 102 lines/mm and is 2 cm in width, so we can find the total number of lines N. The wavelength is $\lambda = 155$ nm, and the grating must resolve $\Delta\lambda = 1$ pm when $m = 12$. We'll solve the grating equation for $\Delta\lambda$ and see if the value obtained is smaller than what's needed.

EVALUATE   The total number of lines is $N = 102$ lines/mm $\times 200$ mm $= 20,400$.

$$\Delta\lambda = \frac{\lambda}{mN} = 0.63 \text{ pm}$$

ASSESS   This is sufficient—in fact the resolution is even higher than necessary.

79.   **INTERPRET**   We use the maximum reflectivity wavelength of an oil film to determine the thickness of that film. We'll start with the principle of interference.

DEVELOP   For maximum constructive interference, the thickness of the film should be such that the reflections from the top and bottom surfaces are shifted by exactly one wavelength. The light reflected from the top surface is shifted by $\delta_1 = \frac{\lambda}{2}$. The light reflected from the bottom surface is shifted on the way through the material and on the way back up, so $\delta_2 = nd + nd = 2nd$. The difference between these two shifts must be 1λ, where $\lambda = 580$ nm. The index of refraction for the oil is $n = 1.38$, and we solve for d.

EVALUATE

$$\delta = \delta_2 - \delta_1 = \lambda = 2nd - \frac{\lambda}{2} \rightarrow d = \frac{3\lambda}{4n} = 315 \text{ nm}$$

ASSESS   We could just use the equation derived for the soap film (Equation 32.7) since the index of refraction of water (1.333) is less than that of the oil and the film then acts at both surfaces exactly like the film with air on both sides. Be careful, though: if the index of this oil were less than 1.333, then there would be another phase shift on reflection at the second surface.

RELATIVITY

EXERCISES

Section 33.2 Matter, Motion, and Ether

13. **INTERPRET**   In this problem we are asked to take wind speed into consideration to calculate the travel time of an airplane.

DEVELOP   Since the velocities are small compared to c, we can use the non-relativistic Galilean transformation of velocities in Equation 3.7, $\vec{u} = \vec{u}' + \vec{v}$, where $\vec{u}$ is the velocity relative to the ground (S), $\vec{u}'$ is that relative to the air (S'), and v is that of S' relative to S (in this case, the wind velocity). We used a notation consistent with that in Equations 33.5a and 33.5b.

EVALUATE   **(a)** If $\vec{v} = 0$ (no wind), then $\vec{u} = \vec{u}'$ (ground speed equals air speed), and the round-trip travel time is

$$t_a = \frac{2d}{u} = \frac{2(1800 \text{ km})}{800 \text{ km/h}} = 4.5 \text{ h}$$

(b) If $\vec{v}$ is perpendicular to $\vec{u}$, then $u'^2 = u^2 + v^2$, or

$$u = \sqrt{u'^2 - v^2} = \sqrt{(800 \text{ km/h})^2 - (130 \text{ km/h})^2} = 789 \text{ km/h}$$

and the round-trip travel time is $t_b = 2d/u = 4.56$ h.

(c) If $\vec{v}$ is parallel or anti-parallel to $\vec{u}$ on alternate legs of the round-trip, then $u = u' \pm v$ and the travel time is (see Equation 33.2, but with c replaced by u')

$$t_c = \frac{d}{u'+v} + \frac{d}{u'-v} = \frac{1800 \text{ km}}{800 \text{ km/h} + 130 \text{ km/h}} + \frac{1800 \text{ km}}{800 \text{ km/h} - 130 \text{ km/h}} = 4.62 \text{ h}$$

ASSESS   We find $t_a < t_b < t_c$, as mentioned in the paragraph following Equation 33.2.

Section 33.4 Space and Time in Relativity

15. **INTERPRET**   This problem is about the distance between two stars, as seen by a spaceship moving at relativistic speed.

DEVELOP   The distance between stars at rest in system S appears Lorentz-contracted in the spaceship's system S', according to Equation 33.4:

$$\Delta x' = \Delta x \sqrt{1 - v^2/c^2}$$

EVALUATE   With $\Delta x = 50$ ly and $v = 0.75c$, we get

$$\Delta x' = \Delta x \sqrt{1 - v^2/c^2} = (50 \text{ ly})\sqrt{1 - (0.75)^2} = 33.1 \text{ ly}$$

ASSESS   The distance appears to be shortened or "contracted" as observed by the spaceship. Note that length contraction occurs only along the direction of motion.

17. **INTERPRET**   This is a problem about the length of the spaceship measured in its rest frame.

DEVELOP   Equation 33.4, $\Delta x' = \Delta x \sqrt{1 - v^2/c^2}$, gives the length Δx, measured in the rest system of the spaceship, in terms of the length $\Delta x' = 35$ m, measured in the system where the spaceship's speed is v (along the x axis).

EVALUATE The above equation gives

$$\Delta x = \frac{\Delta x'}{\sqrt{1 - v^2/c^2}} = \frac{35 \text{ m}}{\sqrt{1 - (1/2)^2}} = 40.4 \text{ m}$$

ASSESS The spaceship is longest in its own rest frame and is shorter to observers for whom it's moving.

19. **INTERPRET** This is a problem about length contraction. The meter stick is measured to be shorter when it appears to be moving relative to you.

    DEVELOP The distance you measure in a system S', moving (in a direction parallel to the length of the meter stick) with speed v, is $\Delta x' = 99$ cm, while the proper length of the meter stick is $\Delta x = 100$ cm (in system S). These are related by Equation 33.4, $\Delta x' = \Delta x \sqrt{1 - v^2/c^2}$, which gives $\sqrt{1 - v^2/c^2} = 0.99$.

    EVALUATE Solving for v, we get $v/c = \sqrt{1 - (0.99)^2} = 0.141$.

    ASSESS In order for the meter stick to be measured 1% shorter, you'd need to be moving at about 14% the speed of light.

Section 33.7 Energy and Momentum in Relativity

21. **INTERPRET** We want to know the change in momentum when we double the speed, both in the non-relativistic and relativistic limits.

    DEVELOP The measure of momentum valid at any speed is given by Equation 33.7:

    $$\vec{p} = \frac{m\vec{u}}{\sqrt{1 - u^2/c^2}} = \gamma m\vec{u}$$

    Thus, doubling the speed ($u \to 2u$) increases the momentum by a factor

    $$2\frac{\sqrt{1 - u^2/c^2}}{\sqrt{1 - 4u^2/c^2}} = 2\sqrt{\frac{1 - u^2/c^2}{1 - 4u^2/c^2}}$$

    EVALUATE (a) When $u = 25$ m/s, $u/c \approx 0$, the above factor is essentially 2.

    (b) If $v/c = 1/3$, the factor is $2\sqrt{1 - (1/9)}/\sqrt{1 - (4/9)} = 2\sqrt{8/5} = 2.53$.

    ASSESS In the non-relativistic limit, momentum $\vec{p}$ is linear in $\vec{u}$ ($\vec{p} \approx m\vec{u}$), but this no longer holds in the relativistic limit.

23. **INTERPRET** The expression $\vec{p} = m\vec{u}$ is valid only when $u \ll c$. We want to know the speed at which the error is 1%.

    DEVELOP The error in the Newtonian expression of momentum is

    $$\frac{\gamma mu - mu}{\gamma mu} = 1 - 1/\gamma = 1 - \sqrt{1 - u^2/c^2}$$

    EVALUATE When the above factor is equal to 0.01, or $\sqrt{1 - u^2/c^2} = 0.99$, the speed is

    $$u/c = \sqrt{1 - (0.99)^2} = 0.141$$

    ASSESS While $\vec{p} = m\vec{u}$ is valid at low velocity, in the relativistic limit where v/c is not negligible, Equation 33.7 should be used instead.

25. **INTERPRET** The electron is moving at a relativistic speed, and we want to know its total energy and kinetic energy.

    DEVELOP The total energy of the electron is given by Equation 33.9:

    $$E = \gamma mc^2 = \frac{mc^2}{\sqrt{1 - u^2/c^2}}$$

    Similarly, its kinetic energy is given by Equation 33.8:

    $$K = \gamma mc^2 - mc^2 = (\gamma - 1)mc^2$$

    For this electron, $v/c = 0.97, \gamma = 4.11$, and $m_e c^2 = 0.511$ MeV.

EVALUATE (a) From the above equation, the total energy is

$$E = \gamma mc^2 = (4.11)(0.511 \text{ MeV}) = 2.10 \text{ MeV}$$

(b) The kinetic energy is $K = (\gamma - 1)mc^2 = (3.11)(0.511 \text{ MeV}) = 1.59 \text{ MeV}$.

ASSESS The total energy and kinetic energy are related by $E = K + mc^2$, where mc^2 is the rest energy of the particle. The expression demonstrates the equivalence between mass and energy.

PROBLEMS

27. **INTERPRET** The problem asks for a comparison of two quantities, t_{parallel} and $t_{\text{perpendicular}}$, given in Equations 33.2 and 33.1.

    DEVELOP The ratio of Equations 33.1 and 33.2 is

    $$\frac{t_{\text{parallel}}}{t_{\text{perpendicular}}} = \frac{2cL/(c^2 - v^2)}{2L/\sqrt{c^2 - v^2}} = \frac{2cL}{c^2 - v^2} \cdot \frac{\sqrt{c^2 - v^2}}{2L} = \frac{1}{\sqrt{1 - v^2/c^2}}$$

    EVALUATE The ratio is greater than 1 for $v/c < 1$.

    ASSESS Since $t_{\text{parallel}} > t_{\text{perpendicular}}$, we conclude that the trip parallel to the ether wind always takes longer.

29. **INTERPRET** This is a problem about travel time measured in different reference frames. Time dilation is involved.

    DEVELOP Note that the distance is given in the system S, where the Earth and the Sun are practically at rest (the orbital speed of the Earth is very small compared to the speed of light or the speed of the spacecraft), but the time interval is given in system S', where the spacecraft is at rest; that is, $\Delta x = 8.3 \text{ c} \cdot \text{min}$, and $\Delta t' = 5 \text{ min}$. (One light-minute, the distance light travels in one minute, equals c multiplied by a minute, so c · min is a convenient notation for this light-unit.) Equations 33.3 and 33.4,

    $$\Delta t' = \Delta t \sqrt{1 - v^2/c^2} \quad \text{and} \quad \Delta x' = v\Delta t \sqrt{1 - v^2/c^2} = \Delta x \sqrt{1 - v^2/c^2}$$

    for time dilation and Lorentz contraction, relate the given quantities to $\Delta x'$ and Δt, so that the spacecraft's speed, $v = \Delta x/\Delta t = \Delta x'/\Delta t'$, can be determined. For example, if we use $\gamma = 1/\sqrt{1 - v^2/c^2}$, the above equations become

    $$\Delta t = \gamma \Delta t', \quad \Delta x = \gamma \Delta x'$$

    EVALUATE (a) From the above equation, we get

    $$v = \frac{\Delta x}{\gamma \Delta t'} = \frac{(8.3 \text{ c} \cdot \text{min})}{\gamma (5 \text{ min})} = \frac{1.66c}{\gamma} \quad \rightarrow \quad \frac{\gamma v}{c} = 1.66$$

    Squaring the equation gives

    $$\gamma^2 (v/c)^2 = \frac{(v/c)^2}{1 - v^2/c^2} = (1.66)^2 \quad \rightarrow \quad (v/c)^2 = \frac{(1.66)^2}{1 + (1.66)^2} = 0.7337$$

    or $v/c = 0.857$.

    (b) The equation $\gamma v/c = 1.66$ is also easily solved for $\gamma = 1.66/(v/c) = 1.66/0.857 = 1.94$. Then $\Delta t = (1.94)(5 \text{ min}) = 9.69 \text{ min}$.

    ASSESS The result demonstrates that clocks (inside the spacecraft) moving relative to the Earth-Sun frame are seen to run more slowly (5 min) as compared to the clocks at rest (9.69 min).

31. **INTERPRET** This is a problem about travel time measured in different reference frames. Time dilation is involved.

    DEVELOP The distance is given in the system S, where the Earth is practically at rest, but the time interval is given in system S', where the spacecraft is at rest; that is, $\Delta x = N \text{ ly}$, and $\Delta t' = N\text{y}$. (One ly or light-year, the distance light travels in one year, equals c multiplied by a year.) Equations 33.3 and 33.4,

    $$\Delta t' = \Delta t \sqrt{1 - v^2/c^2} \quad \text{and} \quad \Delta x' = v\Delta t \sqrt{1 - v^2/c^2} = \Delta x \sqrt{1 - v^2/c^2}$$

    for time dilation and Lorentz contraction, relate the given quantities to $\Delta x'$ and Δt, so that the spacecraft's speed, $v = \Delta x/\Delta t = \Delta x'/\Delta t'$, can be determined. For example, if we use $\gamma = 1/\sqrt{1 - v^2/c^2}$, the above equations become

    $$\Delta t = \gamma \Delta t', \quad \Delta x = \gamma \Delta x'$$

EVALUATE   From the above equation, we get

$$v = \frac{\Delta x}{\gamma \Delta t'} = \frac{N \text{ ly}}{\gamma(N \text{ y})} = \frac{c}{\gamma} \rightarrow \frac{\gamma v}{c} = 1$$

Squaring the equation gives

$$\gamma^2 (v/c)^2 = \frac{(v/c)^2}{1 - v^2/c^2} = 1 \rightarrow (v/c)^2 = \frac{1}{1+1} = \frac{1}{2}$$

or $v/c = 1/\sqrt{2} = 0.707$.

ASSESS   To show consistency of the result, we note that in the reference frame of the spacecraft, the distance is contracted,

$$\Delta x' = \Delta x \sqrt{1 - v^2/c^2} = \Delta x/\gamma = (N \text{ ly})/\sqrt{2} = (N/\sqrt{2}) \text{ ly}$$

So it will take $\Delta t' = \frac{\Delta x'}{v} = (N/\sqrt{2}) \text{ ly}/c\sqrt{2} = N$ y of the traveler's life to get there.

33.   **INTERPRET**   This is the well-known "twin paradox" problem involving time dilation. We want to know the ages of the twins after one undergoes space travel and returns.

DEVELOP   Twin A must wait $\Delta t = \Delta x/v = 2(30 \text{ ly})/(0.95c) = 63.2$ y for twin B to return. On the other hand, because of time dilation, twin B ages for

$$\Delta t' = \Delta t \sqrt{1 - v^2/c^2} = (63.2 \text{ y})\sqrt{1 - (0.95)^2} = 19.7 \text{ y}$$

during the trip.

EVALUATE   Therefore, twin A is 83.2 y old (63.2 + 20), while twin B returns at age 39.7 y (19.7 + 20).

ASSESS   This is an intriguing consequence of time dilation. Note that only twin A's reference frame is inertial.

35.   **INTERPRET**   This is a problem involving relativistic velocity addition.

DEVELOP   Our galaxy (S') is moving with speed $v = 0.75c$ relative to one of the galaxies mentioned in the question (S), and the other galaxy is moving with speed $u' = 0.75c$ relative to us. All velocities are assumed to be along the common x-x' axes.

The speed of one galaxy as measured by an observer at the other galaxy can be obtained by using the relativistic velocity addition formula (Equation 33.5a):

$$u = \frac{u' + v}{1 + u'v/c^2}$$

EVALUATE   Substituting the values given, we get

$$u = \frac{u' + v}{1 + u'v/c^2} = \frac{0.75c + 0.75c}{1 + (0.75)^2} = 0.960c$$

ASSESS   The naïve answer, $0.75c + 0.75c = 1.5c$, is inconsistent with relativity.

37.   **INTERPRET**   In this problem we'd like to show that if the speed of an object is less than c in one inertial reference frame, then the same conclusion holds in another inertial frame.

DEVELOP   The problem is equivalent to showing that in the relativistic velocity addition formula (Equation 33.5a):

$$u = \frac{u' + v}{1 + u'v/c^2}$$

if $u' < c$ and $v < c$, then $u < c$.

EVALUATE   The conclusion follows almost immediately, because if $0 < 1 - u'/c$ and $0 < 1 - v/c$, then

$$0 < (1 - u'/c)(1 - v/c) = 1 - u'/c - v/c + u'v/c^2$$

or $c > (u' + v)/(1 + u'v/c^2) = u$.

ASSESS   Equation 33.5a applies to the special case where all the velocities are collinear, but the conclusion is true in general.

39.   **INTERPRET**   We are given two events (interstellar spacecraft launching) that take place at different instants, as seen by a particular reference frame. We want to know the order of occurrence as observed in another reference frame.

DEVELOP Denote the frame of the galaxy by S. The spacecraft are launched at $t_A = 0$ and $t_B = 5 \times 10^4$ y from $x_A = 0$ and $x_B = 10^5$ ly (we choose the origin of S at civilization A for simplicity). Let S' be the frame of the traveler from C. The Lorentz transformation between S and S' is summarized in Table 33.1, where $v = 0.99c$ (positive x and x' axes from A to B), and

$$\gamma = \frac{1}{\sqrt{1 - v^2/c^2}} = \frac{1}{\sqrt{1 - (0.99)^2}} = 7.089$$

In S', $t'_A = \gamma(t_A - x_A v/c^2) = 0$, and

$$t'_B = \gamma(t_B - x_B v/c^2) = (7.089)[5 \times 10^4 \text{ y} - (10^5 \text{ ly})(0.99c)/c^2] = -3.47 \times 10^5 \text{ y}$$

which is earlier than $t'_A = 0$.

EVALUATE The observer from C assigns priority to civilization B by 3.47×10^5 y.

ASSESS This problem demonstrates that two observers may see the occurrence of two events differently, depending on their frame of reference.

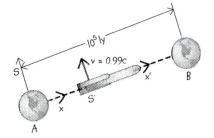

41. **INTERPRET** In special relativity the concept of simultaneity is not absolute but relative; it depends on the reference frame of the observer. We want to know whether or not there exists a reference frame in which the two events are seen to take place simultaneously.

    DEVELOP The light travel time between A and B, in S, is 10^5 ly/$c = 10^5$ y. In Problem 39, t_B is less than this, so the two launchings cannot be causally related. Therefore, there is a frame, S', moving from A to B with speed v relative to S, in which the events are simultaneous.

    EVALUATE Simultaneity requires that (see Table 33.1) $0 = t'_A = t'_B = \gamma(t_B - x_B v/c^2)$, or

    $$\frac{v}{c} = \frac{ct_B}{x_B} = \frac{(1 \text{ ly/y})(5 \times 10^4 \text{ y})}{10^5 \text{ ly}} = 0.5$$

    ASSESS If the launchings are causally related, then they cannot be simultaneous in any frame ($t'_A < t'_B$ always).

43. **INTERPRET** In this problem we want to use the "light box" to derive the time dilation formula given in Equation 33.3.

    DEVELOP The frame of the box, S', is moving with speed v in the x direction relative to the frame, S, at rest in Figure 33.6b. Let the S coordinates of event A be t_A and x_A, and of event B, t_B and $x_B = x_A + v(t_B - t_A)$. To find $\Delta t'$, we apply Lorentz transformation (Table 33.1).

    EVALUATE With $t' = \gamma(t - vx/c^2)$, we get

    $$\Delta t' = t'_B - t'_A = \gamma[t_B - t_A - (x_B - x_A)v/c^2] = \gamma(t_B - t_A)(1 - v^2/c^2) = \Delta t \sqrt{1 - v^2/c^2}$$

    which is Equation 33.3.

    ASSESS The equation shows that $\Delta t' < \Delta t$. That is, the time interval measured in the spaceship frame S' is shorter than that measured in S.

45. **INTERPRET** This problem is about space travel. We want to know the speed one must travel in order to reach a distant star in 75 years.

    DEVELOP The distance is given in the system S, where the Earth and the star are practically at rest, but the time interval is given in system S', where the spacecraft is at rest; that is, $\Delta x = 200$ ly, and $\Delta t' = 75$ y. Equations 33.3 and 33.4,

    $$\Delta t' = \Delta t \sqrt{1 - v^2/c^2} \quad \text{and} \quad \Delta x' = v\Delta t \sqrt{1 - v^2/c^2} = \Delta x \sqrt{1 - v^2/c^2}$$

for time dilation and Lorentz contraction, relate the given quantities to $\Delta x'$ and Δt, so that the spacecraft's speed, $v = \Delta x/\Delta t = \Delta x'/\Delta t'$, can be determined. For example, if we use $\gamma = 1/\sqrt{1 - v^2/c^2}$, the above equations become

$$\Delta t = \gamma \Delta t', \ \Delta x = \gamma \Delta x'$$

EVALUATE From the above equation, we get

$$v = \frac{\Delta x}{\gamma \Delta t'} = \frac{200 \text{ ly}}{\gamma (75 \text{ y})} = \frac{2.67c}{\gamma} \ \rightarrow \ \frac{\gamma v}{c} = 2.67$$

Squaring the equation gives

$$\gamma^2 (v/c)^2 = \frac{(v/c)^2}{1 - v^2/c^2} = (2.67)^2 \ \rightarrow \ (v/c)^2 = \frac{(2.67)^2}{1 + (2.67)^2} = 0.8767$$

or $v/c = 0.936$.

ASSESS The equation $\gamma v/c = 2.67$ is readily solved to give $\gamma = 2.67/(v/c) = 2.67/0.936 = 2.85$. Thus, for an observer on Earth, the time required would have been

$$\Delta t = \gamma \Delta t' = (2.85)(75 \text{ y}) = 214 \text{ y}$$

The result demonstrates that clocks (inside the spacecraft) moving relative to the Earth-star frame are seen to run more slowly (75 y) as compared to the clocks at rest (214 y). The latter far exceeds a human lifetime!

47. **INTERPRET** This is a problem about calculating the distance and time between two events, as measured in different reference frames.

DEVELOP We follow the Problem-solving Strategy 33.1 for Lorentz transformation. In the Earth-star frame (system S), we choose $x_A = 0$ and $t_A = 0$. Events A and B both occur at the spaceship (System S', for which we can choose $x'_A = 0 = x'_B$ and $t'_A = 0$.

EVALUATE (a) In system S, we have $x_B - x_A = 10$ ly (given), and

$$\Delta t = t_B - t_A = \frac{x_B - x_A}{v} = \frac{10 \text{ ly}}{0.8c} = 12.5 \text{ y}$$

(b) In system S', we have $dx'_B - x'_A = 0$. However, $\Delta t' = t'_B - t'_A = (t_B - t_A)/\gamma = \Delta t/\gamma$, from time dilation (Equation 33.3), so $\Delta t' = t'_B - t'_A = 12.5 \text{ y} \sqrt{1 - (0.8)^2} = 7.50 \text{ y}$.

(c) For the space-time interval, one has

$$\Delta s^2 = (c\Delta t)^2 - \Delta x^2 = (12.5 \text{ ly})^2 - (10 \text{ ly})^2 = 56.3 \text{ ly}^2 = (7.50 \text{ ly})^2 = (c\Delta t')^2 - \Delta x'^2 = \Delta s'^2$$

as required by invariance.

ASSESS Our result shows that $\Delta t' < \Delta t$. That is, the time interval measured in the spaceship frame S' is shorter than that measured in S. The space-time interval, however, remains the same in both reference frames, $\Delta s^2 = \Delta s'^2$.

49. **INTERPRET** This is a problem about the spacetime interval between two events.

DEVELOP Choose the x axis along the line separating the positions of the events. Since A and B are connected by the passage of a light beam,

$$|\Delta x| = |x_B - x_A| = c|t_B - t_A| = c|\Delta t|$$

EVALUATE From Equation 33.6, one sees that the spacetime interval between them is zero:

$$(\Delta s)^2 = c^2 (\Delta t)^2 - (\Delta x)^2 = 0$$

ASSESS An event with zero spacetime interval relative to A is said to lie on the light cone of A.

51. **INTERPRET** We're given the time and distance between two distant events observed in a particular reference frame S, and we want to know what they would be in another reference frame S'.

DEVELOP The coordinates of the events in S and S', are related by the Lorentz transformation in Table 33.1, with $v/c = 0.8$ and $\gamma = (1 - v^2/c^2)^{-1/2} = \frac{5}{3}$.

EVALUATE (a) The distance between A and B measured by an observer in S' is

$$x'_B - x'_A = \gamma [x_B - x_A - v(t_B - t_A)] = (5/3)[3.8 \text{ ly} - (0.8 \ c)(1.6 \text{ y})] = (5/3)(3.8 \text{ ly} - 1.28 \text{ ly}) = 4.20 \text{ ly}$$

(b) Similarly, the time between A and B measured by an observer in S' is

$$t'_B - t'_A = \gamma[t_B - t_A - (v/c^2)(x_B - x_A)] = (5/3)[1.6\text{ y} - (0.8/c)(3.8\text{ ly})] = -2.40\text{ y}$$

That is, B occurs before A in S'.

ASSESS Since the light-travel time from the position of A to that of B is greater than the magnitude of the time difference (3.8 y versus 1.6 y in S, or 4.2 y versus 2.4 y is S', the events are not causally connected).

53. **INTERPRET** We're given the kinetic energy of a proton and asked about its speed and momentum.

DEVELOP For the proton, $mc^2 = 938$ MeV, so $K = (\gamma - 1)mc^2 = 500$ MeV implies

$$\gamma = 1 + \frac{K}{mc^2} = 1 + \frac{500\text{ MeV}}{938\text{ MeV}} = 1.53$$

EVALUATE **(a)** Since $\gamma = (1 - v^2/c^2)^{-1/2}$, we find the speed of the proton to be

$$v/c = \sqrt{1 - 1/\gamma^2} = 0.758$$

(b) Using Equation 33.7, we find the relativistic momentum to be

$$p = \gamma mv = 1.53(938\text{ MeV}/c^2)(0.758c) = 1.09\text{ GeV}/c = 5.82 \times 10^{-19}\text{ kg} \cdot \text{m/s}$$

ASSESS Since the kinetic energy is not negligible compared to the rest energy, the Newtonian expression of momentum ($p = mv$) is not applicable.

55. **INTERPRET** This is a problem about mass-energy conversion using $E = mc^2$.

DEVELOP The energy-equivalent of 1 g is

$$E = mc^2 = (10^{-3}\text{ kg})(3 \times 10^8\text{ m/s})^2 = 9 \times 10^{13}\text{ J}$$

EVALUATE This amount of energy could supply a large city, with a power consumption of 10^9 W, for a period of time

$$t = \frac{E}{P} = \frac{9 \times 10^{13}\text{ J}}{10^9\text{ W}} = 9 \times 10^4\text{ s} = 25\text{ h}$$

ASSESS This is an enormous amount of energy harnessed from just 1 g of raisin!

57. **INTERPRET** We are asked about the kinetic energy of an electron, with its speed given. The relativistic formula is needed when v/c is not negligible.

DEVELOP The kinetic energy of the electron is given by Equation 33.8:

$$K = \frac{m_e c^2}{\sqrt{1 - u^2/c^2}} - m_e c^2 = m_e c^2 \left[\frac{1}{\sqrt{1 - u^2/c^2}} - 1 \right]$$

where $m_e c^2 = 0.511$ MeV is the rest energy for an electron.

EVALUATE **(a)** Since $u^2/c^2 = 10^{-6} \ll 1$, we expand the square root and obtain

$$K \approx m_e c^2 \left(\frac{u^2}{2c^2} \right) = \frac{1}{2} m_e u^2 = (0.511\text{ MeV}) \frac{10^{-6}}{2} = 0.256\text{ eV}$$

(b) When $v/c = 0.60$, we have $K = (0.511\text{ MeV}) \left[\frac{1}{\sqrt{1 - (0.60)^2}} - 1 \right] = \frac{1}{4} m_e c^2 = 128$ keV.

(c) Similarly, when $v/c = 0.99$, $K = (0.511\text{ MeV}) \left[\frac{1}{\sqrt{1 - (0.99)^2}} - 1 \right] = 3.11$ MeV.

ASSESS The Newtonian result ($K = m_e u^2/2$) is valid only when $u \ll c$.

59. **INTERPRET** In this problem we want to show that the kinetic energy in Equation 33.8 reduces to the Newtonian result $K = m_e u^2/2$ when $u \ll c$.

DEVELOP The binomial expansion valid for $|x| < 1$ is

$$(1 + x)^p = 1 + px + \frac{p(p-1)}{2!} x^2 + \cdots$$

EVALUATE   For $u/c \ll 1$, Equation 33.8 can be expanded to yield:

$$K = mc^2\left[\frac{1}{\sqrt{1-u^2/c^2}} - 1\right] = mc^2\left[1 + \frac{1}{2}\frac{u^2}{c^2} + \frac{3}{8}\left(\frac{u^2}{c^2}\right)^2 + \cdots - 1\right] = \frac{1}{2}mu^2\left(1 + \frac{3u^2}{4c^2} + \cdots\right) \approx \frac{1}{2}mu^2$$

ASSESS   We indeed recover the Newtonian expression for kinetic energy when $u/c \ll 1$.

61.   **INTERPRET**   In this problem we want to prove that the spacetime interval is relativistically invariant.

DEVELOP   Consider two frames, S and S', related by the Lorentz transformations in Table 33.1. (Since the equations are linear, they apply to differences of coordinates also.) We have

$$c^2\Delta t'^2 - \Delta x'^2 - \Delta y'^2 - \Delta z'^2 = c^2\gamma^2(\Delta t - v\Delta x/c^2)^2 - \gamma^2(\Delta x - v\Delta t)^2 - \Delta y^2 - \Delta z^2$$

$$= \gamma^2(c^2\Delta t^2 - 2v\Delta x\Delta t + v^2\Delta x^2/c^2 - \Delta x^2 + 2v\Delta x\Delta t - v^2\Delta t^2) - \Delta y^2 - \Delta z^2$$

$$= \gamma^2(1 - v^2/c^2)(c^2\Delta t^2 - \Delta x^2) - \Delta y^2 - \Delta z^2 = c^2\Delta t^2 - \Delta x^2 - \Delta y^2 - \Delta z^2$$

EVALUATE   Therefore, $\Delta s'^2 = \Delta s^2$, and the spacetime interval is invariant.

ASSESS   The spacetime interval $(\Delta s)^2 = c^2(\Delta t)^2 - (\Delta x)^2$ describes a relation between two events that's independent of reference frame.

63.   **INTERPRET**   In this problem we are asked to find the speed at which $K = mc^2$.

DEVELOP   The kinetic energy of a particle is given by Equation 33.8:

$$K = \frac{mc^2}{\sqrt{1-u^2/c^2}} - mc^2 = mc^2\left[\frac{1}{\sqrt{1-u^2/c^2}} - 1\right] = mc^2(\gamma - 1)$$

When the kinetic energy is equal to the rest energy, $(\gamma - 1)mc^2 = mc^2$, or $\gamma = 2$.

EVALUATE   Since $\gamma = (1 - v^2/c^2)^{-1/2}$, we find the speed of the particle to be

$$\frac{v}{c} = \sqrt{1 - 1/\gamma^2} = \frac{\sqrt{3}}{2} = 0.866$$

ASSESS   The speed of the particle is about $0.866c$ when $K = mc^2$. This is in the relativistic regime.

65.   **INTERPRET**   This problem explores the change in momentum when the speed of the object is changed.

DEVELOP   The measure of momentum valid at any speed is given by Equation 33.7:

$$\vec{p} = \frac{m\vec{v}}{\sqrt{1-v^2/c^2}} = \gamma m\vec{v}$$

Since $v_2 = 1.05v_1$ and $p_2 = 5p_1$, we write

$$\frac{v_2}{p_2} = \frac{1}{\gamma_2 m} = \frac{1.05v_1}{5p_1} = \frac{1.05}{5}\frac{1}{\gamma_1 m} \rightarrow \frac{1.05}{\gamma_1} = \frac{5}{\gamma_2}$$

EVALUATE   To find the original speed, we rewrite the above equation as

$$(1.05)^2\left(1 - v_1^2/c^2\right) = 5^2\left(1 - v_2^2/c^2\right) = 5^2[1 - (1.05v_1)^2/c^2]$$

which yields $\frac{v}{c} = \sqrt{\frac{25-(1.05)^2}{24(1.05)^2}} = 0.950$.

ASSESS   The particle must be moving at a relativistic speed. In the Newtonian limit where $v \ll c$, $p = mv$, so when the speed is increased by 5%, its corresponding momentum also goes up by 5%.

67.   **INTERPRET**   This problem is about the Doppler shift for light.

DEVELOP   Let S be the rest system of a source of light waves (with frequency and wavelength $\lambda f = c$) which is moving with speed u towards an observer in S' (who measures $\lambda'f' = c$). Suppose that N waves are emitted in S in a time interval Δt. The first wavefront has expanded to a distance $c\,\Delta t$ in S, so the wavelength (i.e., the distance between surfaces of constant phase) is $\lambda = c\Delta t/N$. In S', however, the wavefronts are "piled-up" into a smaller distance, due to the motion of S, so the wavelength is (see sketch)

$$\lambda' = \frac{c\Delta t' - u\Delta t'}{N} = \frac{(c-u)\Delta t'}{N}$$

which gives

$$\frac{\lambda'}{\lambda} = \frac{(1 - u/c)\Delta t'}{\Delta t}$$

Now, Δt (the proper time interval in the source's rest system) is related to $\Delta t'$ (the time interval measured in a system where the source is moving) by time dilation, $\Delta t' = \gamma \Delta t$ (Equation 33.3 with altered notation), so $\lambda'/\lambda = \gamma(1 - u/c)$, or in terms of frequency, $f/f' = \gamma(1 - u/c)$.

EVALUATE Since $\gamma = 1/\sqrt{1 - u^2/c^2}$, this can be written as

$$\frac{f'}{f} = \frac{\sqrt{(1 - u/c)(1 + u/c)}}{1 - u/c} = \sqrt{\frac{1 + u/c}{1 - u/c}}$$

which is the radial Doppler shift (i.e., along the line of sight) in special relativity, with u positive for approach and negative for recession (note the difference in signs with Equation 14.13). For $u/c \ll 1$,

$$\sqrt{\frac{1 + u/c}{1 - u/c}} \approx \left(1 + \frac{1}{2}\frac{u}{c}\right)\left(1 - \frac{1}{2}\frac{-u}{c}\right) \approx 1 + \frac{u}{c}$$

which, allowing for the difference in signs, is the same as the limit of Equation 14.13 for $u/c \ll 1$. *Note:* It is more customary to write this limit as

$$\frac{\Delta f}{f} = \frac{f' - f}{f} = \frac{u}{c}$$

ASSESS Relativistic Doppler effect has been used to measure the shifts in frequency of light emitted by other moving galaxies. The observation of "red shift" suggests that galaxies are receding from us and the Universe is still expanding.

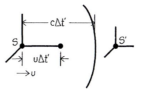

69. **INTERPRET** We use the relativistic momentum, and Newton's second law, to find an equation for the force on a particle in terms of its acceleration. We limit ourselves to acceleration parallel to the velocity.

DEVELOP We know that the relativistic momentum is $p = mu/\sqrt{1 - (u^2/c^2)}$, and Newton's second law is $F = dp/dt$.

EVALUATE

$$F = \frac{dp}{dt} = \frac{dp}{du}\frac{du}{dt} = \left(\frac{m}{\sqrt{1 - \frac{u^2}{c^2}}} + \frac{mu^2}{c^2\left(1 - \frac{u^2}{c^2}\right)^{3/2}}\right)\frac{du}{dt}$$

$$\rightarrow F = \frac{m}{\sqrt{1 - \frac{u^2}{c^2}}}\left(1 + \frac{u^2}{c^2\left(1 - \frac{u^2}{c^2}\right)}\right)\frac{du}{dt} = \gamma m\left(1 + \left(\frac{\gamma u}{c}\right)^2\right)a$$

ASSESS This is quite a bit more complicated than $F = ma$, but the formula reduces to ma for $u = c$.

71. **INTERPRET** Newtonian mechanics is a low-speed approximation to relativistic mechanics. How fast does something have to be going before the relativistic kinetic energy is 50% greater than the Newtonian kinetic energy?

DEVELOP Relativistic kinetic energy is $K_R = E - mc^2$, where $E = \gamma mc^2$. Newtonian kinetic energy is $K_N = \frac{1}{2}mu^2$. We use $K_R = 1.5K_N$, and solve for u.

EVALUATE

$$K_R = \frac{3}{2}K_N \rightarrow (\gamma - 1)mc^2 = \frac{3}{4}mu^2$$

$$\rightarrow 4\left(\sqrt{1 - \frac{u^2}{c^2}}\right)c^2 = 3u^2 \rightarrow 16c^4\left(1 - \frac{u^2}{c^2}\right) = 9u^4$$

$$\rightarrow u^2 = \frac{4}{9}c^2(\sqrt{13} - 2)$$

$$\rightarrow u = 0.714c$$

ASSESS   The kinetic energy actually goes up with speed faster than predicted by Newtonian mechanics.

73.   INTERPRET   We use length contraction to find the apparent length of a spaceship relative to two other frames of reference. One of the reference frames is the Earth, and the other is another moving spaceship so we'll have to use the velocity addition formula as well.

DEVELOP   Commander Krankenfelder's ship has a proper length of $L = 25$ m, and has speed relative to the Earth $u' = -0.5c$. The speed of the Earth relative to Captain Gorgonzoli's ship is $v = -0.65c$. We use $L' = L\sqrt{1 - u'^2/c^2}$ to find the length of Krankenfelder's ship as measured from the Earth. For the length in Gorgonzoli's frame, we do the same, but first find the speed in Gorgonzoli's frame using velocity addition:

$$u = u' + v/1 + \frac{u'v}{c^2}$$

EVALUATE

(a)
$$L' = L\sqrt{1 - \frac{u'^2}{c^2}} = 21.7 \text{ m}$$

(b)
$$u = \frac{u' + v}{1 + \frac{u'v}{c^2}} = 0.868c \rightarrow L' = L\sqrt{1 - \frac{u'^2}{c^2}} = 12.4 \text{ m}$$

ASSESS   Note the reference frames we've used: for part b, we've used Gorgonzoli's ship as the stationary reference frame, Earth as the moving reference frame, and Krankenfelder's ship as the object moving in the moving frame. It's all relative. . .

PARTICLES AND WAVES

EXERCISES

Section 34.2 Blackbody Radiation

15. **INTERPRET** This is a problem about blackbody radiation. We want to explore the connection between temperature and the radiated power.

    DEVELOP From the Stefan-Boltzmann law (Equation 34.1), $P = \sigma A T^4$, we see that the total radiated power, or luminosity, of a blackbody is proportional to T^4.

    EVALUATE Doubling the absolute temperature increases the luminosity by a factor of $2^4 = 16$.

    ASSESS A blackbody is a perfect absorber of electromagnetic radiation. As the temperature of the blackbody increases, its radiated power also goes up.

17. **INTERPRET** We are given the temperature of the blackbody, and asked to find the wavelengths that correspond to peak radiance and median radiance.

    DEVELOP The wavelength at which a blackbody at a given temperature radiates the maximum power is given by Wien's displacement law (Equation 34.2a):

    $$\lambda_{peak} T = 2.898 \text{ mm} \cdot \text{K}$$

    Similarly, the median wavelength, below and above which half the power is radiated, is given by Equation 34.2b:

    $$\lambda_{median} T = 4.11 \text{ mm} \cdot \text{K}$$

    EVALUATE Using the above formulas, we obtain

    $$\lambda_{peak} = (2.898 \text{ mm} \cdot \text{K})/(288 \text{ K}) = 10.06 \text{ } \mu\text{m}$$
    $$\lambda_{median} = (4.11 \text{ mm} \cdot \text{K})/(288 \text{ K}) = 14.27 \text{ } \mu\text{m}$$

    ASSESS The wavelengths are in the infrared. Note that $\lambda_{median} > \lambda_{peak}$.

19. **INTERPRET** We find the wavelength for the peak radiance of solar blackbody radiation, and the median wavelength. In both cases, we'll use the per-unit-wavelength basis, Equations 34.2a and 34.2b.

    DEVELOP Wien's law gives us the peak wavelength: $\lambda_{peak} T = 2.898 \text{ mm} \cdot \text{K}$. The median wavelength is given by $\lambda_{median} T = 4.11 \text{ mm} \cdot \text{K}$. The temperature of the Sun is $T = 5800$ K, so we can use these equations to solve for the desired wavelengths.

    EVALUATE

    (a) $\lambda_{peak} = 2.898 \text{ mm} \cdot \text{K}/5800 \text{ K} = 5.00 \times 10^{-4} \text{ mm} = 500$ nm

    (b) $\lambda_{median} = 4.11 \text{ mm} \cdot \text{K}/5800 \text{ K} = 7.09 \times 10^{-4} \text{ mm} = 709$ nm

    ASSESS The peak wavelength is near the center of the visible spectrum (green) and the median wavelength is just beyond the visible in the near infrared region.

Section 34.3 Photons

21. **INTERPRET** This problem is about the relationship between the wavelength and energy of a photon.

    DEVELOP Using $\lambda = c/f$ and Equation 34.6, the wavelength of a photon can be written as

    $$\lambda = \frac{c}{f} = \frac{hc}{E}$$

EVALUATE   With $E = 6.5$ eV and $hc = 1240$ eV · nm, we find the wavelength to be

$$\lambda = \frac{hc}{E} = \frac{1240 \text{ eV} \cdot \text{nm}}{6.5 \text{ eV}} = 191 \text{ nm}$$

This photon is in the ultraviolet region of the electromagnetic spectrum.

ASSESS   We see that wavelength is inversely proportional to energy; the shorter the wavelength, the more energetic the photon.

23.   **INTERPRET**   The problem asks for a comparison of the power output by a red laser and a blue laser.

DEVELOP   Using $\lambda = c/f$ and Equation 34.6, $E = hf$, the ratio of the photon energies is

$$\frac{E_{\text{blue}}}{E_{\text{red}}} = \frac{f_{\text{blue}}}{f_{\text{red}}} = \frac{\lambda_{\text{red}}}{\lambda_{\text{blue}}}$$

EVALUATE   Using the above equation, the ratio of the energies is

$$\frac{E_{\text{blue}}}{E_{\text{red}}} = \frac{\lambda_{\text{red}}}{\lambda_{\text{blue}}} = \frac{650 \text{ nm}}{450 \text{ nm}} = 1.44$$

Since the lasers emit photons at the same rate, this is also the ratio of their power outputs.

ASSESS   Blue lasers, with shorter wavelength (higher frequency), are more energetic than red lasers.

Section 34.4 Atomic Spectra and the Bohr Atom

25.   **INTERPRET**   This problem is about the energy levels of a hydrogen atom using the Bohr model. We are interested in the wavelengths of the first three lines in the Lyman series.

DEVELOP   The wavelength can be calculated using Equation 34.9:

$$\frac{1}{\lambda} = R_{\text{H}} \left(\frac{1}{n_2^2} - \frac{1}{n_1^2} \right)$$

where $R_{\text{H}} = 1.097 \times 10^7 \text{ m}^{-1}$ is the Rydberg constant and $n_2 = 1$ for the Lyman series.

EVALUATE   The first three lines correspond to $n_1 = 2, 3$, and 4, and the wavelengths are, respectively,

$$\lambda = \frac{1}{R_{\text{H}}} \frac{n_1^2 n_2^2}{n_1^2 - n_2^2} = \frac{1}{R_{\text{H}}} \frac{n_1^2}{n_1^2 - 1} = 122 \text{ nm}, 103 \text{ nm}, \text{ and } 97.2 \text{ nm}$$

Note: $R_{\text{H}}^{-1} = (0.01097)^{-1}$ nm $= 91.2$ nm, which is the Lyman series limit.

ASSESS   The wavelengths are less than 400 nm. Therefore, the Lyman spectral lines are in the ultraviolet regime.

27.   **INTERPRET**   This problem is about the ionization energy of a hydrogen atom at its ground state. We want to find the wavelength of a photon carrying this much energy.

DEVELOP   The energy of the ground state of hydrogen is given by Equation 34.12b (with $n = 1$): $E_1 = -13.6$ eV. Therefore, the ionization energy is $E_I = |E_1| = 13.6$ eV (the subscript "I" for ionization). For a photon whose wavelength is λ, the energy it carries is (Equation 34.6) $E = hf = hc/\lambda$.

EVALUATE   A photon with energy $E_I = 13.6$ eV has wavelength

$$\lambda = \frac{hc}{E_I} = \frac{1240 \text{ eV} \cdot \text{nm}}{13.6 \text{ eV}} = 91.2 \text{ nm}$$

ASSESS   This is the same as the Lyman series limit (Equation 34.9 with $n_2 = 1$ and $n_1 = \infty$) $R_{\text{H}}^{-1} = hc/13.6$ eV, and lies in the ultraviolet.

Section 34.5 Matter Waves

29.   **INTERPRET**   In this problem we are asked to find the de Broglie wavelength of the Earth and an electron, with their speeds given.

DEVELOP   For non-relativistic momentum, Equation 34.14 becomes $\lambda = \frac{h}{p} = \frac{h}{mv}$.

EVALUATE   (a) Using the values given for the Earth, one finds

$$\lambda = \frac{h}{mv} = \frac{6.626 \times 10^{-34} \text{ J} \cdot \text{s}}{(5.97 \times 10^{24} \text{ kg})(30 \text{ km/s})} = 3.70 \times 10^{-63} \text{ m}$$

(b) For the given electron, $\lambda = \frac{h}{mv} = \frac{6.626 \times 10^{-34} \text{ J·s}}{(9.11 \times 10^{-31} \text{ kg})(10 \text{ km/s})} = 72.7$ nm.

ASSESS   The Earth's de Broglie wavelength is much smaller than the smallest physically meaningful distance.

31.   **INTERPRET**   In this problem we compare the speeds of a proton and an electron, both having the same de Broglie wavelength.

DEVELOP   Since $\lambda = h/p$ (Equation 34.14), the same de Broglie wavelength means the same momentum.

EVALUATE   At non-relativistic speeds, $m_p v_p = m_e v_e$, or $v_e / v_p = m_p / m_e = 1836$.

ASSESS   With the same momentum, the speed ratio of two particles is equal to their mass ratio.

Section 34.6 The Uncertainty Principle

33.   **INTERPRET**   We want to find the uncertainty in the velocity of a proton, given its uncertainty in position.

DEVELOP   To find Δv, we use the uncertainty principle, $\Delta x \Delta p \geq \hbar$ (Equation 34.15) with $\Delta p = m \Delta v$ and $\Delta x = 1$ fm.

EVALUATE   The above equation gives

$$\Delta v = \frac{\Delta p}{m} \geq \frac{\hbar}{m \Delta x} = \frac{(197.3 \text{ MeV} \cdot \text{fm}/c)}{(938 \text{ MeV})(1 \text{ fm})} = 0.21c = 6.3 \times 10^7 \text{ m/s}$$

ASSESS   The quantity $\Delta p = \hbar/\Delta x = 197.3$ MeV/c is barely small enough compared to $mc = 938$ MeV/c to justify using the non-relativistic relation $p = mv$, but this is good enough for approximate purposes.

35.   **INTERPRET**   In this problem we want to find the uncertainty in the position of a proton, given its uncertainty in velocity.

DEVELOP   To find Δx, we use the uncertainty principle, $\Delta x \Delta p \geq \hbar$ (Equation 34.15), where $\Delta p = m \Delta v$. We take the uncertainty in velocity to be the full range of variation given; that is, $\Delta v = 0.25$ m/s $-$ (-0.25 m/s) $= 0.5$ m/s.

EVALUATE   The position uncertainty of the proton is

$$\Delta x \geq \frac{\hbar}{m \, \Delta v} = \frac{1.055 \times 10^{-34} \text{ J} \cdot \text{s}}{(1.67 \times 10^{-27} \text{ kg})(0.5 \text{ m/s})} = 126 \text{ nm}$$

ASSESS   The smaller the uncertainty in velocity (Δv), the greater the uncertainty in position (Δx).

37.   **INTERPRET**   The neutron is confined in the uranium nucleus with Δx equal to the diameter of the nucleus. We want to find the minimum energy of the neutron using the uncertainty principle.

DEVELOP   Using the same reasoning as given in Example 34.6, for a neutron ($mc^2 = 940$ MeV) confined to a uranium nucleus ($\Delta x \approx 15$ fm), the uncertainty principle requires that

$$K = \frac{p^2}{2m} \geq \frac{1}{2m} \left(\frac{\hbar}{2 \Delta x} \right)^2$$

EVALUATE   From the above equation, we find the minimum kinetic energy to be

$$K_{min} = \frac{1}{2m} \left(\frac{\hbar}{2 \Delta x} \right)^2 = \frac{1}{2mc^2} \left(\frac{\hbar c}{2 \Delta x} \right)^2 = \frac{(197.3 \text{ MeV} \cdot \text{fm}/30 \text{ fm})^2}{2(940 \text{ MeV})} = 23.0 \text{ keV}$$

ASSESS   This is smaller than the 5 MeV estimated for the nucleon in Example 34.6 by a factor of 15^2, since Δx is 15 times larger. Most estimates of nuclear energies for single-particle states, based on the uncertainty principle, give values of the order 1 MeV, consistent with experimental measurements.

PROBLEMS

39.   **INTERPRET**   We are given the temperature of the Sun which we treat as a blackbody, and asked to compare its radiance at two different wavelengths.

DEVELOP   The radiance of a blackbody is given by Equation 34.4:

$$R(\lambda, T) = \frac{2\pi h c^2}{\lambda^5 (e^{hc/\lambda kT} - 1)}$$

This equation allows us to compare the radiance at two different wavelengths.

EVALUATE From the above equation (also see Example 34.1), the ratio of the blackbody radiances is

$$\frac{R(\lambda_2,T)}{R(\lambda_1,T)} = \left(\frac{\lambda_1}{\lambda_2}\right)^5\left(\frac{e^{hc/\lambda_1 kT}-1}{e^{hc/\lambda_2 kT}-1}\right) = \left(\frac{5}{2}\right)^5\left(\frac{146.9}{2.66\times10^5}\right) = 5.39\times10^{-2}$$

where $\lambda_1 = 500$ nm, $\lambda_2 = 200$ nm, $T = 5800$ K, and $hc/k = 1.449\times10^{-2}$ m·K.

ASSESS The characteristic radiance as a function of wavelength is shown in Figure 34.2. For a given wavelength, the radiance increases with temperature.

41. **INTERPRET** This problem is about blackbody radiation. We are given the wavelengths that correspond to peak radiance and asked to find the temperature of the blackbody. We also want to compare the radiance at two different wavelengths.

    DEVELOP The wavelength at which a blackbody at a given temperature radiates the maximum power is given by Wien's displacement law (Equation 34.2a): $\lambda_{peak}T = 2.898$ mm·K. For part **(b)**, to compare the radiance at two different wavelengths, we use Equation 34.4:

    $$R(\lambda,T) = \frac{2\pi hc^2}{\lambda^5(e^{hc/\lambda kT}-1)}$$

    EVALUATE **(a)** Equation 34.2a gives

    $$T = \frac{2.898 \text{ mm·K}}{\lambda_{peak}} = \frac{2.898 \text{ mm·K}}{660 \text{ nm}} = 4.39\times10^3 \text{ K}$$

    (b) With $hc/kT = 3.30$ μm, the ratio of the radiances is

    $$\frac{R(400 \text{ nm})}{R(700 \text{ nm})} = \left(\frac{700}{400}\right)^5\left(\frac{e^{3.30/0.7}-1}{e^{3.30/0.4}-1}\right) = 0.474$$

    ASSESS The wavelengths considered are in the visible spectrum. Note that the characteristic radiance as a function of wavelength is shown in Figure 34.2. For a given wavelength, the radiance increases with temperature.

43. **INTERPRET** We are given the power output and asked to find the number of photons produced in a time interval.

    DEVELOP The rate of photon emission is the power output (into photons) divided by the photon energy:

    $$\frac{dN}{dt} = \frac{P}{E_\gamma} = \frac{P}{hf} = \frac{P\lambda}{hc}$$

    EVALUATE **(a)** For the antenna, the rate is

    $$\frac{dN}{dt} = \frac{P}{hf} = \frac{1 \text{ kW}}{(6.626\times10^{-34} \text{ J·s})(89.5 \text{ MHz})} = 1.69\times10^{28} \text{ s}^{-1}$$

    (b) For the laser, we have

    $$\frac{dN}{dt} = \frac{P\lambda}{hc} = \frac{(1 \text{ mW})(633 \text{ nm})}{(6.626\times10^{-34} \text{ J·s})(3\times10^8 \text{ m/s})} = 3.18\times10^{15} \text{ s}^{-1}$$

    (c) Similarly, for the X-ray machine, the rate is

    $$\frac{dN}{dt} = \frac{P\lambda}{hc} = \frac{(2.5 \text{ kW})(0.10 \text{ nm})}{(6.626\times10^{-34} \text{ J·s})(3\times10^8 \text{ m/s})} = 1.26\times10^{18} \text{ s}^{-1}$$

    ASSESS For a general device at a given power output, the rate of photon production decreases with the energy of the photon; the more energetic the photons, the smaller the rate of production.

45. **INTERPRET** This problem is about the photoelectric effect of light illuminated on copper. We want to find the cutoff frequency and the maximum energy of the ejected electrons.

    DEVELOP At the cutoff frequency, $K_{max} = 0$, and the photon energy equals the work function, $\phi = hf_{cutoff} = E_\gamma$. The kinetic energy of the ejected electrons is given by Einstein's photoelectric effect equation (Equation 34.7):

    $$K_{max} = hf - \phi = h(f - f_{cutoff})$$

EVALUATE   (a) The work function of copper is (see Table 34.1) $\phi_{Cu} = 4.65$ eV. Therefore, the cutoff frequency is

$$f_{cutoff} = \frac{\phi_{Cu}}{h} = \frac{4.65 \text{ eV}}{4.136 \times 10^{-15} \text{ eV} \cdot \text{s}} = 1.12 \times 10^{15} \text{ Hz}$$

(b) Using Equation 34.7, the maximum kinetic energy of the ejected electrons is

$$K_{max} = h(f - f_{cutoff}) = (4.136 \times 10^{-15} \text{ eV} \cdot \text{s})(1.8 \times 10^{15} \text{ Hz} - 1.12 \times 10^{15} \text{ Hz}) = 2.79 \text{ eV}$$

ASSESS   It takes 4.65 eV to overcome the work function of copper, leaving the electrons with 2.79 eV of kinetic energy.

47.   **INTERPRET**   We're given the cutoff wavelength for the photoelectric effect to take place. Knowing the work function allows us to identify the material(s).

DEVELOP   The energy of a photon of wavelength 275 nm is $E = hf = hc/\lambda = 4.51$ eV. Such photons cannot eject photoelectrons from materials whose work functions are greater than this energy.

EVALUATE   From Table 34.1, we see that the material could be copper, silicon, or nickel.

ASSESS   At the cutoff wavelength, $K_{max} = 0$, and the photon energy equals the work function, $\phi = hf_{cutoff} = hc/\lambda_{cutoff} = E_\gamma$. Therefore, the smaller the value of λ_{cutoff}, the greater the work function.

49.   **INTERPRET**   This problem is about the photoelectric effect. We are given the maximum speed of the ejected electrons, and asked to find the wavelength of the light.

DEVELOP   The maximum speed of the ejected electrons is related to the wavelength of the light by Einstein's photoelectric effect equation (Equation 34.7):

$$K_{max} = \frac{1}{2}mv_{max}^2 = hf - \phi = \frac{hc}{\lambda} - \phi$$

We use this equation to solve for λ.

EVALUATE   The maximum kinetic energy of the electron is

$$K_{max} = \frac{1}{2}mv_{max}^2 = \frac{1}{2}(mc^2)(v_{max}/c)^2 = \frac{1}{2}(511 \text{ keV})(4.2/3000)^2 = 0.501 \text{ eV}$$

From Equation 34.7 and Table 34.1, we find the wavelength to be

$$\lambda = \frac{hc}{K_{max} + \phi} = \frac{1240 \text{ eV} \cdot \text{nm}}{0.501 \text{ eV} + 2.30 \text{ eV}} = 443 \text{ nm}$$

ASSESS   The cutoff wavelength of potassium is

$$\lambda_{cutoff} = hc/\phi = (1240 \text{ eV} \cdot \text{nm})/(2.30 \text{ eV}) = 539 \text{ nm}$$

For the photoelectric effect to take place, we require $\lambda \le \lambda_{cutoff}$.

51.   **INTERPRET**   This problem is about Compton scattering of a photon with an electron. We are interested in the wavelength of the scattered photon and the kinetic energy of the electron.

DEVELOP   The Compton shift of wavelength is given by Equation 34.8:

$$\Delta\lambda = \frac{h}{mc}(1 - \cos\theta) = \lambda_C(1 - \cos\theta)$$

where $\lambda_C = h/mc = 2.43$ pm is the Compton wavelength of the electron. By energy conservation, the kinetic energy of the scattered electron is equal to the energy lost by the photon.

EVALUATE   (a) From Equation 34.8, the wavelength of the scattered photon is

$$\lambda' = \lambda + \Delta\lambda = 150 \text{ pm} + 2.43 \text{ pm} (1 - \cos 135°) = 150 \text{ pm} + 4.15 \text{ pm} = 154 \text{ pm}$$

(b) The kinetic energy of the scattered electron is

$$K = E - E' = \frac{hc}{\lambda} - \frac{hc}{\lambda'} = \frac{hc\Delta\lambda}{\lambda\lambda'} = \frac{(1240 \text{ eV} \cdot \text{nm})(4.15 \text{ pm})}{(150 \text{ pm})(154 \text{ pm})} = 222 \text{ eV}$$

ASSESS   For X rays, the wavelength is in the range 0.01 nm – 10 nm, and therefore, the detection of Compton shift in X rays is difficult.

53. **INTERPRET**   This problem is about whether or not the photoelectric effect can occur with a given wavelength of the light source.

    DEVELOP   To determine whether or not the photoelectric effect can occur, we first find the work function. Einstein's photoelectric effect equation (Equation 34.7), and the data for the blue light, give the work function of the photocathode material as

    $$\phi = E_\gamma - K_{max} = \frac{hc}{\lambda} - K_{max} = \frac{1240 \text{ eV} \cdot \text{nm}}{430 \text{ nm}} - 0.85 \text{ eV} = 2.03 \text{ eV}$$

    EVALUATE   The energy of a photon of the red light is only

    $$\frac{hc}{\lambda} = \frac{1240 \text{ eV} \cdot \text{nm}}{633 \text{ nm}} = 1.96 \text{ eV}$$

    and therefore is insufficient to eject photoelectrons.

    ASSESS   A wavelength of 633 nm is greater than the cutoff wavelength of $\lambda_{cutoff} = hc/\phi = 610$ nm for the photocathode material.

55. **INTERPRET**   This problem is about the wavelength and energy of the photon emitted when the hydrogen atom undergoes a transition.

    DEVELOP   The wavelength of the photon can be calculated using Equation 34.9:

    $$\frac{1}{\lambda} = R_H \left(\frac{1}{n_2^2} - \frac{1}{n_1^2} \right)$$

    where $R_H = 1.097 \times 10^7$ m^{-1} is the Rydberg constant. Once we know the wavelength, the energy of the photon is simply equal to $E_\gamma = hf = hc/\lambda$.

    EVALUATE   (a) The above equation gives

    $$\lambda = \frac{1}{R_H} \frac{n_1^2 n_2^2}{n_1^2 - n_2^2} = \frac{1}{1.097 \times 10^7 \text{ m}^{-1}} \frac{(179)^2 (180)^2}{(180)^2 - (179)^2} = 26.4 \text{ cm}$$

    (b) The energy of the emitted photon is

    $$E_\gamma = \frac{hc}{\lambda} = \frac{1.24 \times 10^{-6} \text{ eV} \cdot \text{m}}{0.264 \text{ m}} = 4.70 \text{ }\mu\text{eV}$$

    ASSESS   The long wavelength corresponds to the radio region of the EM spectrum.

57. **INTERPRET**   The hydrogen atom undergoing a downward transition emits a photon. We are interested in the original state of the atom, given the energy of the photon.

    DEVELOP   The wavelength of the photon can be calculated using Equation 34.9:

    $$\frac{1}{\lambda} = R_H \left(\frac{1}{n_2^2} - \frac{1}{n_1^2} \right)$$

    where $R_H = 1.097 \times 10^7$ m^{-1} is the Rydberg constant. Once we know the wavelength, the energy of the photon is simply equal to

    $$E = hf = \frac{hc}{\lambda} = hcR_H \left(\frac{1}{n_2^2} - \frac{1}{n_1^2} \right)$$

    EVALUATE   Solving for n_1, we get

    $$n_1 = \left[n_2^{-2} - (E/hcR_H) \right]^{-1/2} = [(225)^{-2} - (9.32 \text{ }\mu\text{eV}/13.6 \text{ eV})]^{-1/2} = 229$$

    Note that $hcR_H = 13.6$ eV is the ionization energy.

    ASSESS   The wavelength of the emitted photon is 0.133 m, which falls into the radio wave spectrum.

59. **INTERPRET**   This problem is about the ionization energy of a hydrogen atom in its first excited state.

    DEVELOP   Using Equation 34.12b, the energy of the first excited state $(n = 2)$ is

    $$E_2 = \frac{-13.6 \text{ eV}}{2^2} = -3.40 \text{ eV}$$

    whereas an ionized atom (with zero electron kinetic energy) has energy zero.

EVALUATE   Thus $0 = E_1 + (-3.40 \text{ eV})$, or $E_1 = |E_2| = 3.40 \text{ eV}$.

ASSESS   The ionization energy here is only $\frac{1}{4}$ of the case where the atom is in the ground state. The higher the value of n, the smaller the ionization energy.

61.   **INTERPRET**   He$^+$ is a hydrogen-like atom with a nuclear charge $+2e$. The Bohr model applies in this case.

DEVELOP   Modifying the treatment of the Bohr atom in the text for singly ionized helium (He$^+$), by replacing the nuclear charge with $2e$, one gets $r_n = -2ke^2/2E_n$ and $E_n = -(2ke^2)^2 m/2\hbar^2 n^2$. Thus,

$$E_n = -4\frac{ke^2}{2n^2 a_0} = -2^2\left(\frac{13.6 \text{ eV}}{n^2}\right)$$

and $r_n = \frac{1}{2}n^2 a_0$.

EVALUATE   **(a)** The radius of the ground state of He$^+$ is $r_1 = a_0/2 = 0.0265$ nm.

(b) The energy released in the transition $n = 2$ to $n = 1$ is

$$\Delta E = 4(13.6 \text{ eV})\left(1 - \frac{1}{4}\right) = 40.8 \text{ eV}$$

Note that there is also a small change in m for helium, different from the correction in hydrogen, for the motion of the nucleus.

ASSESS   In general, replacing the nuclear charge with Ze gives results for any one-electron Bohr atom.

63.   **INTERPRET**   The resolution of a microscope depends on the wavelength; a smaller wavelength implies a greater electron speed.

DEVELOP   The electron microscope is better if the wavelength is $\lambda = h/p < 450$ nm. Since $p = mv$, the above condition allows us to obtain the minimum electron speed.

EVALUATE   The above inequality gives

$$v > \frac{h}{m(450 \text{ nm})} = \frac{6.626 \times 10^{-34} \text{ J} \cdot \text{s}}{(9.91 \times 10^{-31} \text{ kg})(450 \text{ nm})} = 1.62 \text{ km/s}$$

So the minimum speed is 1.62 km/s.

ASSESS   Since 450 nm $\gg \lambda_C = 0.00243$ nm (Compton wavelength of electron), our use of the non-relativistic momentum was justified. The electron microscope could provide resolutions down to about 1 nm and magnifications of 10^6.

65.   **INTERPRET**   This problem involves the uncertainty principle. We want to find the minimum velocity of an electron based on its uncertainty in position.

DEVELOP   Using the uncertainty principle given in Equation 34.15 with $\Delta x = 20$ nm (the width of the well), we have

$$\Delta p \geq \frac{\hbar}{\Delta x} = \frac{197.3 \text{ eV} \cdot \text{nm}/c}{20 \text{ nm}} = 9.87 \text{ eV}/c$$

This is small compared to $mc = 511$ keV/c, so one can argue (as in Example 34.6) that $\Delta v = \Delta p/m = 2v$.

EVALUATE   The above conditions lead to

$$v = \frac{\Delta p}{2m} \geq \frac{9.87 \text{ eV}/c}{2(511 \text{ keV}/c^2)} = (9.65 \times 10^{-6})(3 \times 10^8 \text{ m/s}) = 2.90 \text{ km/s}$$

Thus, the minimum speed is $v_{min} = 2.90$ km/s.

ASSESS   Quantum wells have important applications in the field of semiconductor fabrication.

67.   **INTERPRET**   This problem involves energy-time uncertainty. We are interested in the minimum measurement time which sets a minimum uncertainty in energy.

DEVELOP   The electron is non-relativistic ($v/c = 1/300 \ll 1$), and the uncertainty in its (kinetic) energy is

$$\Delta E = 2(0.01\%) \times \left(\frac{1}{2}mv^2\right) = 10^{-4} mc^2 (v/c)^2 = 10^{-4}(511 \text{ keV})(1/300)^2 = 5.68 \times 10^{-4} \text{ eV}$$

The minimum time can then be calculated using Equation 34.16, $\Delta E \Delta t \geq \hbar$.

EVALUATE   An energy measurement of this precision (from Equation 34.16) requires a time

$$\Delta t \geq \frac{\hbar}{\Delta E} = \frac{6.582 \times 10^{-16} \text{ eV} \cdot \text{s}}{5.68 \times 10^{-4} \text{ eV}} = 1.16 \text{ ps}$$

ASSESS   The energy-time uncertainty principle implies that the minimum measurement time must necessarily go up in order to achieve a greater accuracy in energy measurement.

69.   **INTERPRET**   In this problem we want to show that if a photon's wavelength is equal to a particle's Compton wavelength, then the photon's energy is equal to the particle's rest energy.

DEVELOP   The Compton wavelength of a particle is $\lambda_C = h/mc$, and its rest energy is $E = mc^2$. On the other hand, the energy of a photon is $E_\gamma = hf = hc/\lambda$.

EVALUATE When the wavelength of the photon is $\lambda = \lambda_C$, its energy is

$$E_\gamma = \frac{hc}{\lambda_C} = \frac{hc}{h/mc} = mc^2$$

which is the same as the particle's rest energy.

ASSESS We emphasize that photons have zero rest mass and their energy is completely kinetic in nature.

71.   **INTERPRET**   This problem involves Compton scattering of a photon off an electron. The object of interest is the initial photon energy.

DEVELOP   For Compton scattering at $90°$, Equation 34.8 gives $\lambda = \lambda_0 + \lambda_C$. In terms of the photon energies $(\lambda_0 = hc/E_0, \lambda = hc/E)$ and the electron's Compton wavelength $(\lambda_C = hc/m_e c^2)$, this can be written as $1/E = 1/E_0 + 1/m_e c^2$, or $E = E_0 m_e c^2/(E_0 + m_e c^2)$.

The recoil electron's kinetic energy is

$$K_e = (\gamma - 1)m_e c^2 = E_0 - E = E_0 - \frac{E_0 m_e c^2}{E_0 + m_e c^2} = \frac{E_0^2}{E_0 + m_e c^2}$$

This is a quadratic equation in E_0, namely $E_0^2 - (\gamma - 1)mc^2(E_0 + m_e c^2) = 0$. The positive solution is the initial photon energy we seek.

EVALUATE   The positive solution for E_0 is

$$E_0 = \frac{1}{2}\left[(\gamma - 1)m_e c^2 + \sqrt{(\gamma - 1)^2 m_e^2 c^4 + 4(\gamma - 1)m^2 c^4}\right] = \frac{1}{2}m_e c^2\left[(\gamma - 1) + \sqrt{(\gamma - 1)(\gamma + 3)}\right]$$

ASSESS   With some algebra, the kinetic energy of the recoiled electron can be written as

$$K_e = (\gamma - 1)m_e c^2 = \frac{E_0^2}{E_0 + m_e c^2} = \frac{1}{2}m_e c^2 \frac{[(\gamma - 1) + \sqrt{(\gamma - 1)(\gamma + 3)}]^2}{(\gamma + 1) + \sqrt{(\gamma - 1)(\gamma + 3)}}$$

In the non-relativistic limit where $\gamma \approx 1 + (v/c)^2/2$, the above expression reduces to the expected result $K_e \approx \frac{1}{2}m_e v^2$.

73.   **INTERPRET**   We use conservation of energy and conservation of momentum to derive equations related to Compton's equation, and then derive the equation for the Compton shift.

DEVELOP   Relativistic energy for a photon is $E = \frac{hc}{\lambda}$, and for a particle it is $E = \gamma mc^2$. Relativistic momentum is $\vec{p} = \gamma m\vec{u}$ for the electron and $\vec{p} = \frac{h}{\lambda_0}\hat{i}$ for the photon, where $\hat{i}$ is the direction of the photon's motion. We will use conservation of energy to obtain one of the desired equations, and conservation of momentum in two dimensions to obtain the other two.

EVALUATE   The initial energy is the energy of the photon plus the rest energy of the electron: $E_i = \frac{hc}{\lambda_0} + mc^2$. The final energy is the energy of the new photon plus the relativistic energy of the moving electron: $E_f = \frac{hc}{\lambda} + \gamma mc^2$. These two, together, give us the first of the three desired equations: $\frac{hc}{\lambda_0} + mc^2 = \frac{hc}{\lambda} + \lambda mc^2$. The next two equations come from the initial momentum, $\vec{p}_i = \frac{h}{\lambda_0}\hat{i}$, and the components of final momentum $p_{fx} = \frac{h}{\lambda}\cos\theta + \gamma mu\cos\phi$ and $p_{fy} = 0 = \gamma mu\sin\phi - \frac{h}{\lambda}\sin\theta$:

$$p_{ix} = p_{fx} \rightarrow \frac{h}{\lambda_0} = \frac{h}{\lambda}\cos\theta + \gamma mu\cos\phi$$

$$p_{iy} = p_{fy} = 0 \rightarrow 0 = \frac{h}{\lambda}\sin\theta - \gamma mu\sin\phi$$

Solving these three equations for $\lambda_0 - \lambda$ directly is algebraically a lengthy process. An easier approach is to start with the momentum in vector form and use the law of cosines:

$$\vec{p}_\gamma = \vec{p}_{\gamma'} + \vec{p}_{e'} \rightarrow p_{e'}^2 = p_\gamma^2 + p_{\gamma'}^2 - 2p_\gamma p_{\gamma'} \cos\theta$$

$$\rightarrow p_{e'}^2 = \left(\frac{h}{\lambda_0}\right)^2 + \left(\frac{h}{\lambda}\right)^2 - 2\frac{h}{\lambda_0}\frac{h}{\lambda}\cos\theta$$

We now put conservation of energy in the form $\frac{hc}{\lambda_0} + mc^2 = \frac{hc}{\lambda} + \sqrt{m^2c^4 + p_{e'}^2 c^2}$, and solve for $p_{e'}^2$ to obtain

$$p_{e'}^2 = \frac{\left(\frac{hc}{\lambda_0} - \frac{hc}{\lambda} + mc^2\right)^2 - m^2c^4}{c^2}$$

We equate the two equations for $p_{e'}^2$ to obtain

$$\frac{\left(\frac{hc}{\lambda_0} - \frac{hc}{\lambda} + mc^2\right)^2 - m^2c^4}{c^2} = \left(\frac{h}{\lambda_0}\right)^2 + \left(\frac{h}{\lambda}\right)^2 - 2\frac{h}{\lambda_0}\frac{h}{\lambda}\cos\theta$$

$$\rightarrow \left(\frac{hc}{\lambda_0}\right)^2 + \left(\frac{hc}{\lambda}\right)^2 - \frac{2c^3hm}{\lambda} + \frac{2c^3hm}{\lambda_0} - \frac{2c^2h^2}{\lambda\lambda_0} = \left(\frac{hc}{\lambda_0}\right)^2 + \left(\frac{hc}{\lambda}\right)^2 - 2\frac{hc}{\lambda_0}\frac{hc}{\lambda}\cos\theta$$

$$\rightarrow \frac{mc}{\lambda_0} - \frac{mc}{\lambda} = \frac{h}{\lambda\lambda_0}(1 - \cos\theta)$$

$$\rightarrow \lambda - \lambda_0 = \frac{h}{mc}(1 - \cos\theta)$$

ASSESS We've derived the equation for the Compton shift, using conservation of energy and momentum.

75. **INTERPRET** We integrate the radiance equation (Equation 34.3) over all wavelengths and show that the resulting total power radiated per unit area is equivalent to the Stefan-Boltzmann law.

DEVELOP The Stefan-Boltzmann law gives the power per area as $\frac{P}{A} = \sigma T^4$, and the radiance equation is $R(\lambda, T) = \frac{2\pi hc^2}{\lambda^5(e^{hc/\lambda kT}-1)}$. We will integrate $Rd\lambda$ over all λ.

EVALUATE Integrate numerically:

$$\frac{P}{A} = \int_0^\infty R(\lambda, T)d\lambda = \int_0^\infty \frac{2\pi hc^2}{\lambda^5(e^{hc/\lambda kT}-1)}d\lambda = \frac{2k^4\pi^5}{15c^2h^3}T^4$$

ASSESS This is equivalent to the Stefan-Boltzmann law, with $\sigma = \frac{2k^4\pi^5}{15c^2h^3}$.

77. **INTERPRET** We use conservation of momentum to find the recoil angle of the electron in Compton scattering.

DEVELOP The momentum conservation equations from problem 34.73 are $\frac{h}{\lambda_0} = \frac{h}{\lambda}\cos\theta + \gamma mu\cos\phi$ and $0 = \frac{h}{\lambda}\sin\theta - \gamma mu\sin\phi$, and the Compton scattering equation is $\lambda - \lambda_0 = \frac{h}{mc}(1 - \cos\theta)$. We will solve the momentum equations for $\sin\phi$ and $\cos\phi$, take the ratio, and see if we can get the desired result of $\tan\phi = \lambda_0\sin\theta/(\lambda - \lambda_0\cos\theta)$.

Note: The first printing of the text has an incorrect expression for $\tan\phi$. The correct one is given in the problem statement here.

EVALUATE

$$0 = \frac{h}{\lambda}\sin\theta - \gamma mu\sin\phi \rightarrow \sin\phi = \frac{h\sin\theta}{\gamma\lambda mu}$$

$$\frac{h}{\lambda_0} = \frac{h}{\lambda}\cos\theta + \gamma mu\cos\phi \rightarrow \cos\phi = \frac{h}{\gamma mu}\left(\frac{1}{\lambda_0} - \frac{1}{\lambda}\cos\theta\right)$$

$$\rightarrow \tan\phi = \frac{\sin\phi}{\cos\phi} = \frac{\sin\theta}{\lambda\left(\frac{1}{\lambda_0} - \frac{1}{\lambda}\cos\theta\right)}$$

$$\rightarrow \tan\phi = \frac{\lambda_0\sin\theta}{\lambda - \lambda_0\cos\theta}$$

ASSESS This result shows that at wavelengths much longer than λ_0, $\tan\phi$ becomes very small. That makes sense because then the photon's momentum is so small that the electron hardly recoils at all.

79. **INTERPRET** We use the uncertainty principle to estimate the uncertainty in energy measurements for a particular particle detector.

DEVELOP The uncertainty principle tells us that $\Delta E \Delta t \geq \hbar$. The interaction time between the particle and the detector is $t = 12$ fs, so the uncertainty in interaction time Δt must be less than this value. We solve for ΔE.

EVALUATE

$$\Delta E \geq \frac{\hbar}{\Delta t} \rightarrow \Delta E \geq 8.8 \times 10^{-12} \text{ J} = 0.055 \text{ eV}$$

ASSESS The minimum uncertainty in energy measurements is about $\frac{1}{20}$ eV.

QUANTUM MECHANICS

EXERCISES

Section 35.2 The Schrödinger Equation

11. **INTERPRET** We are given the wave function of a particle and asked about its probability distribution.

DEVELOP Since the quantity $\psi^2(x)$ represents the probability density of finding the particle, the particle is most likely to be found at the position where the probability density $\psi^2(x)$ is a maximum.

EVALUATE (a) The maximum of $\psi^2(x) = A^2 e^{-2x^2/a^2}$ is at $x = 0$ (calculate $d(\psi^2)/dx = 0$ for corroboration).

(b) The probability density $\psi^2(x)$ falls to half its maximum value ($\psi^2(0) = A^2$) when $1/2 = e^{-2x^2/a^2}$, or

$$x = \pm a\sqrt{\tfrac{1}{2}\ln 2} = \pm 0.589a.$$

ASSESS The probability distribution is shown below. Note that $\psi^2(x)/A^2 = e^{-2x^2/a^2}$ peaks at $x = 0$ and is halved at $x/a = \pm 0.589$.

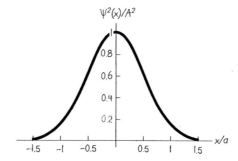

13. **INTERPRET** We use the normalization constant and the wave function from Exercise 12 to find the probability of finding the particle in the region $0 \le x \le \frac{L}{4}$. The probability of finding a particle in a region dx is $\Psi^2 dx$.

DEVELOP We integrate the square of the normalized wave function $\psi(x) = \frac{2\sqrt{3}}{L^{3/2}} x$ from $x = 0$ to $x = \frac{L}{4}$ to find the probability of finding the particle in the region $0 \le x \le \frac{L}{4}$.

EVALUATE

$$P = \frac{12}{L^3} \int_0^{L/4} x^2 dx = \frac{12}{L^3}\left[\frac{1}{3}\frac{L^3}{4^3}\right] = \frac{1}{16}$$

ASSESS There is a one in sixteen chance of finding the particle in the region described. The region is a quarter of the entire region where the particle could be, but the wave function is smaller in this region than elsewhere so the probability is less than one in four.

Section 35.3 Particles and Potentials

15. **INTERPRET** We find the principal quantum number of a particle in an infinite square well, given the particle's energy relative to the ground state.

DEVELOP The energy levels in an infinite square well are $E_n = E_0 n^2$, where $E_0 = \frac{h^2}{8mL^2}$ is the ground state energy. We can immediately see that if the energy is 25 times the ground state energy, then n must be...

EVALUATE $n = 5$.

ASSESS The energy levels go as n squared.

17. **INTERPRET** We have an electron confined in an infinite potential well, and we are interested in its ground-state energy.

DEVELOP The energy levels for an infinite square potential well is given by Equation 35.5:

$$E_n = \frac{n^2 \pi^2 \hbar^2}{2mL^2} = \frac{n^2 h^2}{8mL^2}$$

The ground-state energy corresponds to $n = 1$.

EVALUATE From the above equation, we find the ground-state energy to be

$$E_1 = \frac{h^2}{8mL^2} = \frac{(hc)^2}{8(mc^2)L^2} = \frac{(1240 \text{ eV} \cdot \text{nm})^2}{8(511 \text{ keV})(10 \text{ nm})^2} = 3.76 \times 10^{-3} \text{ eV} = 6.03 \times 10^{-22} \text{ J}$$

ASSESS A non-zero ground-state energy is a common feature of quantum systems. Note that the energy levels are quantized and proportional to n^2.

19. **INTERPRET** In this problem we want to demonstrate the smallness of the quantum effect when dealing with macroscopic objects.

DEVELOP In a one-dimensional infinite square well, the lowest energy of a particle is (see Equation 35.5) $E_1 = h^2/8mL^2$. We set $E_1 = \frac{1}{2}mv^2$ to deduce the "would-be" Planck constant for the quantum effect to be noticeable.

EVALUATE Equating $E_1 = h^2/8mL^2 = \frac{1}{2}mv^2$, we get $h^2 = 8mL^2(\frac{1}{2}mv^2) = 4m^2v^2L^2$, or

$$h = 2mvL = 2(60 \text{ kg})(1 \text{ m/s})(2.6 \text{ m}) = 312 \text{ J} \cdot \text{s}$$

This is 4.71×10^{35} times larger than the actual value of Planck's constant.

ASSESS When dealing with motion of macroscopic objects, one may simply apply classical physics and ignore quantum effect.

21. **INTERPRET** We treat the snail as a particle confined in an infinite potential well. We want to show that classical treatment is adequate for large quantum number n, in accordance with the correspondence principle.

DEVELOP The kinetic energy of the snail is

$$E = \frac{1}{2}mv^2 = \frac{1}{2}(3 \times 10^{-3} \text{ kg})(5 \times 10^{-4} \text{ m/s})^2 = 3.75 \times 10^{-10} \text{ J}$$

If this is regarded as the energy of a 3-g particle confined to a one-dimensional infinite square well of width 15 cm, the energy quantum number can be estimated from Equation 35.5:

$$E_n = \frac{n^2 \pi^2 \hbar^2}{2mL^2} = \frac{n^2 h^2}{8mL^2}$$

EVALUATE Equating the two expressions, we find the quantum number to be

$$n = \frac{\sqrt{8mL^2E}}{h} = \frac{(0.15 \text{ m})\sqrt{8(3 \times 10^{-3} \text{ kg})(3.75 \times 10^{-10} \text{ J})}}{6.63 \times 10^{-34} \text{ J} \cdot \text{s}} = 6.79 \times 10^{26}$$

The correspondence principle implies that classical results ought to be quite adequate for quantum numbers this large.

ASSESS We expect classical physics to be adequate in characterizing the motion of the crawling snail. When we try to quantize the system, we find n to be very large, as expected from the correspondence principle.

23. **INTERPRET** We are given the ground-state energy of a particle in a harmonic oscillator potential, and asked to find the corresponding classical frequency of the oscillator.

DEVELOP The ground-state energy of a one-dimensional harmonic oscillator, as given by Equation 35.7, is $(n = 0)$ $E_0 = \frac{1}{2}\hbar\omega = \frac{1}{2}hf$.

EVALUATE Thus, the classical frequency is

$$f = \frac{2E_0}{h} = \frac{2(0.14 \text{ eV})(1.6 \times 10^{-19} \text{ J/eV})}{6.63 \times 10^{-34} \text{ J} \cdot \text{s}} = 6.76 \times 10^{13} \text{ Hz}$$

ASSESS A non-zero ground-state energy is a common feature of quantum systems. Note that since the spacing between adjacent energy levels is $\Delta E = \hbar\omega = hf$, f represents the frequency of the photon that must be absorbed or emitted for transitions to take place.

25.   **INTERPRET**   In this problem we are asked to find the classical frequency of an oscillator, given the energy of the photon emitted.

DEVELOP   For a one-dimensional harmonic oscillator, the energy spacing is $\Delta E = \hbar \omega = hf$ for transitions between adjacent states (see Equation 35.7).

EVALUATE   From the above equation, we find the frequency to be

$$f = \frac{\Delta E}{h} = \frac{1.1 \text{ eV}}{4.136 \times 10^{-15} \text{ eV} \cdot \text{s}} = 2.66 \times 10^{14} \text{ Hz}$$

ASSESS   For transition to occur, an oscillator must emit (or absorb) photons of this frequency.

27.   **INTERPRET**   We find the separation between energy levels of a harmonic oscillator formed of a macroscopically sized spring and mass.

DEVELOP   The separation between energy levels for the harmonic oscillator is $\Delta E = \hbar \omega$. For a spring-mass system, $\omega = \sqrt{\frac{k}{m}}$. We will find the energy spacing in joules for $k = 80$ N/m and $m = 0.001$ kg.

EVALUATE

$$\Delta E = \hbar \sqrt{\frac{k}{m}} = 2.98 \times 10^{-32} \text{ J}$$

ASSESS   This is a very small spacing between energy levels—much smaller than anything we can measure. This is why we don't notice the quantization of energy in macroscopically sized oscillators.

Section 35.4 Quantum Mechanics in Three Dimensions

29.   **INTERPRET**   This problem is about quantum mechanics in a three-dimensional cubical box. We are interested in the energy difference between the ground state and the first excited state.

DEVELOP   The energy levels of a particle in a three-dimensional box are given by Equation 35.8:

$$E = \frac{h^2}{8mL^2}\left(n_x^2 + n_y^2 + n_z^2\right)$$

The ground state corresponds to $n_x = n_y = n_z = 1$, while the first excited state has one of the quantum numbers equal to 2 (e.g., $n_x = 2$, $n_y = n_z = 1$).

EVALUATE   The difference in energy between the first excited state and the ground-state for a proton (mass 938 MeV/c²) in a cubical box (side length 1 fm) is

$$\Delta E = (2^2 + 1^2 + 1^2 - 1^2 - 1^2 - 1^2)\frac{h^2}{8mL^2} = \frac{3(hc)^2}{8(mc^2)L^2} = \frac{3(1240 \text{ MeV} \cdot \text{fm})^2}{8(938 \text{ MeV})(1 \text{ fm})^2} = 615 \text{ MeV}$$

ASSESS   The typical nuclear energy levels are of order MeV.

PROBLEMS

31.   **INTERPRET**   This problem is about the normalization constant of a given wave function.

DEVELOP   The normalization condition of a wave function is given by Equation 35.3:

$$\int_{-\infty}^{\infty} \psi^2 dx = 1$$

This condition allows us to deduce the form of the normalization constant A.

EVALUATE   With $\psi(x) = \psi(-x)$, we have

$$1 = \int_{-\infty}^{\infty} \psi^2 dx = \int_{-b}^{b} A^2(b^2 - x^2)^2 dx = 2A^2 \int_0^b (b^4 - 2b^2x^2 + x^4)dx = 2A^2\left(b^4 x - \frac{2b^2 x^3}{3} + \frac{x^5}{5}\right)\Bigg|_0^b$$

$$= 2A^2 b^5 \left(1 - \frac{2}{3} + \frac{1}{5}\right) = \frac{16A^2 b^5}{15}$$

Therefore, the normalization constant is $A = \frac{1}{4}\sqrt{15}\ b^{-5/2} = 0.968 b^{-2.5}$.

ASSESS   The normalization condition implies that $\psi(x)$ has dimension (length)$^{-1/2}$. With $\psi(x) = A(b^2 - x^2)$, we expect A to have dimension (length)$^{-5/2}$. Since b also has dimension (length), we see that our result is consistent with dimensional analysis.

33.   INTERPRET   This problem is to show that if two wave functions are solutions of the Schrödinger equation, then their linear combination must also be a solution.

DEVELOP   The time-independent Schrödinger equation is given by Equation 35.1:

$$-\frac{\hbar^2}{2m}\frac{d^2\psi(x)}{dx^2}+U(x)\psi(x)=E\psi(x)$$

We want to show that for any constants a and b, $a\psi_1+b\psi_2$ is a solution if ψ_1 and ψ_2 are.

EVALUATE   Substituting $\psi=a\psi_1+b\psi_2$ into the Schrödinger equation gives

$$\left[-\frac{\hbar^2}{2m}\frac{d^2}{dx^2}+U(x)\right](a\psi_1+b\psi_2)=a\left[-\frac{\hbar^2}{2m}\frac{d^2\psi_1}{dx^2}+U(x)\psi_1\right]+b\left[-\frac{\hbar^2}{2m}\frac{d^2\psi_2}{dx^2}+U(x)\psi_2\right]$$
$$=aE\psi_1+bE\psi_2=E(a\psi_1+b\psi_2)=E\psi$$

ASSESS   The Schrödinger equation is a linear differential equation. Therefore, the result follows directly from the superposition principle.

35.   INTERPRET   This problem is about the energy and wavelength of the photon emitted as the electron trapped in an infinite square well makes a transition to a lower energy level.

DEVELOP   The energy levels for an infinite square potential well are given by Equation 35.5:

$$E_n=\frac{n^2\pi^2\hbar^2}{2mL^2}=\frac{n^2h^2}{8mL^2}$$

Thus, the energy of the photon emitted when the electron drops from n_i to $n_f<n_i$ is

$$E_\gamma=\Delta E=\left(n_i^2-n_f^2\right)\frac{h^2}{8mL^2}$$

The wavelength of the photon is $\lambda=hc/E_\gamma$.

EVALUATE   (a) Substituting the values given, we get

$$E_\gamma=\left(n_i^2-n_f^2\right)\frac{h^2}{8mL^2}=\left(n_i^2-n_f^2\right)\frac{(hc)^2}{8(mc^2)L^2}=(7^2-6^2)\frac{(1240\text{ eV}\cdot\text{nm})^2}{8(511\text{ keV})(1.5\text{ nm})^2}=2.17\text{ eV}$$

(b) The wavelength of the photon is $\lambda=\frac{hc}{E_\gamma}=\frac{8mc^2L^2}{(n_i^2-n_f^2)hc}=571$ nm.

ASSESS   The wavelength is in the visible region of the EM spectrum.

37.   INTERPRET   An electron must absorb a photon in order to make a transition from the ground state to an excited state. We want to know the maximum wavelength associated with the transition.

DEVELOP   The smallest transition energy is to the first excited state ($n=1$ to $n=2$), so

$$E_\gamma=\left(n_i^2-n_f^2\right)\frac{h^2}{8mL^2}=\left(n_i^2-n_f^2\right)\frac{(hc)^2}{8(mc^2)L^2}=(2^2-1^2)\frac{(1240\text{ eV}\cdot\text{nm})^2}{8(511\text{ keV})(4.4\text{ nm})^2}=0.0583\text{ eV}$$

EVALUATE   Thus, the maximum wavelength that can be absorbed is

$$\lambda=\frac{hc}{E_\gamma}=\frac{1240\text{ eV}\cdot\text{nm}}{0.0583\text{ eV}}=21.3\ \mu\text{m}$$

ASSESS   The wavelength corresponds to the infrared region of the EM spectrum.

39.   INTERPRET   There are various ways for the electron which is initially in the $n=4$ state to make a transition to the ground state. We want to find the wavelengths associated with all possible spectral lines in this process.

DEVELOP   The energy levels for an infinite square potential well are given by Equation 35.5:

$$E_n=\frac{n^2\pi^2\hbar^2}{2mL^2}=\frac{n^2h^2}{8mL^2}$$

Thus, the energy of the photon emitted when the electron drops from n_i to $n_f<n_i$ is

$$E_\gamma=\Delta E=\left(n_i^2-n_f^2\right)\frac{h^2}{8mL^2}$$

The possible photon wavelengths are

$$\lambda = \frac{hc}{E_\gamma} = \frac{8(mc^2)L^2}{\left(n_i^2 - n_f^2\right)hc} = \frac{8(511\ \text{keV})(0.1\ \text{nm})^2}{\left(n_i^2 - n_f^2\right)(1240\ \text{eV}\cdot\text{nm})} = \frac{33.0\ \text{nm}}{n_i^2 - n_f^2}$$

EVALUATE Starting from the $n = 4$ state, the possible transitions to the ground state are $4 \to 1$, $4 \to 2$, $4 \to 3$, $3 \to 1$, $3 \to 2$, and $2 \to 1$, with corresponding wavelengths

$$\lambda_{4\to1} = 33.0\ \text{nm}/(4^2 - 1^2) = 2.20\ \text{nm}, \quad \lambda_{4\to2} = 33.0\ \text{nm}/(4^2 - 2^2) = 2.75\ \text{nm}$$

$$\lambda_{4\to3} = 33.0\ \text{nm}/(4^2 - 3^2) = 4.71\ \text{nm}, \quad \lambda_{3\to1} = 33.0\ \text{nm}/(3^2 - 1^2) = 4.12\ \text{nm}$$

$$\lambda_{3\to2} = 33.0\ \text{nm}/(3^2 - 2^2) = 6.59\ \text{nm}, \quad \lambda_{2\to1} = 33.0\ \text{nm}/(2^2 - 1^2) = 11.0\ \text{nm}$$

ASSESS Only photons of these discrete wavelengths will be emitted during the transition. The wavelengths correspond to the ultraviolet region of the EM spectrum.

41. **INTERPRET** The potential is the same as that in Figure 35.5, except the origin of coordinates is at the center of the well. We want to find the normalized wave function and the energy levels.

DEVELOP The wave function for this well can be found by using Equation 35.6 but replace x by $x' + \frac{1}{2}L$, where $-\frac{1}{2}L \le x' \le \frac{1}{2}L$. The normalized wave function is

$$\psi_n(x') = \sqrt{\frac{2}{L}} \sin\frac{n\pi}{L}\left(x' + \frac{L}{2}\right) = \sqrt{\frac{2}{L}}\left(\sin\frac{n\pi x'}{L}\cos\frac{n\pi}{2} + \cos\frac{n\pi x'}{L}\sin\frac{n\pi}{2}\right)$$

But we need to distinguish between even and odd values of n.

EVALUATE (a) If n is odd, then $\cos\frac{1}{2}n\pi = 0$ and $\sin\frac{1}{2}n\pi = \pm1$, so the wave function is

$$\psi_{n\text{-odd}}(x') = \sqrt{\frac{2}{L}}\cos\left(\frac{n\pi x'}{L}\right)$$

The probability density, ψ_n^2, is unaffected by the overall sign of ψ_n, which we choose to be positive. If n is even, then $\cos\frac{1}{2}n\pi = \pm1$ and $\sin\frac{1}{2}n\pi = 0$, and the wave function is

$$\psi_{n\text{-even}}(x') = \sqrt{\frac{2}{L}}\sin\left(\frac{n\pi x'}{L}\right)$$

(b) The energy levels are the same, $E_n = n^2h^2/8mL^2$, regardless of how the potential is parameterized.

ASSESS These results can be confirmed by direct solution of the Schrödinger equation. For $-\frac{1}{2}L \le x \le \frac{1}{2}L$, Equation 35.4 has two solutions, $A\sin kx$ or $A\cos kx$. In order for ψ to vanish at $x = \pm\frac{1}{2}L$, one must use the cosine solution for odd quantum numbers $(\cos(\pm kL/2)$ vanishes for kL equal to odd multiples of π) and the sine solution for even quantum numbers $(\sin(\pm kL/2)$ vanishes for kL equal to even multiples of π). Since the average of $\sin^2$ or $\cos^2$ over an integer number of half-cycles is 1/2, the normalization constant is $\sqrt{2/L}$ for either wave function (or use integrals in Appendix A). (Note that these wave functions have even or odd parity about the center of the potential well; see Section 39.2.)

43. **INTERPRET** We're given the energy of the photon emitted in a transition and asked about the width of the infinite potential well.

DEVELOP The energy levels for an infinite square potential well are given by Equation 35.5:

$$E_n = \frac{n^2\pi^2\hbar^2}{2mL^2} = \frac{n^2h^2}{8mL^2}$$

Thus, the energy of the photon emitted when the electron drops from n_i to $n_f < n_i$ is

$$\Delta E = \left(n_i^2 - n_f^2\right)\frac{h^2}{8mL^2}$$

EVALUATE   From the above equation, the energy difference between $n_i = 2$ and $n_f = 1$ is $\Delta E = 3h^2/8mL^2$, so the width of the potential well is

$$L = \frac{\sqrt{3}hc}{\sqrt{8mc^2 \Delta E}} = \frac{\sqrt{3}(1240 \text{ eV} \cdot \text{nm})}{\sqrt{8(511 \text{ keV})(1.96 \text{ eV})}} = 0.759 \text{ nm}$$

ASSESS   The result is consistent with the typical quantum well width (about the size of an atom).

45.   **INTERPRET**   This problem is about quantizing a classical mass-spring system and finding its corresponding quantum number.

DEVELOP   The total energy of the mass-spring system (kinetic plus potential) is $E = \frac{1}{2}kA^2$. We equate this to the quantum energy (Equation 35.7), $E_n = (n + \frac{1}{2})\hbar\omega$ to find n.

EVALUATE   Solving for n, we get

$$n = \frac{1}{2}\left(\frac{kA^2}{\hbar\omega} - 1\right) \approx \frac{A^2\sqrt{mk}}{2\hbar} = \frac{(0.12 \text{ m})^2 \sqrt{(0.01 \text{ kg})(150 \text{ N/m})}}{2(1.055 \times 10^{-34} \text{ J} \cdot \text{s})} = 8.36 \times 10^{31}$$

ASSESS   This huge value, via the correspondence principle, justifies the classical treatment of a mass on a spring.

47.   **INTERPRET**   This problem involves comparison of energy spacing between adjacent levels in a harmonic oscillator with the actual energy.

DEVELOP   The energy levels of a harmonic oscillator are given by Equation 35.7:

$$E_n = \left(n + \frac{1}{2}\right)\hbar\omega$$

so the spacing of adjacent levels is $\Delta E = \hbar\omega$. The problem is to find a condition for n such that $\Delta E/E_n < 0.01$.

EVALUATE   The above condition implies that

$$\frac{\Delta E}{E_n} = \frac{\hbar\omega}{(n + 1/2)\hbar\omega} = \frac{1}{n + 1/2} < 0.01 \quad \rightarrow \quad n + \frac{1}{2} > 100$$

or $n > 99$.

ASSESS   As n increases, the ratio $\Delta E/E_n$ becomes vanishingly small, in accordance with Bohr's correspondence principle.

49.   **INTERPRET**   We calculate the probability of finding a particle in the central one-fourth of the infinite square well for various energy states, and compare our answers with the classical value.

DEVELOP   The wave function for the infinite square well is $\psi_n = \sqrt{\frac{2}{L}} \sin(\frac{n\pi x}{L})$. The region in which we are interested is the central fourth: $\frac{3}{8}L \le x \le \frac{5}{8}L$. We will calculate the probability by integrating $\psi^2 dx$ from $x = \frac{3}{8}L$ to $x = \frac{5}{8}L$.

EVALUATE   First, we find the genereral solution for any value of n:

$$P(n) = \frac{2}{L}\int_{\frac{3}{8}L}^{\frac{5}{8}L} \sin^2\left(\frac{n\pi x}{L}\right)dx = \frac{1}{4} + \frac{1}{2n\pi}\left[\sin\left(\frac{3n\pi}{4}\right) - \sin\left(\frac{5n\pi}{4}\right)\right]$$

(a) $n = 1 \rightarrow P(1) = \frac{2\sqrt{2}+\pi}{4\pi} = 0.475$

(b) $n = 2 \rightarrow P(2) = 0.0908$

(c) $n = 5 \rightarrow P(5) = 0.205$

(d) $n = 20 \rightarrow P(20) = 0.250$

(e) The classical model predicts that the particle would be anywhere in the box with equal probability, so the total probability of being in the central $\frac{1}{4}$ of the box would be 0.25.

ASSESS   Note that as n becomes higher, the quantum probability becomes closer to the classical probability.

51.   **INTERPRET**   This problem is about quantum mechanics in a three-dimensional cubical box. Some excited-state energy levels exhibit degeneracy.

DEVELOP   The energy levels of a particle in a three-dimensional box are given by Equation 35.8:

$$E = \frac{h^2}{8mL^2}\left(n_x^2 + n_y^2 + n_z^2\right)$$

The ground state corresponds to $n_x = n_y = n_z = 1$, while the first excited state has one of the quantum numbers equal to 2 (e.g., $n_x = 2$, $n_y = n_z = 1$).

EVALUATE   The quantum numbers, energy, and degeneracy of the first six levels in the three-dimensional infinite square well are summarized below:

Energy level	(n_x, n_y, n_z)	$E/(h^2/8mL^2)$	Degeneracy factor
1	(1,1,1)	3	1
2	(2,1,1), (1,2,1), (1,1,2)	6	3
3	(2,2,1), (1,2,2), (2,1,2)	9	3
4	(3,1,1), (1,3,1), (1,1,3)	11	3
5	(2,2,2)	12	1
6	(1,2,3), (1,3,2), (2,1,3) (2,3,1), (3,1,2), (3,2,1)	14	6

The energy-level diagram looks similar to that shown in Figure 35.6.

ASSESS   In general, degeneracy arises from symmetry in the system. In our case, the symmetry is due to the fact that we have a cube with $L_x = L_y = L_z = L$.

53.   **INTERPRET**   Since this is a question about probability, we analyze ψ^2 which represents the probability density.

DEVELOP   A straightforward generalization of Example 35.2 shows that the probability of finding a particle, in the quantum state n, in the left-hand quarter of a one-dimensional infinite square well is just

$$P = \int_0^{L/4} \psi_n^2(x)\, dx = \frac{2}{L}\left(\frac{x}{2} - \frac{\sin(2n\pi x/L)}{(4n\pi/L)}\right)\Bigg|_0^{L/4} = \frac{1}{4} - \frac{\sin(n\pi/2)}{2\pi n}$$

The probability is equal to 1/4 for any even n, and is greater than 1/4 for $n = 3 + 4n'$ (or less than 1/4 for $n = 1 + 4n'$) where $n' = 0, 1, 2, \ldots$ (This follows from the fact that $\sin(n\pi/2)$ alternates between ± 1 for integer n.)

EVALUATE   In this problem, the probability is greater than 1/4, so

$$0.303 - 0.25 = \frac{1}{2}\pi n \;\rightarrow\; n = \frac{1}{2}\pi(0.053) = 3.00$$

for the third quantum state.

ASSESS   Classically we expect the probability to be 1/4. From the above equation, we see that this limit is approached for large n, as expected from the correspondence principle.

55.   **INTERPRET**   Electrons emit photons to undergo transitions to lower states. We are interested in the photon wavelengths associated with the spectral lines of all possible transitions.

DEVELOP   The energy levels for an infinite square potential well are given by Equation 35.5:

$$E_n = \frac{n^2\pi^2\hbar^2}{2mL^2} = \frac{n^2 h^2}{8mL^2}$$

Thus, the energy of the photon emitted when the electron drops from initial state n_i to final state $n_f < n_i$ is

$$E_\gamma = \Delta E = \left(n_i^2 - n_f^2\right)\frac{h^2}{8mL^2}$$

The possible photon wavelengths are

$$\lambda = \frac{hc}{E_\gamma} = \frac{8mc^2 L^2}{\left(n_i^2 - n_f^2\right)hc} = \frac{8(511\ \text{keV})(1.2\ \text{nm})^2}{\left(n_i^2 - n_f^2\right)(1240\ \text{eV}\cdot\text{nm})} = \frac{4.75\ \mu\text{m}}{n_i^2 - n_f^2}$$

EVALUATE   (a) For visible photons, $0.4\ \mu\text{m} < \lambda < 0.7\ \mu\text{m}$, or

$$4.75/0.7 = 6.78 < n_i^2 - n_f^2 < 11.9 = 4.75/0.4$$

There are four transitions satisfying this condition: $3 \rightarrow 1$, $4 \rightarrow 3$, $5 \rightarrow 4$, and $6 \rightarrow 5$.

(b) Infrared photons, with $\lambda > 0.7\ \mu\text{m}$, occur for transitions with

$$n_i^2 - n_f^2 < 4.75/0.7 = 6.78$$

The only positive integers $n_i > n_f \geq 1$ satisfying this condition are for $2 \rightarrow 1$ and $3 \rightarrow 2$ transitions, so there are only two infrared photons in the spectrum. (*Note:* The spectrum includes only infrared and shorter wavelengths, so we did not need to consider the upper limit of the infrared region.)

ASSESS   Since energy is quantized, only photons with wavelengths (4 in the visible spectrum and 2 in the infrared) that satisfy the above condition will be emitted during the transitions.

57. **INTERPRET**   This problem is about a potential well of finite depth. We substitute the wave functions given in the problem statement to show that they satisfy the Schrödinger equation.

   DEVELOP   The time-independent Schrödinger equation is given by Equation 35.1:

$$-\frac{\hbar^2}{2m}\frac{d^2\psi(x)}{dx^2} + U(x)\psi(x) = E\psi(x)$$

   In the interior of the well, $U(x) = 0$, while in the exterior, $U(x) = U_0$. After multiplying by $2m/\hbar$ and rearranging terms, Equation 35.1 becomes:

$$\frac{d^2\psi}{dx^2} = \begin{cases} -\dfrac{2mE}{\hbar^2}\psi & \text{for}\ \ 0 \leq x \leq L \\ \dfrac{2m(U_0 - E)}{\hbar^2}\psi & \text{for}\ \ x \geq L \end{cases}$$

   Solving the equation separately allows us to obtain two wave functions, one for $0 \leq x \leq L$, and the other for $x \geq L$.

   EVALUATE   Since $E > 0$, ψ is oscillatory inside the well. The solution which satisfies the condition $\psi = 0$ at $x = 0$ (where $U(x) = \infty$) is

$$\psi_1 = A\sin(\sqrt{2mE}x/\hbar) = A\sin(\sqrt{\varepsilon}x/L)$$

   where $\varepsilon = 2mL^2 E/\hbar^2$. Outside the well, the type of solution (exponential or oscillatory) depends on whether $U_0 - E$ is positive or negative. The former corresponds to a bound state and the latter to an unbound or scattering state. Here, we have assumed that $E < U_0$ (exponential ψ), so the normalizable solution outside is

$$\psi_2 = Be^{\sqrt{2m(U_0-E)}x/\hbar} = Be^{\sqrt{\mu-\varepsilon}x/L}$$

   where $\mu = 2mU_0 L^2/\hbar^2$.

   ASSESS   Straightforward substitution would show that ψ_1 and ψ_2 indeed satisfy the Schrödinger equation.

59. **INTERPRET**   In this problem we continue our investigation of the potential well of finite depth. We use a graphical procedure to find solutions to the Schrödinger equation.

   DEVELOP   In Problem 58, by imposing continuity of ψ and $d\psi/dx$ at $x = L$, we obtained

$$A\sin(\sqrt{\varepsilon}) = Be^{-\sqrt{\mu-\varepsilon}}$$
$$\sqrt{\varepsilon}A\cos\sqrt{\varepsilon} = -\sqrt{\mu-\varepsilon}Be^{-\sqrt{\mu-\varepsilon}}$$

   We then showed that the two equations can be combined to give

$$\tan\sqrt{\varepsilon} = -\sqrt{\frac{\varepsilon}{\mu-\varepsilon}}$$

   Before a graphical solution is attempted, it is convenient to rewrite the above condition for the bound-state energies

$$-\sqrt{\varepsilon}\cot\sqrt{\varepsilon} = \sqrt{\mu-\varepsilon} \qquad (0 < \varepsilon < \mu)$$

The right-hand side, as a function of $\sqrt{\varepsilon}$, represents a quarter of a circle of radius $\sqrt{\mu}$. The left-hand side is a function which ranges from 0 to ∞ in the intervals $\frac{1}{2}\pi \le \sqrt{\varepsilon} \le \pi$, $\frac{3}{2}\pi \le \sqrt{\varepsilon} \le 2\pi$, $\frac{5}{2}\pi \le \sqrt{\varepsilon} \le 3\pi$, etc., for which physical solutions are possible.

EVALUATE   From the sketch, it can be seen that there are no bound states for $\sqrt{\mu} < \frac{1}{2}\pi$ (or $\mu < 2.47$), there is one bound state for $\frac{1}{2}\pi < \sqrt{\mu} < \frac{3}{2}\pi$ (or $2.47 < \mu < 22.2$), there are two bound states for $\frac{3}{2}\pi < \sqrt{\mu} < \frac{5}{2}\pi$ (or $22.2 < \mu < 61.7$), etc. Values of ε can be found from the graph or by numerical computation. For the specified cases, they are: **(a)** none for $\mu = 2$, **(b)** $\varepsilon = 6.44$ for $\mu = 20$, and **(c)** $\varepsilon = 7.52$ and 29.3 for $\mu = 50$.

ASSESS   From our graphical solution, we see that when the potential well is too shallow ($\mu < 2.47$), it cannot trap electrons, and hence there's no bound-state solution. However, the number of bound states increases with μ. This is to be expected since a deeper well can accommodate more quantum states.

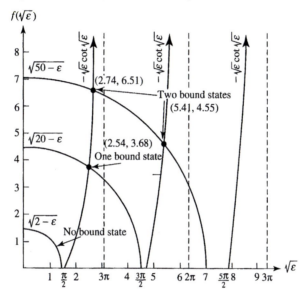

61.   **INTERPRET**   We show by direct solution that a given wave function is a solution to the time-independent Schrödinger equation for the simple harmonic oscillator, then normalize the wave function and show that the probability density is greatest in the center.

DEVELOP   The time-independent Schrödinger equation is $-\hbar^2/2m\, d^2\psi/dx^2 + U(x)\psi = E\psi$, where for the simple harmonic oscillator $U(x) = \frac{1}{2}kx^2 = \frac{1}{2}\omega^2 m x^2$. The potential solution we're given is $\psi_1 = A_1 x e^{-\alpha^2 x^2/2}$, where $\alpha^2 = m\omega/\hbar$. For part **(a)** we substitute this ψ and show that the solution corresponds to $n = 1$ in $E_n = (n + \frac{1}{2})\hbar\omega$. For part **(b)**, we find the normalization constant A_1 by setting the integral of ψ_1^2 over all space equal to one. For **(c)**, we show that ψ_1^2 has a minimum at $x = 0$ and maxima at $x = \pm\frac{1}{\alpha}$.

EVALUATE

(a) $\frac{d\psi_1^2}{dx^2} = A_1 e^{-\alpha^2 x^2/2}[x^3\alpha^4 - 3x\alpha^2]$. The time-independent Schrödinger equation then becomes, with this substitution, $-\frac{\hbar^2}{2m} A_1 e^{-\alpha^2 x^2/2}[x^3\alpha^4 - 3x\alpha^2] + \frac{1}{2}\omega^2 m x^2 x\, A_1 e^{-\alpha^2 x^2/2} = E x\, A_1 e^{-\alpha^2 x^2/2}$. Further simplification gives us

$$-\frac{\hbar^2}{2m}x^3\left(\frac{m\omega}{\hbar}\right)^2 + \frac{\hbar^2}{2m}3x\left(\frac{m\omega}{\hbar}\right) + \frac{1}{2}\omega^2 m x^3 = Ex$$

$$\rightarrow -\frac{x^3 m\omega^2}{2} + \frac{3\hbar x\omega}{2} + \frac{x^3 m\omega^2}{2} = Ex \rightarrow E = \frac{3}{2}\hbar\omega$$

So the wave function given is a solution, corresponding to the $n = 1$ energy state for the harmonic oscillator.

(b)
$$1 = \int_{\infty}^{\infty} \psi_1\, dx = A_1 \int_{\infty}^{\infty} x^2 e^{-\alpha^2 x^2}\, dx = A_1 \frac{\sqrt{\pi}}{2\alpha^3} \rightarrow A_0 = \frac{2\alpha^3}{\sqrt{\pi}}$$

(c) The quantum probability density is a minimum when $\frac{d}{dx}[\psi^2(x)] = 0$ and $\frac{d^2}{dx}[\psi^2(x)] > 0$, so

$\frac{d}{dx}[A_0^2 x^2 e^{-\alpha^2 x^2}] = 0 = 2A_1^2 e^{-\alpha^2 x^2}[x - x^3 \alpha^2] \rightarrow x - x^3\alpha^2 = 0 \rightarrow x = \{0, \pm\frac{1}{\alpha}\}$. To find which of these solutions are

maxima and which are minima, take the second derivative: $d^2/dx^2[A_0^2 x^2 e^{-\alpha^2 x^2}] = 2A_1^2 e^{-\alpha^2 x^2}[1 - 5x^2\alpha^2 + 2x^4\alpha^4]$.

At $x = 0$, this is a positive value, so $x = 0$ is a minimum and $x = \pm\frac{1}{\alpha}$ are maxima.

ASSESS In part c, we have neglected possible solutions to $d/dz[\psi^2(x)] = 0$ at $x = \pm\infty$. The function goes to zero there as well, but that's not really what we're interested in.

63. **INTERPRET** We use the boundary conditions on the wave function to find the normalization constants for a wave function on either side of a potential step.

   DEVELOP For the region $x < 0$, $\psi_1(x) = Ae^{ik_1 x} + Be^{-ik_1 x}$. For $x > 0$, $\psi_2(x) = Ce^{ik_2 x}$. We know that the wave function is continuous, so $\psi_1(0) = \psi_2(0)$. The first derivative of the wave function is also continuous, so $d\psi_1/dx(0) = d\psi_2/dx(0)$. We use these two boundary conditions to find B and C in terms of k_1, k_2, and A.

   EVALUATE $\psi_1(0) = \psi_2(0) \rightarrow A + B = C$, and $\frac{d\psi_1}{dx}(0) = \frac{d\psi_2}{dx}(0) \rightarrow ik_1 A - ik_1 B = ik_2 C$. Solving for B and C gives us

$$B = C - A \rightarrow k_1 A - k_1(C - A) = k_2 C$$

$$\rightarrow 2k_1 A = C(k_1 + k_2) \rightarrow C = A\left(\frac{2k_1}{k_1 + k_2}\right)$$

$$\rightarrow B = C - A = A\left(\frac{2k_1}{k_1 + k_2} - 1\right) = A\left(\frac{k_1 - k_2}{k_1 + k_2}\right)$$

   ASSESS These coefficients can be used to find the probability that a particle is reflected or transmitted at the barrier.

65. **INTERPRET** We find the width of an infinite square well from the energy of a photon that is emitted when an electron changes levels in that well.

   DEVELOP The photon energy must equal the electron energy difference between the two energy levels in the square well, so $E_\gamma = 1.4 \text{ eV} = h^2/8mL^2(n_1^2 - n_2^2)$ and we solve for L with $n_1 = 5$ and $n_2 = 2$. The electron mass is $m = 9.11 \times 10^{-31}$ kg.

   EVALUATE

$$L = \sqrt{\frac{h^2\left(n_1^2 - n_2^2\right)}{8mE_\gamma}} = 2.4 \text{ nm}$$

   ASSESS The professor's estimate of "a few" nanometers is about right.

67. **INTERPRET** What temperature would be required for proton-antiproton pair production? We use the thermal energy $E = kT$ and the mass energy $E = mc^2$.

   DEVELOP In order for thermal energy to create a proton-antiproton pair, the thermal energy $E_T = kT$ would have to equal the mass energy of the pair, $E = 2mc^2$. We set these equal and solve for T.

   EVALUATE $kT = 2m_p c^2 \rightarrow T = 2m_p c^2/k = 2.2 \times 10^{13}$ K.

   ASSESS That's a bit hotter than a barbeque grill... or a supernova, for that matter!

ATOMIC PHYSICS

Note: For the problems in this chapter, useful numerical values of Planck's constant, in SI and atomic units, are: $h = 6.626 \times 10^{-34}$ J·s $= 4.136 \times 10^{-15}$ eV·s $= 1240$ eV·nm/c, and $\hbar = h/2\pi = 1.055 \times 10^{-34}$ J·s $= 6.582 \times 10^{-16}$ eV·s $= 197.3$ MeV·fm/c.

Useful constants and combinations, in SI and atomic units, are:

$$h = 6.626 \times 10^{-34} \text{ J·s} = 4.136 \times 10^{-15} \text{ eV·s} = 1240 \text{ eV·nm}/c$$

$$\hbar = h/2\pi = 1.055 \times 10^{-34} \text{ J·s} = 6.582 \times 10^{-16} \text{ eV·s} = 197.3 \text{ MeV·fm}/c$$

$$c = 2.998 \times 10^8 \text{ m/s}, \ ke^2 = 1.440 \text{ eV·nm},$$

$$1u = 1.661 \times 10^{-27} \text{ kg} = 931.5 \text{ MeV}/c^2$$

$$k_B = 1.381 \times 10^{-23} \text{ J/K} = 8.617 \times 10^{-5} \text{ eV/K}$$

EXERCISES

Section 36.1 The Hydrogen Atom

15. **INTERPRET** Our system consists of a group of hydrogen atoms in the same excited state characterized by the quantum number n. The 1.5 eV minimum energy is the ionization energy.

DEVELOP The hydrogen energy levels are given by Equation 36.6:

$$E_n = \frac{E_1}{n^2} = \frac{-13.6 \text{ eV}}{n^2}$$

where n is the principal quantum number. The ionization energy for this state is the difference between the zero of energy and the energy of the state.

EVALUATE So $1.5 \text{ eV} = -E_n = 13.6 \text{ eV}/n^2$, which yields $n = \sqrt{13.6/1.5} = 3$.

ASSESS The result makes sense since principal quantum must be a positive integer.

17. **INTERPRET** This problem is about the allowed values of the magnitude of angular momentum.

DEVELOP The magnitude of orbital angular momentum is given by Equation 36.9:

$$L = \sqrt{l(l+1)}\hbar, \quad l = 0,1,2,\dots$$

EVALUATE Successive values of $\sqrt{l(l+1)}$, starting with $l = 3$, are: $\sqrt{3 \times 4} = \sqrt{12}$, $\sqrt{4 \times 5} = \sqrt{20}$, $\sqrt{5 \times 6} = \sqrt{30}$, $\sqrt{6 \times 7} = \sqrt{42}$, etc. Evidently, **(d)** is an erroneous possibility.

ASSESS Since orbital angular momentum is quantized, only certain discrete values are allowed.

19. **INTERPRET** We're given the magnitude of the orbital angular momentum of an electron and asked to find its corresponding orbital quantum number.

DEVELOP The magnitude of orbital angular momentum is given by Equation 36.9:

$$L = \sqrt{l(l+1)}\hbar, \quad l = 0,1,2,\dots$$

EVALUATE From the above equation, we have $l(l+1) = 30$, or $l = 5$.

ASSESS In this case, the l-value is easily found by inspection. In general, l is a nonnegative integer solution of the quadratic formula.

21.   **INTERPRET**   We're given the energy and orbital angular momentum of an electron and asked about its state. We need to find both principal and orbital quantum numbers.

   DEVELOP   The hydrogen energy levels are given by Equation 36.6:

$$E_n = \frac{E_1}{n^2} = \frac{-13.6 \text{ eV}}{n^2}$$

   where n is the principal quantum number. Similarly, the magnitude of orbital angular momentum is given by Equation 36.9:

$$L = \sqrt{l(l+1)}\hbar, \quad l = 0,1,2,\ldots$$

   EVALUATE   From Equation 36.6, we get $n = \sqrt{13.6/1.51} = 3$, and from Equation 36.9, we have $l(l+1) = 6 = 2(3)$. Thus, $n = 3$ and $l = 2$, and this is a $3d$ state.

   ASSESS   Our result makes sense since orbital quantum number l can only take on values between 0 and $n-1$.

Section 36.2 Electron Spin

23.   **INTERPRET**   We're given the spin quantum number and asked about the magnitude of spin angular momentum.

   DEVELOP   The magnitude of spin angular momentum is given by Equation 36.11:

$$S = \sqrt{s(s+1)}\hbar$$

   EVALUATE   With $s = 2$, the above equation gives $S = \sqrt{2 \times 3}\hbar = \sqrt{6}\hbar = 2.58 \times 10^{-34}$ J·s for the graviton.

   ASSESS   Spin angular momentum is quantized, just like the orbital angular momentum, so only a discrete set of values is allowed.

25.   **INTERPRET**   We have a hydrogen atom in the $3d$ state, and we want to know all the possible values the quantum number j can take.

   DEVELOP   The total angular momentum of the hydrogen atom is (see Equation 36.15) $\vec{J} = \vec{L} + \vec{S}$. For the $3d$ state, $l = 2$ and $s = \frac{1}{2}$ (hydrogen has one electron).

   EVALUATE   The possible j-values are $j = l - \frac{1}{2} = \frac{3}{2}$ and $j = l + \frac{1}{2} = \frac{5}{2}$.

   ASSESS   For an atom with a single electron, the quantum number j takes the values (see Equation 36.17)

$$j = \begin{cases} l + \frac{1}{2}, l \neq 0 \\ \frac{1}{2}, l = 0 \end{cases}$$

Section 36.3 The Exclusion Principle

27.   **INTERPRET**   We find the energy of the highest of several electrons in a harmonic oscillator potential. We will assume that the electrons are in the lowest states possible, and use the exclusion principle.

   DEVELOP   The exclusion principle states that no more than one electron may be in any given state. Electrons have spin $\frac{1}{2}$, so two electrons may exist in each n energy state, one with spin $+\frac{1}{2}$ and one with spin $-\frac{1}{2}$. We will find the energy of the highest-energy electron by "filling" the energy levels with pairs of electrons, and finding the lowest energy level available to the ninth electron. The energy levels are given by $E_n = (n + \frac{1}{2})\hbar\omega$.

   EVALUATE   The first 20 electrons fill levels 0–9, so the 21st electron must be in level $n = 10$ and

   $E = (10 + \frac{1}{2})\hbar\omega = \frac{21\hbar\omega}{2}$.

   ASSESS   Electrons are fermions, so they follow the exclusion principle. If the particles in question were bosons, they could all be in the lowest energy state.

Section 36.4 Multielectron Atoms and the Periodic Table

29.   **INTERPRET**   We're asked for the symbolic configuration like those in Table 36.2.

   DEVELOP   Scandium ($Z = 21$), the first of the transition metals, has one $3d$-electron in addition to the configuration of the preceding element, calcium ($Z = 20$). (See explanation in the text just before Example 36.4.)

   EVALUATE   Thus, $1s^2 2s^2 2p^6 3s^2 3p^6 4s^2 3d^1$ is the full electronic configuration for scandium.

   ASSESS   As a useful check, we note that the sum of all the superscripts is 21.

Section 36.5 Transitions and Atomic Spectra

31. **INTERPRET** This problem is about the relationship between the wavelength and energy of a photon, with wavelength expressed in nm and energy given in eV.

DEVELOP Using Equation 34.6, the energy of a photon can be written as

$$E = hf = \frac{hc}{\lambda}$$

We complete the proof by expressing the Planck's constant and the speed of light in the appropriate units.

EVALUATE Writing $h = 6.626 \times 10^{-34}$ J·s $= 4.136 \times 10^{-15}$ eV·s and $c = 3 \times 10^{17}$ nm/s, we obtain

$$\lambda = \frac{hc}{E} = \frac{(4.136 \times 10^{-15} \text{ eV·s})(3 \times 10^{17} \text{ nm/s})}{E} = \frac{1240 \text{ eV·nm}}{E}$$

ASSESS The units eV and nm are more appropriate for photons. The above formula allows us to calculate the photon energy (in units of eV) easily, once its wavelength (in nm) is given.

33. **INTERPRET** In this problem we are asked about the energy splitting between two $4p$ states of sodium.

DEVELOP Since the $3s$-level is not split, the fine structure splitting of the $4p$-levels is equal to the difference in the photon energies.

EVALUATE Using Equation 34.6, we find the energy difference to be

$$\Delta E = hc \left(\frac{1}{\lambda_1} - \frac{1}{\lambda_2} \right) = (1240 \text{ eV·nm}) \left(\frac{1}{330.2 \text{ nm}} - \frac{1}{330.3 \text{ nm}} \right) = 1.14 \times 10^{-3} \text{ eV}$$

ASSESS The two states correspond to $4p_{3/2}$ and $4p_{1/2}$, with $4p_{3/2}$ having a slightly higher energy compared to $4p_{1/2}$.

PROBLEMS

35. **INTERPRET** We're given the energy and orbital angular momentum of an electron and asked about the principal quantum number and orbital quantum number.

DEVELOP The hydrogen energy levels are given by Equation 36.6:

$$E_n = \frac{E_1}{n^2} = \frac{-13.6 \text{ eV}}{n^2}$$

where n is the principal quantum number. Similarly, the magnitude of orbital angular momentum is given by Equation 36.9:

$$L = \sqrt{l(l+1)}\hbar, \quad l = 0, 1, 2, \ldots$$

EVALUATE From Equation 36.6, we get $n = \sqrt{13.6/0.85} = 4$, and from Equation 36.9, we have $l(l+1) = 12 = 3(4)$. Thus, $n = 4$ and $l = 3$.

ASSESS This is a $4f$ state. Our result makes sense since orbital quantum number l can only take on values between 0 and $n-1$.

37. **INTERPRET** We find the orbital quantum number l for the Moon. We would expect that for a macroscopic object such as this, the number would be rather high—so high that the difference between angular momentum states is smaller than anything we could measure.

DEVELOP We know that the angular momentum is quantized by $L = \sqrt{l(l+1)}\hbar$. We will find the classical angular momentum of the Moon using $L = mvr = mr^2\omega$, where $r = 384{,}400$ km and $m = 7.35 \times 10^{22}$ kg. ω is the angular velocity of the moon, $\omega = \frac{2\pi \text{ radians}}{28 \text{ days}} = 2.60 \times 10^{-6}$ rad/s. We will equate these values of L, and solve for l.

EVALUATE

$$\sqrt{l(l+1)}\hbar = mr^2\omega \rightarrow l^2 + l - \frac{m^2 r^4 \omega^2}{\hbar^2} = 0 \rightarrow l = 2.68 \times 10^{68}$$

ASSESS This is such a large value of l that there is no way that we could measure the difference between angular momentum states, and the angular momentum appears to be a continuous function as classical theory predicts.

39.   **INTERPRET**   This problem is about the possible orientations of the angular momentum vector in the $l = 2$ state.

DEVELOP   For $l = 2$, the magnetic orbital angular momentum quantum number has $2l + 1 = 5$ values, namely $m_l = 0, \pm 1, \pm 2$. The corresponding "angles" $\vec{L}$ makes with the z axis are given by

$$\theta = \cos^{-1}\left(\frac{L_z}{L}\right) = \cos^{-1}\left(\frac{m_l}{\sqrt{l(l+1)}}\right) = \cos^{-1}\left(\frac{m_l}{\sqrt{6}}\right)$$

EVALUATE   Substituting the possible values for m_l, we find the angles to be $90°, 65.9°,$ or $114°, 35.3°,$ or $145°$ respectively.

ASSESS   The angular momentum is quantized, not only in magnitude but also in direction. Note that the "angle" has meaning only in the vector model for angular momentum.

41.   **INTERPRET**   We're given the state of the electron, and asked about the possible values of L_z when measurements are performed.

DEVELOP   An f state has $l = 3$; thus $m_l = 0, \pm 1, \pm 2,$ and ± 3.

EVALUATE   Since $L_z = m_l \hbar$ (Equation 36.10), when a measurement is performed, the possible results are (in units of $\hbar$) $m_l = 0, \pm 1, \pm 2,$ and ± 3.

ASSESS   These are the possible components of the orbital angular momentum in units of $\hbar$. Note that L_z can only take on certain discrete values. This is a consequence of space quantization in quantum mechanics.

43.   **INTERPRET**   This problem is about the radial probability distribution of the electron in the hydrogen ground state. We want to show that the probability density has a maximum at $r = a_0$, where a_0 is the Bohr radius.

DEVELOP   Using Equations 36.3 and 36.5, the radial probability in the ground state can be written as

$$P(r) = 4\pi r^2 \psi^2 = 4\pi r^2 A^2 e^{-2r/a_0}$$

EVALUATE   We differentiate $P(r)$ with respect to r and get

$$\frac{dP(r)}{dr} = 4\pi A^2 e^{-2r/a_0}(2r - 2r^2/a_0) = 0$$

The result shows that $r = a_0$ is a maximum of $P(r)$. Physically, this means that the electron is most likely to be found at $r = a_0$.

ASSESS   The radial probability density function is plotted in Figure 36.4. From the figure, we see that $P(r)$ indeed is a maximum at $r = a_0$.

45.   **INTERPRET**   We have a hydrogen atom in the $n = 3$ state, and we want to know all the possible values the quantum number j can take.

DEVELOP   The total angular momentum of the hydrogen atom is (see Equation 36.15) $\vec{J} = \vec{L} + \vec{S}$. For $n = 3$, $l = 2, 1,$ or 0 and $s = \frac{1}{2}$ (hydrogen has one electron).

EVALUATE   For $l = 0$, there is only $j = s = \frac{1}{2}$, while for $l \neq 0$, $j = l \pm \frac{1}{2}$ (as in Exercise 25). Then $j = \frac{1}{2}$ and $\frac{3}{2}$ for $l = 1$, and $\frac{3}{2}$ and $\frac{5}{2}$ for $l = 2$. All the j-values are $\frac{1}{2}, \frac{1}{2}, \frac{3}{2}, \frac{3}{2},$ and $\frac{5}{2}$. (Note that some j-values occur twice.)

ASSESS   The vector diagrams for spin-orbit coupling with $l = 1$ and $l = 2$ are shown in Figures 36.10 and 36.11, respectively.

47.   **INTERPRET**   The state of the atom written in spectroscopic notation contains information about the principal quantum number, the orbital quantum number, and the total angular momentum.

DEVELOP   The quantum numbers of the $4F_{5/2}$ state are $n = 4$, $l = 3$, and $j = \frac{5}{2}$. The hydrogen energy levels are given by Equation 36.6: $E_n = E_1/n^2$. The magnitude of the electron's orbital angular momentum is (see Equation 36.9) $L = \sqrt{l(l+1)}\hbar$. Similarly, for the total angular momentum, we have (see Equation 37.1) $J = \sqrt{j(j+1)}\hbar$.

EVALUATE   (a) With $n = 2$, we get $E_4 = E_1/16$.

(b) Since $l = 3$, $L = \sqrt{3(3+1)}\hbar = \sqrt{12}\ \hbar$.

(c) With $j = 5/2$, $J = \sqrt{(5/2)(7/2)}\hbar = \sqrt{35}\ \hbar/2$.

(d) Since this state has $j = l - \frac{1}{2}$, the magnitude of L is greater than that of J. The orbital angular momentum and electronic spin angular momentum are anti-parallel.

ASSESS The vector diagram for the spin-orbit coupling with $l = 3$ looks similar to that shown on the left of Figure 36.11 (but with $j = 5/2$).

49. **INTERPRET** We have eight particles in the ground state of a harmonic oscillator potential, and we want to know the energy of the system when the particles are electrons and when they are spin-1 particles.

DEVELOP Each one-dimensional harmonic oscillator state can be occupied by, at most, two electrons, as required by the Pauli exclusion principle for spin $-\frac{1}{2}$ fermions. The exclusion principle does not apply to spin-1 bosons,

EVALUATE (a) Thus, the lowest energy configuration of 8 electrons has 2 electrons in the 4 lowest states. With $E_n = (n + 1/2)\hbar\omega$ for one electron, the total energy is

$$E = 2(E_0 + E_1 + E_2 + E_3) = (1 + 3 + 5 + 7)2E_0 = 16\hbar\omega$$

(b) Since the exclusion principle does not apply in this case, all 8 particles may occupy the ground state. In this case, lowest energy is $8E_0 = 4\hbar\omega$.

ASSESS The exclusion principle makes the ground-state energy of the N-electron (fermion) system higher than the energy of a system of N bosons.

51. **INTERPRET** The atomic number of copper is 29. We want to find its electronic configuration.

DEVELOP Generally electrons fill in order of increasing shell number n, and within each shell with increasing l ($s, p, d, f, \ldots$). In the case of copper which is a transition element, we actually run into an exception.

EVALUATE As mentioned in the text just before Example 36.4, the electronic configuration of copper is $1s^2 2s^2 2p^6 3s^2 3p^6 4s^1 3d^{10}$, instead of $\ldots 4s^2 3d^9$. The closed $3d$-subshell (with total angular momentum zero) has enough of a lower energy (in spin-orbit, orbit-orbit, and spin-spin interactions) to compensate for the $4s - 3d$ difference.

ASSESS As a useful check, we note that the sum of all the superscripts is 29.

53. **INTERPRET** In this problem all the photons have the same energy, so the power output is proportional to the number of photons emitted per second.

DEVELOP The energy released in the emission of a photon of wavelength 30 μm is

$$E = \frac{hc}{\lambda} = \frac{(6.626 \times 10^{-34} \text{ J} \cdot \text{s})(3 \times 10^8 \text{ m/s})}{3 \times 10^{-5} \text{ m}} = 6.626 \times 10^{-21} \text{ J}$$

The power output to be supplied is $P = d(NE)/dt = E(dN/dt)$.

EVALUATE Thus, photons would have to be emitted at a rate of

$$\frac{dN}{dt} = \frac{P}{E} = \frac{2 \times 10^{-3} \text{ J/s}}{6.626 \times 10^{-21} \text{ J}} = 3.02 \times 10^{17}$$

transitions per second.

ASSESS The laser is of "far-infrared" nature since it emits far-infrared photons.

55. **INTERPRET** This problem is about the radial probability distribution of the electron in the hydrogen ground state. We want to know the probability that the electron in the hydrogen ground state is found in $r = a_0 \pm 0.1a_0$, where a_0 is the Bohr radius.

DEVELOP Using Equations 36.3 and 36.5, the radial probability in the ground state can be written as

$$P(r) = 4\pi r^2 \psi^2 = 4\pi r^2 A^2 e^{-2r/a_0}$$

where $A^2 = 1/\pi a_0^3$. The probability can be found by integrating $P(r)$ from $0.9a_0$ to $1.1a_0$.

EVALUATE Integrating over the range, we find the probability to be

$$P = \int_{0.9a_0}^{1.1a_0} 4\pi r^2 \psi^2 dr = 4\pi \left(\frac{1}{\pi a_0^3}\right) \left\{ \frac{r^2 e^{-2r/a_0}}{(-2/a_0)} - \frac{2}{(-2/a_0)} \left[\frac{e^{-2r/a_0}}{(-2/a_0)^2} \left(-\frac{2r}{a_0} - 1 \right) \right] \right\} \Bigg|_{0.9a_0}^{1.1a_0}$$

$$= 2 \left\{ (0.9)^2 e^{-1.8} - (1.1)^2 e^{-2.2} + \frac{1}{2}[-(3.2)e^{-2.2} + (2.8)e^{-1.8}] \right\} = 0.108$$

ASSESS In Problem 43, we have shown that the radial probability density function (plotted in Figure 36.4) has a maximum at $r = a_0$. Our result indicates that the probability of finding the electron in $r = a_0 \pm 0.1a_0$ is about 10.8%.

57.   **INTERPRET**   We want to add orbital angular momentum to a 6d-electron with the principal quantum number n kept fixed at 6.

DEVELOP   If the electron stays in an $n = 6$ state, then the maximum l-value is 5.

EVALUATE   The new state corresponds to the 6h-state. The angular momentum is $L = \sqrt{l(l+1)}\hbar$, so the difference between the 6h ($l = 5$) and 6d states ($l = 2$) is $\sqrt{5(5+1)}\hbar - \sqrt{2(2+1)}\hbar = 3.03\hbar$. (Such numerical magnitudes are rarely used.)

ASSESS   Since orbital quantum number l can only take on values between 0 and $n-1$, $l_{max} = n-1$, which is 5 in this case. So, the transition is $6d \to 6h$.

59.   **INTERPRET**   Given a hydrogen atom in the F state, we want to know what are the possible values of j, and the number of m_j associated with j_{max}.

DEVELOP   An F state has $l = 3$. For hydrogen with one electron, $s = \frac{1}{2}$. We can then use Equation 36.17a to find all possible values of j.

EVALUATE   (a) With $j = l \pm \frac{1}{2}$, we obtain $j = l \pm \frac{1}{2} = \frac{5}{2}$ or $\frac{7}{2}$. These j-values correspond to magnitudes $J = \sqrt{j(j+1)}\ \hbar = \frac{1}{2}\sqrt{35}\ \hbar$ or $\frac{3}{2}\sqrt{7}\hbar$.

(b) For any j, there are $2j+1$ values of m_j (from $-j$ to $+j$), so for $j = j_{max} = \frac{7}{2}$ there are 8 m_j-values.

ASSESS   The vector diagrams for the spin-orbit coupling with $l = 3$ and $s = \frac{1}{2}$ look similar to that shown in Figure 36.11 (but with $j = \frac{5}{2}$ and $j = \frac{7}{2}$).

61.   **INTERPRET**   In this problem we want to show that the nth shell of an atom can accommodate a maximum number of $2n^2$ electrons.

DEVELOP   To find the maximum possible number of electrons in the nth shell, we note that for a given value of n (or shell), there are n l-values from 0 to $n-1$. For each l-value, there are $2l+1$ m_l-values, and for each state n, l, m_l there are 2 m_s-values for electron spin allowed by the Pauli exclusion principle.

EVALUATE   Therefore, the total number of states in a given shell is:

$$2\sum_{l=0}^{n-1}(2l+1) = 4\sum_{l=0}^{n-1}l + 2\sum_{l=0}^{n-1}1 = \frac{4(n-1)n}{2} + 2n = 2n^2$$

ASSESS   Let's check a few cases to see how the formula works. The $n = 1$ shell only has an s sub-shell and therefore can only accommodate 2 electrons. The $n = 2$ shell has an s sub-shell and a p sub-shell and can take $2 + 6 = 8$ electrons. For $n = 3$, the $s, p,$ and d sub-shells can accommodate a maximum of $2 + 6 + 10 = 18$ electrons.

63.   **INTERPRET**   In this problem we want to verify that ψ_2 satisfies the Schrödinger equation with energy $E = E_2$.

DEVELOP   Before making a substitution, we first find the derivatives of $\psi_{2s} = Ae^{-r/2a_0}(2 - r/a_0)$, where $A = 1/4\sqrt{2\pi a_0^3}$, using the product rule and collecting terms:

$$\frac{d\psi_{2s}}{dr} = -Ae^{-r/2a_0}(2 - r/2a_0)/a_0$$

$$\frac{d^2\psi_{2s}}{dr^2} = Ae^{-r/2a_0}(3 - r/2a_0)/2a_0^2$$

EVALUATE   We expand the derivatives on the left-hand side of Equation 36.4, substitute and factorize:

$$-\frac{\hbar^2}{2m}\left(\frac{d^2\psi_{2s}}{dr^2} + \frac{2}{r}\frac{d\psi_{2s}}{dr}\right) = -\frac{\hbar^2}{2m}Ae^{-r/2a_0}\left[\frac{1}{2a_0^2}\left(3 - \frac{r}{2a_0}\right) - \frac{2}{ra_0}\left(2 - \frac{r}{2a_0}\right)\right]$$

$$= \frac{\hbar^2}{2m}Ae^{-r/2a_0}\frac{1}{ra_0}\left(4 - \frac{r}{2a_0}\right)\left(1 - \frac{r}{2a_0}\right) = \frac{\hbar^2}{4mra_0}\left(4 - \frac{r}{2a_0}\right)\psi_{2s}$$

When this is substituted into the full Equation 36.4, we can cancel a common factor of ψ_{2s} and use $\hbar^2/ma_0 = ke^2$:

$$\frac{\hbar^2}{4mra_0}\left(4 - \frac{r}{2a_0}\right) - \frac{ke^2}{r} = \frac{\hbar^2}{mra_0} - \frac{\hbar^2}{8ma_0^2} - \left(\frac{\hbar^2}{ma_0}\right)\frac{1}{r} = \frac{-\hbar^2}{8ma_0^2} = E$$

This is just the left-hand part of Equation 36.6 with $n = 2$.

ASSESS   As expected, ψ_{2s} satisfies the Schrödinger equation:

$$-\frac{\hbar^2}{2m}\frac{d}{dr}\left(r^2\frac{d\psi_{2s}}{dr}\right)-\frac{ke^2}{r}\psi_{2s}=E_2\psi_{2s}$$

The wave function ψ_{2s} corresponds to a spherically symmetric state with energy $E_2 = E_1/4$. Its radial probability density is plotted in Figure 36.5.

65.   **INTERPRET**   This problem is about transitions between different states of an infinite square well potential allowed by a selection rule.

DEVELOP   The energy levels for an electron in this one-dimensional infinite square well (Equation 35.5) are

$$E_n=\frac{n^2h^2}{8mL^2}=\frac{n^2(hc/L)^2}{8mc^2}=\frac{n^2(1240\ \text{eV}\cdot\text{nm}/0.2\ \text{nm})^2}{8(511\ \text{keV})}=n^2(9.40\ \text{eV})$$

The energies associated with these transitions is

$$\Delta E_{i\to f}=\left(n_i^2-n_f^2\right)(9.40\ \text{eV})$$

EVALUATE   (a) The allowed transitions (Δn odd) from the $n = 4$ level are $4 \to 1$, $4 \to 3$, $3 \to 2$, and $2 \to 1$, as shown on the right.

(b) The photon energies associated with these transitions are

$$\Delta E_{4\to1}=(4^2-1^2)(9.40\ \text{eV})=15(9.40\ \text{eV})=141\ \text{eV}$$
$$\Delta E_{4\to3}=(4^2-3^2)(9.40\ \text{eV})=7(9.40\ \text{eV})=65.8\ \text{eV}$$
$$\Delta E_{3\to2}=(3^2-2^2)(9.40\ \text{eV})=5(9.40\ \text{eV})=47.0\ \text{eV}$$
$$\Delta E_{2\to1}=(2^2-1^2)(9.40\ \text{eV})=3(9.40\ \text{eV})=28.2\ \text{eV}$$

ASSESS   The wavelengths of the photons (using the equation given in Exercise 31) range from 8.8 nm to 44 nm and they are in the far ultraviolet region.

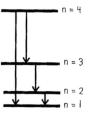

67.   **INTERPRET**   We find the average radius of an electron in the ground state of hydrogen, using the probability density from Equation 36.5 with the wave function from Equation 36.3 and Example 36.1.

DEVELOP   Given a probability density $P(r)$, the average value of r is $\langle r\rangle=\int_0^\infty rP\,dr$. We will use the square of the wave function given in Equation 36.3 and normalized in Example 36.1, $P(r)=4\pi r^2\psi^2$, with $\psi(r)=\frac{1}{\sqrt{\pi a_0^3}}e^{-r/a_0}$.

EVALUATE

$$P(r)=4\pi r^2\left(\frac{1}{\pi a_0^3}e^{-2r/a_0}\right)=\frac{4r^2}{a_0^3}e^{-2r/a_0}$$
$$\langle r\rangle=\int_0^\infty rP\,dr=\frac{4}{a_0^3}r^3e^{-2r/a_0}=\frac{3}{2}a_0$$

ASSESS   The probability density is not symmetric, so the *average* value of r is (in this case) greater than the *most probable* value of r.

69.   **INTERPRET**   We find the g-factor for a classical electron orbit and for electron spin. G-factor is defined as the ratio of the magnetic moment in units of μ_B to the angular momentum in units of $\hbar$.

DEVELOP   We first find the classical orbital g-factor, in which we find the orbital magnetic moment by treating the electron orbit as a current loop. The magnetic moment of a current loop is $\mu = IA$, where the current is $I = -e/T$, the area is $A = \pi r^2 = \pi a_0^2$, and the period can be found from the speed $v = n\hbar/mr$ of the electron. We will express this magnetic moment in terms of the Bohr magneton $\mu_B = e\hbar/2m$, and divide by the angular momentum $mvr = n\hbar$. For the second part of the problem, we use $\mu = -e/mS$, where $S = \frac{1}{2}\hbar$, and divide by the spin angular momentum $\frac{1}{2}\hbar$.

EVALUATE

(a) The classical electron speed is $v = n\hbar/mr$, so the orbital period is $T = 2\pi a_0/v = 2\pi ma_0^2/n\hbar$. The current in this orbital loop is then $I = -e/T = -ne\hbar/2\pi ma_0^2$. The magnetic moment of a loop is

$$\mu = IA = -\frac{ne\hbar}{2\pi ma_0^2}\pi a_0^2 = -n\left(\frac{e\hbar}{2m}\right) = -n\mu_B$$

The classical orbital angular momentum is $L = mvr = n\hbar$, so the g-factor is $g = \frac{\mu/\mu_B}{L/\hbar} = 1$.

(b) $\mu = -e/mS$, but $S = 1/2\hbar$ so $\mu = -e\hbar/2m = -\mu_B$. The angular momentum is $S = \pm\frac{1}{2}\hbar$, so $g = 2$.

ASSESS We have shown what was required.

71. **INTERPRET** We calculate the maximum efficiency of a HeNe laser, given the energy level diagram.

DEVELOP The efficiency of light production for this system is the energy of a single photon divided by the energy required to reach the highest state in the production cycle for that photon. The highest state reached is given as $E_2 = 20.66$ eV, and the wavelength of the photon emitted is $\lambda = 632.8$ nm. The energy of a photon is $E = hf = hc/\lambda$. We will calculate the efficiency given these values.

EVALUATE Converting E_2 to joules gives us $E_2 = 3.31\times10^{-18}$ J, so the maximum efficiency is

$E = E_\gamma/E_2 = hc/\lambda E_2 = 9.484\%$.

ASSESS Make sure you use E_2 rather than E_1 in calculating the efficiency, otherwise you'll get an answer of $E = 9.507\%$. That extra bit of thermal energy to move from 20.61 eV to 20.66 eV has to come from somewhere!

EXERCISES

Section 37.2 Molecular Energy Levels

17. **INTERPRET** The oxygen molecule must absorb a photon in order to make a transition to the excited rotational energy state. We are interested in the wavelength of the photon.

DEVELOP Using Equation 37.2, the difference in energy between the $l = 1$ and $l = 0$ states is

$$\Delta E_{rot} = \frac{\hbar^2}{2I}1(1+1) - 0 = \frac{\hbar^2}{I}$$

The photon wavelength corresponding to this transition is $\lambda = hc/\Delta E_{rot}$.

EVALUATE Substituting the value of I given in the problem, we get

$$\lambda = \frac{hc}{\Delta E} = \frac{hc}{\hbar^2/I} = \frac{2\pi cI}{\hbar} = \frac{2\pi(3 \times 10^8 \text{ m/s})(1.95 \times 10^{-46} \text{ kg} \cdot \text{m}^2)}{1.055 \times 10^{-34} \text{ J} \cdot \text{s}} = 3.48 \text{ mm}$$

ASSESS Transition between adjacent rotational energy levels requires absorbing a photon in the microwave region (frequency $f \sim 10^{11}$ Hz).

19. **INTERPRET** The gas molecules must absorb a photon in order to make a transition to the excited rotational state. We are given the wavelength of the photon, and asked to find the rotational inertia of the molecule.

DEVELOP Using Equation 37.2, the difference in energy between the $l = 1$ and $l = 0$ states is

$$\Delta E_{1 \to 0} = \frac{\hbar^2}{2I}1(1+1) - 0 = \frac{\hbar^2}{I}$$

The energy of the absorbed photon equals the difference in energy: $\Delta E_{rot} = hf = hc/\lambda$. Thus,

$$\frac{\hbar^2}{I} = \Delta E_{1 \to 0} = \frac{hc}{\lambda} = \frac{1240 \text{ eV} \cdot \text{nm}}{1.68 \text{ cm}} = 7.38 \times 10^{-5} \text{ eV} = 1.18 \times 10^{-23} \text{ J}$$

We can find I readily from the above equation.

EVALUATE The rotational inertia is

$$I = \frac{\hbar^2}{\Delta E_{1 \to 0}} = \frac{(1.055 \times 10^{-34} \text{ J} \cdot \text{s})^2}{1.18 \times 10^{-23} \text{ J}} = 9.41 \times 10^{-46} \text{ kg} \cdot \text{m}^2$$

ASSESS The value of I is reasonable for a molecule. We can estimate the bond length of the molecule using $I = mR^2$. With $m \sim 10^{-26}$ kg, we get $R \sim 0.3$ nm, which is also a reasonable value.

21. **INTERPRET** This problem is about the vibrational motion of the N_2 molecule. We're given the energy difference between the adjacent levels, and asked about its corresponding classical vibrational frequency.

DEVELOP The quantized vibrational energy levels are given by Equation 37.3:

$$E_{vib} = (n + 1/2)\hbar\omega$$

Therefore, the energy difference between the adjacent levels is $\Delta E_{vib} = \hbar\omega = hf$.

EVALUATE The classical vibrational frequency is

$$f = \frac{\Delta E_{vib}}{h} = \frac{0.293 \text{ eV}}{4.136 \times 10^{-15} \text{ eV} \cdot \text{s}} = 7.08 \times 10^{13} \text{ Hz}$$

ASSESS The frequency associated with vibrational motion is $f_{vib} \sim 10^{13}$ Hz, which is higher than that associated with rotation, $f_{rot} \sim 10^{11}$ Hz. This means that a more energetic photon must be absorbed by the molecule in order to excite the vibrational modes.

Section 37.3 Solids

23. **INTERPRET** This problem is about expressing the ionic cohesive energy of NaCl in units of kcal/mol.

DEVELOP To convert eV to kcal/mol, we use the following conversion factors (see Appendix C):

$$1 \text{ eV} = 1.602 \times 10^{-19} \text{ J}$$
$$1 \text{ kcal} = 4184 \text{ J}$$
$$1 \text{mol} = 6.022 \times 10^{23}$$

EVALUATE Using the above conversion factors, we find:

$$7.84 \text{ eV} = (7.84 \text{ eV})(1.602 \times 10^{-19} \text{ J/eV})(1 \text{ kcal}/4184 \text{ J})(6.022 \times 10^{23}/\text{mol}) = 181 \text{ kcal/mol}$$

ASSESS The result means that it takes 181 kcal to break one mole of NaCl into its constituent ions.

25. **INTERPRET** Gallium phosphide is a semiconductor with a band-gap energy of $E_g = 2.26$ eV (see Table 37.1).

DEVELOP A photon of energy corresponding to the band gap would have a wavelength of $\lambda = hc/E_g$.

EVALUATE The wavelength is $\lambda = \frac{hc}{E_g} = \frac{1240 \text{ eV·nm}}{2.26 \text{ eV}} = 548.7$ nm.

ASSESS The light is green in color. (The mercury green line has wavelength 546.0 nm, for comparison.)

27. **INTERPRET** From the list of materials given in Table 37.1, we want to know which one would emit the longest wavelength.

DEVELOP The wavelength emitted depends on the energy gap, since $\lambda = hc/E_g$. Thus, the maximum wavelength for the materials in Table 37.1 (corresponding to the smallest gap) is for InAs.

EVALUATE The wavelength is $\lambda = hc/E_g = 1240$ eV·nm/0.35 eV = 3.54 μm.

ASSESS The wavelength is in the infrared of the EM spectrum.

PROBLEMS

29. **INTERPRET** We have a molecule that emits a photon to make a transition to a lower rotational energy state. We want to know the energy of the emitted photon for the $l = 1 \rightarrow l = 0$ transition.

DEVELOP Using Equation 37.2, the difference in energy between the $l = 2$ and $l = 1$ states is

$$\Delta E_{2 \rightarrow 1} = \frac{\hbar^2}{2I}\left[2(2+1) - 1(1+1)\right] = \frac{2\hbar^2}{I}$$

This means that $\hbar^2/I = \Delta E_{2 \rightarrow 1}/2$. For the $l = 1 \rightarrow l = 0$ transition,

$$\Delta E_{1 \rightarrow 0} = \frac{\hbar^2}{2I}\left[1(1+1) - 0\right] = \frac{\hbar^2}{I} = \frac{1}{2}\Delta E_{2 \rightarrow 1}$$

EVALUATE Since $\Delta E_{2 \rightarrow 1} = 2.50$ meV, we have $\Delta E_{1 \rightarrow 0} = (2.50 \text{ meV})/2 = 1.25$ meV.

ASSESS The result shows that the energy levels are not evenly spaced. In general, the energy of a photon emitted in a transition between rotational levels with $|\Delta l| = 1$ is $\Delta E_{l \rightarrow (l-1)} = l\hbar^2/I$.

31. **INTERPRET** A molecule must absorb a photon to make a transition to a higher rotational energy state. We want to know the energy of the photon associated with the $l - 1 \rightarrow l$ transition.

DEVELOP The quantized rotational energy of a molecule is given by Equation 37.2:

$$E_{rot} = \frac{\hbar^2}{2I}l(l+1)$$

The energy of the photon is equal to the energy difference between levels l and $l - 1$.

EVALUATE Using the above equation, we get

$$\Delta E_{l \rightarrow (l-1)} = E_{rot} = \frac{\hbar^2}{2I}\left[l(l+1) - (l-1)l\right] = \frac{l\hbar^2}{I}$$

ASSESS The energy difference between two adjacent rotational levels is proportional to the upper l-value.

33. **INTERPRET** In this problem we are asked to find the atomic separation in the O_2 molecule.

DEVELOP The separation of the rotational spectral lines in energy is $\Delta(\Delta E) = \hbar^2/I$ (see Example 37.1), or $\hbar^2/I = 0.356$ meV for O_2. In a diatomic molecule, with equal-mass atoms and atomic separation R, each atom rotates about the center of mass at a distance of $R/2$, so $I = 2(m_O)(R/2)^2 = (8\text{ u})R^2$, where the mass of an oxygen atom is about $m_O = 16$ u.

EVALUATE The above expressions can be simplified to give

$$R^2 = \frac{I}{8\text{ u}} = \frac{\hbar^2}{(8\text{ u})(0.356\text{ meV})}$$

$$\text{or } R = \frac{\hbar c}{\sqrt{(8\text{ uc}^2)(0.356\text{ meV})}} = \frac{197.3\text{ eV}\cdot\text{nm}}{\sqrt{(8\times931.5\text{ MeV})(0.356\text{ meV})}} = 0.121\text{ nm}$$

ASSESS Our result is in good agreement with the experimental value of 0.146 nm.

35. **INTERPRET** This problem is about the vibrational energy levels of the HCl molecule.

DEVELOP The quantized vibrational energy levels are given by Equation 37.3:

$$E_{vib} = (n+1/2)\hbar\omega$$

Therefore, the ground state energy is $(n=0)$ $E_{vib,0} = \hbar\omega/2$. In addition, energy difference between the adjacent levels is $\Delta E_{vib} = \hbar\omega = hf$.

EVALUATE (a) The vibrational ground-state energy is

$$E_{vib,0} = \frac{1}{2}\hbar\omega = \frac{1}{2}hf = \frac{1}{2}(4.136\times10^{-15}\text{ eV}\cdot\text{s})(8.66\times10^{13}\text{ Hz}) = 0.179\text{ eV}$$

(b) The photon energy for allowed transitions $(\Delta n = 1)$ is $\hbar\omega$, which is twice the zero point energy, or 0.358 eV.

ASSESS Since $f \sim 10^{14}$ Hz, in the infrared region of the spectrum, study of molecular vibrations typically involves infrared spectroscopy.

37. **INTERPRET** This problem involves both rotational and vibrational transitions of an oxygen molecule which is initially in the ground state with $n = 0$ and $l = 0$.

DEVELOP The quantized vibrational energy levels are given by Equation 37.3:

$$E_{vib} = (n+1/2)\hbar\omega$$

Therefore, the energy difference between the adjacent levels is $\Delta E_{vib} = \hbar\omega = hf$. Similarly, using Equation 37.2, the difference in energy between the $l = 1$ and $l = 0$ states is

$$\Delta E_{rot} = \frac{\hbar^2}{2I}1(1+1) - 0 = \frac{\hbar^2}{I}$$

Thus, for the $n = 0, l = 0$ to $n = 1, l = 1$ transition, the energy difference is

$$\Delta E = \Delta E_{vib} + \Delta E_{rot} = hf + \frac{\hbar^2}{I} = 0.19653\text{ eV}$$

and for the $n = 1, l = 1$ to $n = 0, l = 2$ transition, $0.19546\text{ eV} = hf - 2(\hbar^2/I)$. We can solve these equations simultaneously for hf and $\hbar^2/I$ to find f and I.

EVALUATE (a) Multiplying the first equation by 2 and adding to the second, we get

$$2(0.19653\text{ eV}) + (0.19546\text{ eV}) = 3hf \rightarrow f = 4.74\times10^{13}\text{ Hz}$$

(b) Similarly, subtracting the second equation from the first one gives

$$(0.19653 - 0.19546)\text{ eV} = 3(\hbar^2/I) \rightarrow I = 1.95\times10^{-46}\text{ kg}\cdot\text{m}^2$$

ASSESS Since $f \sim 10^{13}$ Hz, in the infrared region of the spectrum, study of molecular vibrations typically involves infrared spectroscopy. A rotational inertia $\sim 10^{-46}$ kg$\cdot$m^2 is typical of diatomic molecules.

39. **INTERPRET** This problem involves both rotational and vibrational transitions of a KCl molecule.

DEVELOP The quantized vibrational energy levels are given by Equation 37.3:

$$E_{vib} = (n+1/2)\hbar\omega$$

Therefore, the energy difference between the adjacent levels is $\Delta E_{vib} = \hbar\omega = hf$. Similarly, using Equation 37.2, the difference in energy between the $l = 1$ and $l = 0$ states is

$$\Delta E_{rot} = \frac{\hbar^2}{2I}1(1+1) - 0 = \frac{\hbar^2}{I}$$

Thus, the difference in energy between these vibrational-rotational levels is

$$\Delta E = \Delta E_{vib} + \Delta E_{rot} = hf - \frac{3\hbar^2}{I}$$

$$= (4.136 \times 10^{-15} \text{ eV} \cdot \text{s})(8.40 \times 10^{12} \text{ Hz}) - \frac{3(6.582 \times 10^{-16} \text{ eV} \cdot \text{s})^2 (1.602 \times 10^{-19} \text{ J/eV})}{2.43 \times 10^{-45} \text{ J} \cdot \text{s}^2}$$

$$= 34.7 \text{ meV}$$

EVALUATE The energy corresponds to a photon wavelength of $\lambda = hc/\Delta E = 35.8 \ \mu\text{m}$.

ASSESS Since $\Delta E > 0$, the final state has higher energy than the initial state, so the transition involves absorbing a photon in the infrared.

41. **INTERPRET** This problem is about the ionic cohesive energy of KCl. We are asked to solve for the constant n in Equation 37.4.

DEVELOP As shown in Example 37.3, the constant is given by

$$n = \left(1 + \frac{U_0 r_0}{\alpha k e^2}\right)^{-1}$$

Since the crystal structures of KCl and NaCl are the same, $\alpha = 1.748$.

EVALUATE Substituting the values given, we get

$$n = \left[1 + \frac{(-7.21 \text{ eV})(0.315 \text{ nm})}{(1.748)(1.44 \text{ eV} \cdot \text{nm})}\right]^{-1} = 10.2$$

where we used a convenient value of ke^2 in atomic units:

$$ke^2 = (9 \times 10^9 \text{ N} \cdot \text{m}^2/\text{C}^2)(1.6 \times 10^{-19} \text{ C})^2 (1 \text{ eV}/1.6 \times 10^{-19} \text{ J}) = 1.44 \text{ eV} \cdot \text{nm}$$

ASSESS The result can be compared with $n = 8.22$ for NaCl (see Example 37.3). The large value of the exponent implies that KCl is strongly resistant to compression.

43. **INTERPRET** This problem is about the ionic cohesive energy of LiCl which has the same structure as NaCl.

DEVELOP As calculated in Example 37.3, the ionic cohesive energy for LiCl is

$$U_0 = -\frac{\alpha k e^2}{r_0}\left(1 - \frac{1}{n}\right)$$

EVALUATE With $n = 7$ and $r_0 = 0.257$ nm, the cohesive energy is

$$U_0 = -\frac{\alpha k e^2}{r_0}\left(1 - \frac{1}{n}\right) = -(1.748)(1.44 \text{ eV} \cdot \text{nm}/0.257 \text{ nm})\left(1 - \frac{1}{7}\right) = -8.40 \text{ eV}$$

ASSESS The result can be compared with $U_0 = -7.84$ eV for NaCl (see Example 37.3). The large value of the exponent implies that LiCl is strongly resistant to compression.

45. **INTERPRET** We integrate the density of states for a metal over all occupied states to find the number of conduction electrons per unit volume.

DEVELOP Equation 37.5 tells us that the density of states is $g(E) = \left(\frac{2^{7/2}\pi m^{3/2}}{h^3}\right)\sqrt{E}$. We will integrate this density from $E = 0$ to $E = E_F$ to find the electron number density.

EVALUATE

$$n = \int_0^{E_F} g(E)dE = \frac{2^{7/2}\pi m^{3/2}}{h^3}\int_0^{E_F} E^{1/2} dE = \frac{2^{7/2}\pi m^{3/2}}{h^3}\left[\frac{2}{3}E^{3/2}\right]_0^{E_F}$$

$$\rightarrow n = \frac{2^{9/2}\pi m^{3/2}}{3h^3}E_F^{3/2}$$

ASSESS We have shown what was required.

47. **INTERPRET**   We use the results of Problem 45 to calculate the Fermi energy for calcium, given the number density.

DEVELOP   The number density of conduction electrons for calcium is $n = 4.6 \times 10^{28}$ m^{-3}. We use this, and the electron mass $m = 9.109 \times 10^{-31}$ kg, to solve $n = (2^{9/2} \pi m^{3/2}/3h^3) E_F^{3/2}$ for E_F.

EVALUATE

$$E_F = \left(\frac{3nh^3}{2^{9/2} \pi m^{3/2}} \right)^{2/3} = 4.68 \text{ eV}$$

ASSESS   This is a reasonable value for the Fermi energy in an alkali metal.

49. **INTERPRET**   This problem is about the Fermi temperature for silver.

DEVELOP   From the problem statement, we see that the Fermi temperature can be expressed mathematically as $T_F = E_F/k_B$.

EVALUATE   With $E_F = 5.48$ eV, we find the Fermi temperature of silver to be

$$T_F = \frac{E_F}{k_B} = \frac{5.48 \text{ eV}}{8.617 \times 10^{-5} \text{ eV} \cdot \text{K}^{-1}} = 6.36 \times 10^4 \text{ K}$$

This is about 212 times room temperature (300 K).

ASSESS   The characteristic Fermi temperature is on the order of 10^5 K for a metal, and the Fermi energy is about 1 – 10 eV, much higher than the thermal energy at typical temperatures (0.025 eV at room temperatures).

51. **INTERPRET**   We want to know whether zinc selenide would make a good photovoltaic cell. We find this by treating the Sun as a blackbody and finding the median wavelength of solar emission, then comparing that energy at that wavelength with the band gap energy for ZnSe.

DEVELOP   We approximate the Sun as a $T = 5800$ K blackbody, and use $\lambda_{median} T = 4.11$ mm $\cdot$ K to find the median wavelength. Next, we convert that wavelength to an energy using $E = \frac{hc}{\lambda}$ and compare the energy of the median photon with the band gap energy of ZnSe, $E_g = 3.6$ eV.

EVALUATE $\lambda_{median} = \frac{4.11 \text{ mm·K}}{5800 \text{ K}} = 709$ nm, so the energy of the median photon is

$$E = \frac{hc}{\lambda_{median}} = 2.80 \times 10^{-19} \text{ J} = 1.75 \text{ eV}$$

ASSESS   The energy of this photon is less than the energy of the band gap for ZnSe, so ZnSe would not make a good photovoltaic cell.

53. **INTERPRET**   In this problem we're asked about the current required in a solenoid in order to achieve the critical magnetic field strength of a superconductor.

DEVELOP   The magnetic field inside a long thin solenoid is given by Equation 26.20, $B = \mu_0 n I$.

EVALUATE   Substituting the values given, we find the current to be

$$I = \frac{B}{\mu_0 n} = \frac{15 \text{ T}}{(4\pi \times 10^{-7} \text{ T} \cdot \text{m/A})(5000/0.75 \text{ m})} = 1.79 \text{ kA}$$

ASSESS   The current required is enormous! Note that niobium-titanium is a Type-II superconductor with a critical temperature of about 10 K.

55. **INTERPRET**   We're given the mass density and ionic cohesive energy of RbI, and asked to solve for the equilibrium separation between Rb and I and the constant n in Equation 37.4.

DEVELOP   The equilibrium ionic separation, r_0, can be estimated from the density, ρ. The constant n is given by (see Example 37.3)

$$n = \left(1 + \frac{U_0 r_0}{\alpha k e^2} \right)^{-1}$$

EVALUATE   (a) If we assume that the crystal structure of RbI is cubic, like NaCl in Example 37.3, then each cubic volume, r_0^3, contains one ion of average mass $(m_{Rb} + m_I)/2$. Thus,

$$r_0 = \left(\frac{m_{Rb} + m_I}{2\rho}\right)^{1/3} = \left(\frac{85.47 \text{ g} + 126.9 \text{ g}}{2(6.022 \times 10^{23})(3.55 \text{ g/cm}^3)}\right)^{1/3} = 0.368 \text{ nm}$$

where we used Appendix D for the atomic weights.

(b) The ionic cohesive energy per ion pair is,

$$U_0 = (-145 \text{ kcal/mol})(4184 \text{ J/kcal})(1 \text{ eV}/1.602 \times 10^{-19} \text{ J})(1 \text{ mol}/6.022 \times 10^{23}) = -6.29 \text{ eV}$$

where we have used the conversion factors given in Appendix C. Equation 37.4 evaluated at $r = r_0$, can be solved for n (as in Example 37.3) with the following result:

$$n = \left(1 + \frac{U_0 r_0}{\alpha k e^2}\right)^{-1} = \left[1 + \frac{(-6.29 \text{ eV})(0.368 \text{ nm})}{(1.748)(1.44 \text{ eV} \cdot \text{nm})}\right]^{-1} = 12.3$$

ASSESS   The results can be compared with $n = 8.22$ and $r_0 = 0.282$ nm for NaCl (see Example 37.3). The large value of the exponent implies that RbI is strongly resistant to compression.

57.   INTERPRET   This problem is about the electron energy distribution at $T = 0$ K.

DEVELOP   The density of states of electrons is given by Equation 37.5:

$$g(E) = \left(\frac{2^{7/2} \pi m^{3/2}}{h^3}\right)\sqrt{E}$$

Thus, the number of electrons with $E \leq E_F/2$ at $T = 0$ K is simply equal to

$$n\left(E \leq \frac{1}{2}E_F\right) = \int_0^{E_F/2} g(E)\,dE$$

EVALUATE   Carrying out the integration, we obtain

$$n\left(E \leq \frac{1}{2}E_F\right) = \left(\frac{2^{7/2} \pi m^{3/2}}{h^3}\right)\int_0^{E_F/2} E^{1/2}\,dE = \frac{2^{9/2} \pi m^{3/2}}{3h^3}\left(\frac{E_F}{2}\right)^{3/2} = \left(\frac{1}{2}\right)^{3/2} n(E \leq E_F)$$

where $n(E \leq E_F)$ is the density (number per unit volume) of conduction electrons. Thus, the fraction is $(1/2)^{3/2} = 35.4\%$.

ASSESS   The density of states at $T = 0$ K is plotted in Figure 37.16a. We see that $n(E \leq E_F/2)$ is less than half $n(E \leq E_F)$.

59.   INTERPRET   This question deals with a model for the energy of a diatomic molecule. We determine that the model has a minimum, and we determine what that minimum is.

DEVELOP   The Morse potential is $U(r) = U_0(1 - e^{-a(r-r_0)})^2$. We will use the first and second derivatives to find the minimum: the first derivative must be zero and the second must be positive for some value of r. Part **(a)** asks for the minimum energy U_{min}, and part **(b)** asks for the radius r_{min} at which this minimum occurs, but in order to find the minimum energy we must know the radius so we'll do part **(b)** first.

EVALUATE

(b)

$$\frac{dU}{dr} = 0 = 2aU_0(1 - e^{-a(r-r_0)}) \rightarrow e^{-a(r-r_0)} = 1 \rightarrow r = r_0$$

$$\frac{d^2U}{dr^2} = 2a^2 U_0 e^{-a(r-r_0)} \rightarrow \left.\frac{d^2U}{dr^2}\right|_{r=r_0} = 2a^2 U_0$$

There is a local maximum or minimum at $r = r_0$, and the second derivative of the function at $r = r_0$ is greater than zero so it must be a minimum.

(a) $U_{min} = U(r_0) = U_0(1 - e^0) = 0$

ASSESS   Unlike the harmonic oscillator potential, the Morse potential is asymmetric. This asymmetry means that the average spacing between the atoms increases with energy level, which leads to a "prediction" that materials would expand with increasing temperature.

61. **INTERPRET** We find the average energy of a conduction electron at very low temperature, using the density of states $g(E)$.

DEVELOP We find the average by dividing the total energy $\int_0^\infty E g(E) dE$ by the number of energy states $\int_0^\infty g(E) dE$. Since at $T = 0$ all states above E_F are empty, we can integrate to E_F rather than to ∞, $g(E) = \left(\frac{2^{7/2} \pi m^{3/2}}{h^3} \right) \sqrt{E}$.

EVALUATE

$$\bar{E} = \frac{\int_0^{E_F} E g(E) dE}{\int_0^{E_F} g(E) dE} = \frac{\int_0^{E_F} E^{3/2}}{\int_0^{E_F} E^{1/2}} = \frac{\frac{2}{5} E_F^{5/2}}{\frac{2}{3} E_F^{3/2}} = \frac{3}{5} E_F$$

ASSESS It is not actually possible to reach $T = 0$, but the information gained by calculations such as this is applicable at other low temperatures, often including such "low" temperatures as $T = 300$ K.

63. **INTERPRET** We find the Fermi energy for an imaginary material in which the charge carriers are protons instead of electrons.

DEVELOP Ignoring the implausibility of the premise, we will use $n = (2^{9/2} \pi m^{3/2} / 3h^3) E_F^{3/2}$, where $n = 10^{28}$ m^{-3}, and solve for E_F.

EVALUATE

$$E_F = \left(\frac{3 n h^3}{2^{9/2} \pi m^{3/2}} \right)^{2/3} = 0.00255 \text{ eV}$$

ASSESS A proton in this material would have thermal energy equal to the Fermi energy at a temperature of merely 29.6 Kelvin!

65. **INTERPRET** Would a semiconductor with a band gap of 2.77 eV work for a blue-green laser? We compare the energy of the photon and the band gap, and see if it the material is a possible candidate for a laser medium.

DEVELOP The wavelength desired for the laser is $\lambda = 447$ nm, and the band gap of the material is $E_g = 2.77$ eV. Given a wavelength λ of a photon, the energy of that photon is $E_\lambda = hc/\lambda$. The energy of the photon must be less than or equal to the band gap, which represents the maximum energy of a photon emitted by the material.

EVALUATE $E_\lambda = hc/\lambda = 2.77$ eV

ASSESS This material might work.

EXERCISES

Section 38.1 Elements, Isotopes, and Nuclear Structure

13. **INTERPRET** This problem is about writing the conventional symbols for the isotopes of radon.

DEVELOP The conventional symbol for a nucleus X is $^A_Z X$, where A is the mass number and Z is the atomic number.

EVALUATE With the number of protons ($Z = 86$ for all radon isotopes) and neutrons ($N = A - Z$) given, the mass numbers of the three isotopes are, respectively, $A = Z + N = 86 + 125 = 211$, 220, and 222. Therefore, the nuclear symbols are $^{211}_{86} Ra, ^{220}_{86} Ra$, and $^{222}_{86} Ra$.

ASSESS Isotopes of a given element have the same number of protons (and hence Z) but different number of neutrons (and hence A).

15. **INTERPRET** This problem asks for a comparison of the number of nucleons and charges between two nuclei.

DEVELOP The comparison can be made by noting that the conventional symbol for a nucleus X is $^A_Z X$, where A is the mass number and Z is the atomic number.

EVALUATE (a) The mass number (number of nucleons) is $A = 35$ for both.

(b) The charge, Ze, of a potassium nucleus, $Z = 19$, is two electronic charge units greater than that for a chlorine nucleus, $Z = 17$.

ASSESS Equality in mass number A does not imply equality in atomic number Z. Two nuclei have the same Z only when they are isotopes.

17. **INTERPRET** This problem is about the size of the fission products of $^{235}_{92} U$.

DEVELOP The nuclear radius can be estimated using Equation 38.1:

$$R = R_0 A^{1/3} = (1.2 \text{ fm}) A^{1/3}$$

EVALUATE Two fission products as equal as possible would have $A = 117$ or 118, and radii of about $R = (1.2 \text{ fm}) A^{1/3} \approx 5.9$ fm.

ASSESS Equation 38.1 is a good approximation for R since nucleons are packed tightly into the nucleus. The tight packing also suggests that all nuclei have roughly the same density.

Section 38.2 Radioactivity

19. **INTERPRET** In this problem we are asked to write down all possible beta-decay processes for $^{64}_{29} Cu$.

DEVELOP Beta decay in $^{64}_{29} Cu$ can involve positron-neutrino or electron-anti-neutrino emission, or electron capture.

EVALUATE The reactions are:

$$^{64}_{29} Cu \rightarrow ^{64}_{30} Zn + \beta^- + \bar{\nu} \quad (40\%)$$
$$^{64}_{29} Cu \rightarrow ^{64}_{28} Ni + \beta^+ + \nu \quad (19\%)$$
$$^{64}_{29} Cu + e^- \rightarrow ^{64}_{28} Ni + \nu \quad (41\%)$$

ASSESS In each decay mode, charge and mass number are conserved.

21. **INTERPRET**   This problem is about the decay process of $^{10}_6\text{C}$ and its activity after a period of time.

DEVELOP   The process is a β^+-decay, in which A remains constant and Z decreases by one, and a positron-neutrino pair is created. To answer **(b)**, we note that in radioactive decay, the activity decays with the same half-life as the number of nuclei.

EVALUATE   **(a)** The decay process can be written as $^{10}_6\text{C} \rightarrow ^{10}_5\text{B} + e^+ + \bar{\nu}_e$.

(b) With a half-life of $t_{1/2} = 19.3$ s, after $t = 1\,\text{min}$, the activity is

$$(\lambda N) = (\lambda N)_0 \left(\frac{1}{2}\right)^{t/t_{1/2}} = (48 \text{ MBq})\left(\frac{1}{2}\right)^{60/19.3} = 5.56 \text{ MBq}$$

ASSESS   A quick check shows that our answer is reasonable. One minute is about $t = 60 \text{ s} \approx 3t_{1/2}$, so $(1/2)^3 = 1/8$, so the activity decreases by roughly a factor of 8.

23. **INTERPRET**   This is a problem about the decay of ^{90}Sr. From Table 38.1, we find the half-life of ^{90}Sr to be 29 years.

DEVELOP   Using Equation 38.3b, $N = N_0 2^{-t/t_{1/2}}$, the time elapsed since the bomb tests is

$$t = t_{1/2} \frac{\ln(N_0/N)}{\ln 2}$$

The fraction of ^{90}Sr that remains, N/N_0, is one minus the fraction that decays, which is given.

EVALUATE   **(a)** For 99% of the radioactive contaminant to decay, or 1% to remain, the time required is

$$t = t_{1/2} \frac{\ln(N_0/N)}{\ln 2} = (29 \text{ y})\frac{\ln(1/0.01)}{\ln 2} = 193 \text{ y}$$

(b) Similarly, for 99.9% of the radioactive contaminant to decay, or 0.1% to remain, the time required is

$t = (29 \text{ y})(\ln 10^3/\ln 2) = 289$ y.

ASSESS   Since $(1/2)^7 = 0.0078 < 0.01 < (1/2)^6 = 0.0156$, for 1% of contaminant to remain, $t/t_{1/2}$ must be between 6 and 7, so our result of $193/29 = 6.66$ is reasonable. Similarly reasoning also shows that our result for **(b)** is about right.

Section 38.3 Binding Energy and Nucleosynthesis

25. **INTERPRET**   We're given the binding energy per nucleon for $^{60}_{28}\text{Ni}$, and asked about its atomic mass.

DEVELOP   Using Equation 38.7, the binding energy of a nucleus can be written as

$$E_b = Z m_p c^2 + (A - Z) m_n c^2 - m_N c^2$$

where m_p, m_n, and m_N are the masses of the protons, neutrons, and the nucleus, respectively. The total nuclear binding energy of $^{60}_{28}\text{Ni}$ is the number of nucleons $(A = 60)$ times the given binding energy per nucleon, or

$$E_b = 60(8.8 \text{ MeV})(1 \text{ u} \cdot c^2/931.5 \text{ MeV}) = 0.567 \text{ u} \cdot c^2$$

The result can be used to solve for the atomic mass of $^{60}_{28}\text{Ni}$.

EVALUATE   If we express Equation 38.7 in terms of atomic masses, by adding $Z = 28$ electron rest energies $(m_e c^2)$ to both sides, and neglect atomic binding energies (as mentioned in the text following Example 38.4), we obtain:

$$M(^{60}_{28}\text{Ni}) = 28M(^1_1\text{H}) + (60 - 28)m_n - E_b/c^2 = 28(1.00783 \text{ u}) + 32(1.00867 \text{ u}) - 0.567 \text{ u}$$
$$= 59.930 \text{ u}$$

ASSESS   The actual binding energy of ^{60}Ni is so close to 8.8 MeV/nucleon that the accuracy of the atomic mass just calculated is better than one might expect from data given to two figures.

27. **INTERPRET**   In this problem we are asked to find the binding energy per nucleon for ^7_3Li.

DEVELOP   Using Equation 38.7, the binding energy per nucleon of a nucleus can be written as

$$\frac{E_b}{A} = \frac{1}{A}[Z m_p c^2 + (A - Z)m_n c^2 - m_N c^2]$$

where m_p, m_n, and m_N are the masses of the protons, neutrons, and the nucleus, respectively.

EVALUATE   Substituting the values given, we get

$$\frac{E_b}{A} = \frac{1}{7}[3(1.00728 \text{ u}) + (7 - 3)(1.00867 \text{ u}) - 7.01435 \text{ u}](931.5 \text{ MeV/u})$$
$$= 5.61 \text{ MeV/nucleon}$$

ASSESS   The binding energy per nucleon as a function of A is plotted in Figure 38.9. For very light nuclides such as $^{7}_{3}\text{Li}$, the energy is low because the nuclear force is not yet saturated for so few nucleons.

Section 38.4 Nuclear Fission

29.   **INTERPRET**   This problem is about the number of neutrons released in the fission of ^{235}U.

DEVELOP   In this reaction, the conservation of charge (atomic number Z) is the same as the conservation of the number of protons, so the conservation of the number of nucleons (mass number A) is equivalent to the conservation of the number of neutrons.

EVALUATE   The numbers of neutrons in the nuclei ^{235}U, ^{139}I, and ^{95}Y are $A-Z = 235 - 92 = 143$, $139 - 53 = 86$, and $95 - 39 = 56$, respectively, so the number of neutrons released in the reaction is $1 + 143 - 86 - 56 = 2$.

ASSESS   As expected, the fission process produces two middle-weight products with notably unequal masses. The number of neutrons released is in accordance with conservation of the number of nucleons.

31.   **INTERPRET**   We are given the power output and energy released per fission and asked about the fission rate.

DEVELOP   We assume that all of the energy released in fissions goes into thermal power output. Thus,

$$P = \frac{dE}{dt} = \frac{d(NE_1)}{dt} = E_1 \frac{dN}{dt}$$

where $E_1 = 200$ MeV.

EVALUATE   The fission rate is

$$\frac{dN}{dt} = \frac{P}{E_1} = \frac{3200 \text{ MW}}{(200 \text{ MeV/fission})(1.6 \times 10^{-13} \text{ J/MeV})} = 1.00 \times 10^{20} \text{ fissions/s}$$

ASSESS   A very high fission rate is required to achieve this power output.

Section 38.5 Nuclear Fusion

33.   **INTERPRET**   This problem is about the Lawson criterion for D-T fusion. We are interested in the density of the nuclei.

DEVELOP   The Lawson criterion for D-T fusion is $n\tau > 10^{20}$ s/m^3 (Equation 38.11). This condition allows us to solve for n.

EVALUATE   For $\tau = 0.5$ s, a particle density of $n > (10^{20}$ s/m$^3)/(0.5$ s$) = 2 \times 10^{20}$ m^{-3} would be required.

ASSESS   This is a very high density. From the Lawson criterion, we see that in order to achieve self-sustaining fusion, we must have a high nuclei density and confine them long enough such that the product $n\tau$ exceeds 10^{20} s/m^3. So far, the approach has not yet produced a sustained energy yield.

35.   **INTERPRET**   We find the required confinement time for D-T fusion, given the plasma density. We will use the Lawson criterion.

DEVELOP   The D-T Lawson criterion is $n\tau > 10^{22}$ s/m^3, and the number density given in the problem is $n = 10^{19}$ particles/m^3. We will solve for the confinement time τ.

EVALUATE $\tau > \frac{10^{22} \text{ s·m}^{-3}}{10^{19} \text{ m}^{-3}} = 10^3$ s.

ASSESS   This confinement time is inconveniently long, but creating a greater particle density so as to decrease the time is also difficult.

PROBLEMS

37.   **INTERPRET**   In this problem we want to know the energy required to flip the spin state of a proton in the Earth's magnetic field.

DEVELOP   As discussed in Example 38.1, a proton acts like a magnetic dipole whose component along the magnetic field is $\mu_p = 1.41 \times 10^{-26}$ J/T. The energy needed to flip the spin is $\Delta U = \mu_p B - (-\mu_p B) = 2\mu_p B$.

EVALUATE   With $B = 30\ \mu$T, we get

$$\Delta U = 2\mu_p B = 2(30\ \mu\text{T})(1.41 \times 10^{-26}\ \text{J/T}) = 8.46 \times 10^{-31}\ \text{J} = 5.28 \times 10^{-12}\ \text{eV}$$

ASSESS   The frequency of a photon with this energy, 1.28 kHz, is in the audible range!

39.   **INTERPRET**   In this problem we are asked to find the binding energy per nucleon for $^{56}_{26}$Fe.

DEVELOP   Using Equation 38.7, the binding energy of a nucleus can be written as

$$E_b = Zm_p c^2 + (A - Z)m_n c^2 - m_N c^2$$

where m_p, m_n, and m_N are the masses of the protons, neutrons, and the nucleus, respectively.

EVALUATE   Since the nuclear mass is given, the above equation gives directly

$$E_b(^{56}_{26}\text{Fe}) = [26(1.00728\ \text{u}) + 30(1.00867\ \text{u}) - 55.9206\ \text{u}]c^2 = (0.5288\ \text{u})(931.5\ \text{MeV/u})$$
$$= 493\ \text{MeV}$$

Therefore, $E_b/A = 8.80$ MeV/nucleon. The result is in good agreement with the curve given in Figure 38.9.

ASSESS   The value $E_b/A = 8.80$ MeV/nucleon is very close to the peak of the curve in Figure 38.9. Nuclei with mass numbers around $A = 60$ are most tightly bound.

41.   **INTERPRET**   This problem is about the age of the Earth in half-lives of the isotopes specified.

DEVELOP   From Table 38.1, the half-lives are 5730 y for ^{14}C, 4.46×10^9 y for ^{238}U, and 1.25×10^9 y for ^{40}K. The number of nuclei remaining can be found by using Equation 38.3b: $N = N_0 2^{-t/t_{1/2}}$.

EVALUATE   (a) For ^{14}C, we have $n = t/t_{1/2} = 4.5 \times 10^9/5730 = 7.85 \times 10^5$, and $N = 10^6/2^{7.85 \times 10^5} = 10^{-2.36 \times 10^5}$, which is practically zero.

(b) For ^{238}U, we get $n = t/t_{1/2} = 4.5 \times 10^9/4.46 \times 10^9 = 1.01$, and $N = 10^6/2^{1.01} = 4.97 \times 10^5$, about half of the original amount.

(c) Similarly, for ^{40}K, we have $n = t/t_{1/2} = 4.5 \times 10^9/1.25 \times 10^9 = 3.60$, and $N = 10^6/2^{3.60} = 8.25 \times 10^4$.

ASSESS   The longer the half-life $t_{1/2}$, the greater the number of nuclei remaining.

43.   **INTERPRET**   This is a problem about radioactive decay of radon. We want to know the fraction of nuclei remaining after time t.

DEVELOP   Equation 38.3b gives the fraction of a sample remaining after time t as

$$f(t) = \frac{N}{N_0} = 2^{-t/t_{1/2}}$$

For $^{222}_{86}$Rn, the half-life is $t_{1/2} = 3.82$ d.

EVALUATE   (a) After $t = 1$ d, the fraction remaining is $f(1\ \text{d}) = 2^{-1/3.82} = 0.834$.

(b) After one week or 7 days, $f(7\ \text{d}) = f(1\ \text{d})^7 = (0.834)^7 = 0.281$.

(c) Similarly, after a month or 30 days,

$$f(30\ \text{d}) = f(1\ \text{d})^{30} = (0.834)^{30} = 4.32 \times 10^{-3} = 0.432\%$$

ASSESS   As a quick check, we note that one half-life reduces the number to 50% of the original, two half-lives to 25%, three to 12.5%, and so on. In the $^{222}_{86}$Rn case, one week is slightly less than two half-lives, so our result of 28.1% makes sense. On the other hand, one month is just under 8 half-lives, where only $(1/2)^8 = 0.0039 = 3.9\%$ remains. So our answer in (c) is also reasonable.

45.   **INTERPRET**   This problem is about the decay of $^{232}_{90}$Th that results in a series of short-lived nuclei.

DEVELOP   In α-decay, the numbers of neutrons and protons both decrease by two, while in β^--decay, the number of neutrons decreases and that of protons increases by one (see Equations 38.4 and 38.5a). Thus, after one α- and two β^--decays, the number of neutrons is four less and the number of protons is the same.

EVALUATE   From the reasoning above, we conclude that the third daughter nucleus in the $^{232}_{90}$Th -series is $^{228}_{90}$Th, which is another thorium isotope.

(b) The chart is shown below. The half-lives are given in the problem statement.

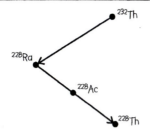

ASSESS We have shown what was required.

47. **INTERPRET** This is a problem about ^{14}C dating. We want to deduce the age of the bone.

    DEVELOP Using Equation 38.3b, $N = N_0 2^{-t/t_{1/2}}$, the age of the bone is given by

    $$t = t_{1/2} \frac{\ln(N_0/N)}{\ln 2}$$

    where N is the number of nuclei remaining. The half-life of ^{14}C is $t_{1/2} = 5730$ y.

    EVALUATE From the above equation, we determine the age of the bone to be

    $$t = t_{1/2} \frac{\ln(N_0/N)}{\ln 2} = (5730 \text{ y}) \frac{\ln(1/0.34)}{\ln 2} = 8.92 \times 10^3 \text{ y}$$

    ASSESS As a quick check, we note that one half-life reduces the number to 50% of the original, two half-lives to 25%, three to 12.5%, and so on. In our ^{14}C dating case, we have $t = (8920 \text{ y})/(5730 \text{ y}) = 1.56 t_{1/2}$. Therefore, we expect the fraction remaining to be between 25% and 50%. This is consistent with the 34% given in the problem statement.

49. **INTERPRET** This problem is about the decay of ^{131}I in milk. We're given the initial activity per liter of milk, and we want to know how long it takes for the activity to decay to the level provided by the safety guideline.

    DEVELOP Using Equation 38.3b, $N = N_0 2^{-t/t_{1/2}}$, the waiting time required can be written as

    $$t = t_{1/2} \frac{\ln(N_0/N)}{\ln 2}$$

    where N is the number of nuclei remaining. The half-life of ^{131}I is $t_{1/2} = 8.04$ d.

    EVALUATE For Poland, we get

    $$t = t_{1/2} \frac{\ln (\text{reported level/safety guideline})}{\ln 2} = (8.04 \text{ d}) \frac{\ln(2000/1000)}{\ln 2} = 8.04 \text{ d}$$

    Similar calculations yield 16.2 d for Austria, and 10.0 d for Germany.

    ASSESS As a quick check, we note that one half-life reduces the number to 50% of the original, two half-lives to 25%, three to 12.5%, and so on. For Poland, the waiting time is exactly one half-life since the initial reported level is twice the safety guideline. For Austria, $N/N_0 \approx 1/4 = 0.25$, so about two half-lives (16 days) of waiting time are required. Similar reasoning can be applied to the Germany case.

51. **INTERPRET** This problem is about using the decay of ^{40}K to deduce the age of a rock.

    DEVELOP Using Equation 38.3b, $N = N_0 2^{-t/t_{1/2}}$, the age of the rock is given by

    $$t = t_{1/2} \frac{\ln(N_0/N)}{\ln 2}$$

    where N is the number of nuclei remaining. The half-life of ^{40}K is $t_{1/2} = 1.2 \times 10^9$ y. If 82% of the original ^{40}K decayed, then 18% remains in a rock of age t.

    EVALUATE From the above equation and the given half-life, we get

    $$t = t_{1/2} \frac{\ln(N_0/N)}{\ln 2} = (1.2 \times 10^9 \text{ y}) \frac{\ln(1/0.18)}{\ln 2} = 2.97 \times 10^9 \text{ y}$$

    Note: A type of lunar highlands rock rich in potassium (K), rare earth elements (REE), and phosphorus (P), is called KREEP norite.

    ASSESS The age of the rock is on the same order as the age of the Earth, which is about 4.5 billion years old.

53.   **INTERPRET**   Uranium-235 and uranium-238 have different half lives. We are given their relative abundance at the present time and asked to find their relative abundance 4.5 billion years ago.

DEVELOP   Suppose that when the Earth formed ($t = 0$), natural uranium consisted of just the two longest-lived isotopes in Table 38.1 (we ignore ^{234}U which has an abundance of 0.0057%). Then the percentage of ^{235}U today ($t = 4.5 \times 10^9$ y) is

$$0.0072 = \frac{N^{235}}{N^{235} + N^{238}} \rightarrow N^{238} = 138\, N^{235}$$

We next find the original amounts of the two isotopes are given by Equation 38.3b, $N = N_0 2^{-t/t_{1/2}}$, with half-lives from Table 38.1. The results are

$$N_0^{235} = N^{235} 2^{4.5/0.704} \quad \text{and} \quad N_0^{238} = N^{238} 2^{4.5/4.46}$$

EVALUATE   So that the original percentage must have been

$$\frac{N_0^{235}}{N_0^{235} + N_0^{238}} = \frac{1}{1 + (N_0^{238}/N_0^{235})} = \frac{1}{1 + (N^{238}/N^{235})2^{(4.5/4.46)-(4.5/0.704)}} = \frac{1}{1 + (138)2^{-5.38}} = \frac{1}{1 + 3.30}$$
$$= 23.2\%$$

ASSESS   Since ^{235}U has a shorter half-life than ^{238}U, we expect the ratio $N_0^{235}/(N_0^{235} + N_0^{238})$ to decrease with time. Note that from current data on nuclear reactions and models of nucleosynthesis in supernova explosions, one can predict the isotopic abundances of U^{235} and U^{238} when they were produced. By reversing the above argument, one can then estimate the age of the elements in the nebula from which the solar system formed.

55.   **INTERPRET**   This problem is about the decay process of ^{15}O. We are given its activity at time t and asked about its initial activity.

DEVELOP   Multiplying Equation 38.3b by the decay rate, λ, the initial activity is

$$(\lambda N_0) = (\lambda N)2^{t/t_{1/2}}$$

EVALUATE   With $t_{1/2} = 2.0$ min, we get

$$(\lambda N_0) = (\lambda N)2^{t/t_{1/2}} = (0.5 \text{ mCi/L})2^{3.5/2.0} = 1.68 \text{ mCi/L}$$

ASSESS   As a quick check, we note that one half-life reduces the activity to 50% of the original, two half-lives to 25%, three to 12.5%, and so on. In our case, $t/t_{1/2} = 3.5/2 = 1.75$, which is less than 2. So we expect the activity to be between 25% and 50% of the original. Indeed, we have (0.5 mCi/L)/(1.68 mCi/L) = 0.298.

57.   **INTERPRET**   This problem is about the amount of uranium-235 present in a fission bomb with a 20-kt explosive yield.

DEVELOP   Using the data in Appendix C, we find that a 1 kt explosive yield is about 4.18×10^{12} J. On the other hand, the energy content of a ^{235}U is 8.2×10^{13} J/kg. Therefore, 1 kt is equivalent to the energy released in the fission of about

$$\frac{4.18 \times 10^{12} \text{ J/kt}}{8.21 \times 10^{13} \text{ J/kg of } ^{235}\text{U}} = 50.9 \text{ g of } ^{235}\text{U}$$

EVALUATE   A 20 kt yield is equivalent to the fission energy of 20×50.9 g $= 1.02$ kg of ^{235}U.

ASSESS   The bomb "Little Boy" was a uranium-235 fission bomb which had a yield of about 12–15 kt.

59.   **INTERPRET**   This problem is about a reactor in subcritical state with multiplication factor $k < 1$. We're asked to calculate the time it takes for the power to be halved.

DEVELOP   A multiplication factor k means that the fission rate is multiplied by a factor k with each generation time. The reactor power changes by a factor of $\frac{1}{2} = k^n$ in n generations. The time elapsed is $t = n\tau$.

EVALUATE   Taking the logarithm on both sides of $\frac{1}{2} = k^n$, we get

$$n = \ln\left(\frac{1}{2}\right)/\ln(0.992) = 86.3$$

If the generation time is $\tau = 0.1$ s, this process consumes $t = n\tau = 8.63$ s.

ASSESS   The power output decreases rather rapidly. A fission reactor with $k < 1$ produces fission without achieving criticality, and hence no sustaining chain reaction.

61.   INTERPRET   This problem is about the power output and efficiency of a nuclear plant.

DEVELOP   To compute the power output, we note that the energy content of ^{235}U is 8.2×10^{13} J/kg (see Appendix C).

EVALUATE   (a) If all the ^{235}U consumed fissions to produce heat, the thermal power output would be

$$P = (1311 \text{ kg/y})(8.2 \times 10^{13} \text{ J/kg})(1 \text{ y}/3.156 \times 10^7 \text{s}) = 3.41 \text{ GW}$$

(b) The efficiency is $W/Q = 1.2 \text{ GW}/3.41 \text{ GW} = 35.2\%$.

ASSESS   An alternative way to calculate the power output is to follow the procedure used in Example 38.6. Since each fission produces about 200 MeV of energy, the power output is

$$P = \frac{(1311 \text{ kg/y})(200 \text{ MeV/fission})(1.602 \times 10^{-13} \text{ J/MeV})(1 \text{ y}/3.156 \times 10^7 \text{s})}{(235 \text{ u/atom})(1.66 \times 10^{-27} \text{ kg/u})} = 3.41 \text{ GW}$$

As illustrated in Example 38.6, only 0.35g of ^{235}U is required to provide the same amount of energy as 1000 kg of coal. That's why a nuclear power plant is more efficient.

63.   INTERPRET   This problem is about the amount of heavy water required for a given energy output in D-D fusion.

DEVELOP   In one year, the amount of energy produced by a 1 GW power plant is

$$E = Pt = 1 \text{ GW} \cdot \text{y} = (10^9 \text{ J/s})(3.156 \times 10^7 \text{ s})(1 \text{ MeV}/1.602 \times 10^{-13} \text{ J}) = 1.97 \times 10^{29} \text{ MeV}$$

We shall use 7.2 MeV/deuteron as the average energy release in a D-D reactor (note that the D-D reactions of Equations 38.10b and 38.10c produce $^3_2$He and $^3_1$H that can undergo further fusion reaction). Thus, $N = (1.97 \times 10^{29} \text{ MeV})/(7.2 \text{ MeV}) = 2.74 \times 10^{28}$ deuterons are required.

EVALUATE   The molecular weight of D$_2$O is about 20 u, so about

$$m = \frac{1}{2}(20 \text{ u})(1.66 \times 10^{-27} \text{ kg/u})(2.74 \times 10^{28}) = 454 \text{ kg}$$

of heavy water (each molecule of which contains two deuterium atoms) would be needed.

ASSESS   About 0.015% of hydrogen nuclei in seawater molecules are deuterium. They can be used to produce heavy water. Note that heavy water plays the role of moderator and coolant in certain reactors (e.g., the Canadian CANDU design).

65.   INTERPRET   We want to know the energy released if all the deuterium in a gallon of water undergoes fusion.

DEVELOP   The number of D-atoms in a gallon of water (mostly H$_2$O) is

$$N = \frac{(1 \text{ gal} \times 3.786 \times 10^{-3} \text{ m}^3/\text{gal})(10^3 \text{ kg/m}^3)(2 \text{ atoms/molecule})(0.00015 \text{ D/H})}{(18 \times 1.66 \times 10^{-27} \text{ kg/molecule})} = 3.80 \times 10^{22}$$

Each deuterium on the average yields $E_1 = 7.2$ MeV of energy.

EVALUATE   (a) If all of these release energy via D-D fusion reactions, an average of

$$E = NE_1 = (3.80 \times 10^{22})(7.2 \text{ MeV})(1.602 \times 10^{-13} \text{ J/MeV}) = 4.38 \times 10^{10} \text{ J}$$

would be produced.

(b) This is equivalent to the energy content of $(4.38 \times 10^{10} \text{ J})/(36 \times 10^3 \times 3600 \text{ J/gal}) = 338$ gallons of gasoline.

ASSESS   The amount of energy released by D-D fusion is enormous compared to the energy content of gasoline. Controlled fusion devices could play a key role in future energy production.

67.   INTERPRET   We're asked about the energy released in the α-decay of ^{238}U.

DEVELOP   The energy released in the decay is the difference in the mass-energy of the initial nucleus and that of the decay products, or

$$E_\alpha/c^2 = M(^A_Z X) - M(^{A-4}_{Z-2} Y) - M(^4_2 \text{He})$$

Note that atomic or nuclear masses can be used to calculate E_α, since in α-decay, the mass of the Z-electrons cancels. The neglect of atomic binding energies is also less serious here than in Equation 38.7 because only differences enter. (For $^4_2$He, from Table 38.2, we have 4.001506 u + 2(0.000548579 u) = 4.002603 u; the atomic masses of $^{238}_{92}$U and $^{234}_{90}$Th are given.)

EVALUATE   Substituting the values obtained, we get

$$E_\alpha/c^2 = M(^{238}_{92}\text{U}) - M(^{234}_{90}\text{Th}) - M(^4_2\text{He}) = 238.050784 \text{ u} - 234.043593 \text{ u} - 4.002603 \text{ u}$$
$$= 0.004588 \text{ u} = 4.27 \text{ MeV}/c^2$$

Thus, the energy released during the α-decay is 4.27 MeV. Note that the half-life of this decay process is about $t_{1/2} = 4.47 \times 10^9$ y.

ASSESS   The energy released in the decay is analogous to the binding energy for a stable nucleus, which has the opposite sign. In nuclear reactions, the difference in mass-energy between the initial and final reactants is called the Q-value, which is positive for exothermic (energy releasing) and negative for endothermic (energy absorbing) reactions.

69. **INTERPRET**   This problem is about producing a gold nucleus from bombarding mercury with neutrons.

    DEVELOP   If we assume only one unknown product, the reaction is of the form

    $$^1_0 n + {}^{198}_{80}\text{Hg} \rightarrow {}^{197}_{79}\text{Au} + {}^A_Z\text{X}$$

    Conservation of charge and mass number implies $80 = 79 + Z$ and $1 + 198 = 197 + A$, or $Z = 1$ and $A = 2$.

    EVALUATE   Thus, the unknown product is a deuteron, ^2_1H. The reaction can be written as

    $$^1_0 n + {}^{198}_{80}\text{Hg} \rightarrow {}^{197}_{79}\text{Au} + {}^2_1\text{H}$$

    or symbolized by $^{198}\text{Hg}(n,d)^{197}\text{Au}$.

    ASSESS   The production of gold is an age-old dream of alchemists. While it is possible in nuclear reactors, the expense of the procedure far exceeds the financial gain!!

71. **INTERPRET**   This problem is about the energy released in D-T fusion. We also want to compare the fusion energy with the energy produced by a coal-burning power plant.

    DEVELOP   Half of the number of deuterons in one fuel pellet is

    $$N = \frac{1}{2}(2.5 \text{ mg})/(2 \text{ u} \times 1.66 \times 10^{-27} \text{ kg/u}) = 3.77 \times 10^{20}$$

    (since there is 2.5 mg of deuterium in a pellet and the mass of a deuteron is approximately 2 u). As given in Equation 38.10a, each D-T reaction releases $E_1 = 17.6$ MeV of energy.

    EVALUATE   **(a)** The total amount of energy released in the D-T fusion reactions is

    $$E = NE_1 = (3.77 \times 10^{20})(17.6 \text{ MeV/fusion})(1.602 \times 10^{-13} \text{ J/MeV}) = 1.06 \text{ GJ}$$

    (b) The thermal power output, 3 GW, is equal to the energy release per pellet times the rate that pellets are consumed, hence this rate is 3 GW/1.06 GJ = 2.83 s^{-1}.

    (c) 5 mg pellets, consumed at this rate for a year, have a total mass of (5 mg)(2.83 s^{-1})(3.156 × 10^7 s) = 446 kg. A comparable coal-burning power plant uses more than 7 million times this mass for its fuel.

    ASSESS   The amount of energy released by D-T fusion is enormous compared to the energy content of coal fuel. Controlled fusion devices could play a key role in future energy production.

73. **INTERPRET**   We find the value of Z in the liquid-drop model that gives the minimum nuclear mass for a value of A. This will be the most stable of the nuclei for that atomic mass.

    DEVELOP   The formula given us for the liquid-drop model is $M(A,Z) = c_1 A - c_2 Z + (c_2 A^{-1} + c_3 A^{-1/3})Z^2$. We will take the partial derivative $\frac{\partial M}{\partial Z}$ and set it equal to zero to find Z_{min}. We hope to show that

    $$Z_{min} = \frac{A}{2\left[1 + \frac{c_3}{c_2}A^{2/3}\right]}$$

    EVALUATE

    $$\frac{\partial M}{\partial Z} = 0 = -c_2 + 2\left(\frac{c_2}{A} + \frac{c_3}{A^{1/3}}\right)Z \rightarrow Z_{min} = \frac{c_2}{2\left(\frac{c_2}{A} + \frac{c_3}{A^{1/3}}\right)} = \frac{A}{2\left(1 + \frac{c_3}{c_2}A^{2/3}\right)}$$

    ASSESS   We have shown what was required.

75. **INTERPRET** When one radionuclide decays into another radioactive nucleus, the total activity of the sample becomes somewhat complicated. Here we will calculate the total activity in such a situation, starting with the differential equations that define activity.

DEVELOP The activity of a sample is $R(t) = \lambda N(t)$, so if we find $N(t)$ for each radionuclide, we can find the activity $R(t) = \lambda_A N_A + \lambda_B N_B$. The equation for the "parent" nuclide is found in the usual way: it does not depend on the daughter at all, so $dN_A/dt = -\lambda_A N_A \rightarrow N_A = N_0 e^{-\lambda_A t}$. The differential equation for the "daughter" has the usual decay term, but it also has a growth term since each decay of a parent nucleus creates a daughter nucleus: so $dN_B/dt = -\lambda_B N_B + (-dN_A/dt)$, where we *subtract* the change in the parent since a decrease in parent count is an increase in daughter count. We will solve this second differential equation for $N_B(t)$ and then find the activity R.

EVALUATE

$$\frac{dN_B}{dt} = -\lambda_B N_B + \left(-\frac{dN_A}{dt}\right) = -\lambda_B N_B + \lambda_A N_A = -\lambda_B N_B + \lambda_A N_0 e^{-\lambda_A t}$$

Let's guess at a solution in the form $N_B(t) = A(e^{-\lambda_A t} - e^{-\lambda_B t})$, and substitute to find A.

$$A(-\lambda_A e^{-\lambda_A t} + \lambda_B e^{-\lambda_B t}) = -\lambda_B A(e^{-\lambda_A t} - e^{-\lambda_B t}) + \lambda_A N_0 e^{-\lambda_A t}$$
$$\rightarrow -A\lambda_A e^{-\lambda_A t} = -A\lambda_B e^{-\lambda_A t} + \lambda_A N_0 e^{-\lambda_A t}$$
$$\rightarrow A(\lambda_B - \lambda_A) = \lambda_A N_0 \rightarrow A = \frac{\lambda_A N_0}{\lambda_B - \lambda_A}$$
$$\rightarrow N_B(t) = \frac{\lambda_A N_0}{\lambda_B - \lambda_A}(e^{-\lambda_A t} - e^{-\lambda_B t})$$

Now we have both $N_A(t)$ and $N_B(t)$ so we can find the activity:

$$R(t) = \lambda_A N_A(t) + \lambda_B N_B(t) = \lambda_A N_0 e^{-\lambda_A t} + \frac{\lambda_B \lambda_A N_0}{\lambda_B - \lambda_A}(e^{-\lambda_A t} - e^{-\lambda_B t})$$

ASSESS Interestingly, the activity can actually *increase* with time! This is a serious problem right now with some of the radioactive holding tanks left over from the production of bomb-making material during WWII—their activity is increasing, and is expected to continue increasing for several more decades.

77. **INTERPRET** We find the binding energy per nucleon for gold. To do this, we find the mass of the particles that go into gold, and compare it with the actual mass of gold.

DEVELOP Gold consists of 79 hydrogen atoms and 118 neutrons. The mass of hydrogen is $m_H = 1.00794$ AMU, and the mass of a neutron is $m_N = 1.00866$ AMU. The actual mass of the gold atom is $M = 196.96655$ AMU. The difference between the parts and the whole, Δm, gives us the total binding energy $E = \Delta mc^2$, which we divide by the nucleon number to get the binding energy per nucleon.

EVALUATE

$$\Delta m = 79 m_H + 118 m_N - M = 1.68306 \text{ AMU}$$
$$\rightarrow E = \Delta mc^2 = 1568 \text{ MeV}$$

so the binding energy per nucleon is $\frac{E}{A} = 7.96$ MeV.

ASSESS This answer is consistent with the data shown in Figure 38.3. Counting the protons and electrons together as if they were hydrogen atoms is a common shortcut for this type of problem. You can also add protons and electrons separately.

79. **INTERPRET** Will a given amount of a sample of a radioisotope remain after a given time? We use the equation for radioactive decay.

DEVELOP We want to know whether $N/N_0 = 10/15 = 2/3$ of an isotope will remain after $t = 3.5$ hours. The half-life of the isotope is $t_{1/2} = 6.01$ hours, so $\lambda = \ln 2/t_{1/2} = 0.1153$ hour^{-1}. We use $N = N_0 e^{-\lambda t}$.

EVALUATE $N/N_0 = e^{-\lambda t} = 0.891$.

ASSESS This is more than 2/3, so the initial 15 mg is more than sufficient for the job. Note that we don't have to go through all the gory details of finding the number of atoms and so on—just use ratios!

FROM QUARKS TO THE COSMOS

EXERCISES

Section 39.1 Particles and Forces

17. **INTERPRET** This problem is about finding the lifetime of a virtual photon by applying the uncertainty principle.

    DEVELOP In order to test the conservation of energy in a process involving one virtual photon, a measurement of energy with uncertainty less than the photon's energy ($\Delta E < hc/\lambda$), must be performed in a time interval less than the virtual photon's lifetime ($\Delta t < \tau$). Thus, $\Delta E \Delta t < hc\tau/\lambda$. But Heisenberg's uncertainty principle limits the product of these uncertainties to $\Delta E \, \Delta t \le \hbar$, so $hc\tau/\lambda > \hbar$. Knowing the wavelength allows us to find the upper bound on the lifetime of the virtual photon.

    EVALUATE With $\lambda = 633$ nm, we get

    $$\tau > \frac{\lambda}{2\pi c} = \frac{633 \times 10^{-9} \text{ m}}{2\pi (3 \times 10^8 \text{ m/s})} \approx 3 \times 10^{-16} \text{ s}$$

    for the given wavelength.

    ASSESS If the lifetime of a virtual photon of wavelength 633 nm were less than 3×10^{-16} s, no measurement showing a violation of conservation of energy would be possible.

Section 39.2 Particles and More Particles

19. **INTERPRET** We're asked about the decay of a positive pion to a muon and a neutrino.

    DEVELOP The decay process must conserve both charge and muon-lepton number L_μ (see Table 39.1).

    EVALUATE The only decay of the positive pion consistent with conservation laws is

    $$\pi^+ \to \mu^+ + \nu_\mu$$

    ASSESS The muon-lepton numbers of π^+, μ^+, and ν_μ are 0, −1, and +1, respectively. Therefore, we see that L_μ is conserved. Similarly, the charges of π^+, μ^+, and ν_μ are +1, +1, and 0, respectively, so we see that charge is also conserved.

21. **INTERPRET** We're asked to verify the conservation laws for the decay of η^0 into a positive, a negative, and a neutral pion.

    DEVELOP The decay process $\eta \to \pi^+ + \pi^- + \pi^0$ must conserve charge, strangeness, and baryon number (see Table 39.1).

    EVALUATE The decay conserves charge ($0 = 1 - 1 + 0$), and all the particles have zero baryon number and strangeness.

    ASSESS The problem illustrates how conservation laws restrict the possible decay modes of a particle.

23. **INTERPRET** In this problem we're asked to apply conservation laws to decide whether or not the interaction $p + p \to p + \pi^+$ is possible.

    DEVELOP The relevant properties to be considered here are: charge, baryon number, and spin (given in Table 39.1).

    EVALUATE The reaction $p + p \to p + \pi^+$ violates the conservation of baryon number ($0 + 0 \ne 0 + 1$), and also angular momentum, since the proton's spin is $\frac{1}{2}$ and the spin of the pion is 0. Therefore, the process is not allowed in the standard model with electro-weak unification.

    ASSESS The problem illustrates how conservation laws restrict the possible outcomes of particle interactions.

Section 39.3 Quarks and the Standard Model

25.   **INTERPRET**   We want to know the quark composition of a baryon with strangeness –3.

DEVELOP   A baryon is a particle that consists of three quarks. A strange quark has strangeness $s = -1$ and charge $-e/3$ (see Table 39.2).

EVALUATE   The baryon which is composed of three strange quarks with strangeness $s = -3$ and charge $-e$ is the $\Omega^- = sss$.

ASSESS   Gell-Mann's prediction of the existence of Ω^- was confirmed experimentally, and he received the Nobel Prize in 1969 for his work on sub-atomic particles.

Section 39.4 Unification

27.   **INTERPRET**   The Kamiokande experiment consists of 50,000 tons of water and is designed to detect rare nuclear reactions such as neutrino interactions and hypothetical proton decays. We want to estimate the volume of the water.

DEVELOP   The mass of 50,000 tons of water is (see Appendix C for the conversion factors)

$$M = (5 \times 10^4 \text{ tons})(2000 \text{ lb/ton})(0.454 \text{ kg/lb}) = 4.54 \times 10^7 \text{ kg}$$

The volume is $V = M/\rho$, where ρ is the density of water.

EVALUATE   At the ordinary density of $\rho = 10^3$ kg/m^3, this amount of water occupies a volume of

$$V = M/\rho = 4.54 \times 10^4 \text{ m}^3 = 1.20 \times 10^7 \text{ gal}$$

ASSESS   This is the volume of a cube of side length

$$L = (4.54 \times 10^4)^{1/3} \text{ m} = 35.7 \text{ m} = 117 \text{ ft}$$

29.   **INTERPRET**   In this problem we want to estimate the temperature in a gas of particles such that the thermal energy kT is about 10^{15} GeV, the energy of grand unification.

DEVELOP   The temperature corresponding to the energy where the strong and electro-weak forces unify is given by $U_{GUT} = 10^{15}$ GeV $= 10^{24}$ eV $= kT$, where $k = 8.617 \times 10^{-5}$ eV $\cdot$ K^{-1} is the Boltzmann's constant.

EVALUATE   The temperature is $T_{GUT} = \frac{U_{GUT}}{k} = \frac{10^{24} \text{ eV}}{8.617 \times 10^{-5} \text{ eV} \cdot \text{K}^{-1}} \approx 10^{28}$ K.

ASSESS   This is an extremely high temperature! According to the inflationary Big Bang Theory, at about 10^{-35} s after the Big Bang, grand-unified interaction breaks up into strong and electro-weak interactions.

Section 39.5 The Evolving Universe

31.   **INTERPRET**   We use Hubble's law to calculate the distance to a galaxy that is receding at a given speed.

DEVELOP   Hubble's law is $v = H_0 d$. The Hubble constant is $H_0 = 22$ km/s/Mly and the recessional speed is $v = 2 \times 10^4$ km/s. We solve for d.

EVALUATE $d = \frac{v}{H_0} = 6.61 \times 10^8$ ly $= 661$ Mly.

ASSESS   The galaxy is hundreds of millions of light years away. Looking at distant galaxies such as this gives us a chance to see what things were like 600 million years ago.

33.   **INTERPRET**   We are asked to find the age of the universe based on a given Hubble constant.

DEVELOP   As discussed in Example 39.3, the age of the universe is given by $t = 1/H_0$, where H_0 is the Hubble constant.

EVALUATE   With $H_0 = 25$ km/s/Mly, we find the age of the universe to be

$$t = \frac{1}{H_0} = \frac{1}{(25 \text{ km/s/Mly})/[3 \times 10^5 \text{ km/s/(ly/y)}]} = 12 \text{ Gy}$$

ASSESS   This is uncomfortably close to the age of stars in old globular clusters.

PROBLEMS

35.   **INTERPRET**   We're asked to apply conservation laws to decide whether or not the interactions given in the problem are possible.

DEVELOP   The relevant properties to be considered here are: charge, baryon number, and spin (see Table 39.1).

EVALUATE (a) The decay, $\Lambda^0 \to \pi^+ + \pi^-$, while conserving charge $(0 = 1 + (-1))$, violates the conservation of baryon number $(1 \neq 0 + 0)$, and angular momentum, since the spin of Λ^0 is $\frac{1}{2}$ and that of the pions is 0.

(b) The decay $K^0 \to \pi^+ + \pi^-$ is an observed weak interaction.

ASSESS The problem illustrates how conservation laws restrict the possible outcomes of particle interactions.

37. **INTERPRET** This problem is about the conservation laws in the hypothetical proton decay suggested by the grand unification theory.

DEVELOP The relevant properties we wish to consider in the hypothetical decay $p \to \pi^0 + e^+$ are charge and baryon number (given in Table 39.1).

EVALUATE (a) The hypothetical decay $p \to \pi^0 + e^+$ does not conserve baryon number (which is 1 for the proton and 0 for mesons and leptons), nor does it conserve lepton number.

(b) The hypothetical decay does conserve charge $(1 = 0 + 1)$.

ASSESS Baryon number is conserved in the Standard Model but not in the grand unification theory.

39. **INTERPRET** The composition of the J/ψ particle includes charm quarks but has charm $c = 0$.

DEVELOP A meson is a particle that consists of one quark and one anti-quark, so its baryon number is zero. A charmed quark (c) has charm $c = +1$, while an anti-charmed quark $(\bar{c})$ has charm $c = -1$.

EVALUATE The J/ψ must have quark content $c\bar{c}$ in order to have zero net charm.

ASSESS Charmed is one of the six flavors of quarks (up, down, strange, charmed, bottom, and top).

41. **INTERPRET** This problem is about the Tevatron accelerator at the Fermilab.

DEVELOP The conversion can be done by noting that one TeV is equal to 10^{12} eV, and 1 eV $= 1.602 \times 10^{-19}$ J. To answer (b), we note that the energy of a particle of mass m falling a distance y in Earth's gravitational field is $U = mgy$.

EVALUATE (a) 1 TeV $= (10^{12}$ eV$)(1.602 \times 10^{-19}$ J/eV$) \approx 0.16$ μJ.

(b) Setting $mgy = 1$ TeV with $m = 1$ g, the distance traveled by the mass is

$$y = \frac{U}{mg} = \frac{0.16\ \mu\text{J}}{(10^{-3}\,\text{kg})(9.8\ \text{m/s}^2)} = 16.3\ \mu\text{m}$$

ASSESS This is a very small distance. But remember that 1 g of protons contains about 6×10^{23} particles, compared to each single proton in the beam of the Tevatron.

43. **INTERPRET** We want to estimate the time it takes for a 7-TeV proton to travel a circular path with a circumference of 27 km.

DEVELOP With 7 TeV energy, the speed of a proton $(m_p = 938$ MeV/$c^2)$ is very close to the speed of light. So the time it takes to complete one circuit is simply given by $t = d/c$.

EVALUATE Writing $c = 3 \times 10^5$ km/s, we find the time to be

$$t = \frac{d}{c} = \frac{27\ \text{km}}{3 \times 10^5\ \text{km}} = 90\ \mu\text{s}$$

ASSESS Light travels about 3×10^5 km in one second. Since the speed of the proton is so close to the speed of light, the time it takes to travel 27 km would be very short.

45. **INTERPRET** What would be the diameter of the Sun if it had the same density as the critical density of the universe? We will assume that the mass of the Sun is the same, and use the value for ρ_c obtained in the Problem 44.

DEVELOP The mass of the Sun is $M = 1.99 \times 10^{30}$ kg, and the critical density (from Problem 39.44) is $\rho_c = 1.83 \times 10^{-26}$ kg/m^3. The volume of a sphere is $V = \frac{4}{3}\pi r^3$, and $\rho = \frac{m}{V}$. We will solve for r.

EVALUATE

$$\rho = \frac{M}{\frac{4}{3}\pi r^3} = \rho_c \to r^3 = \frac{3M}{4\pi\rho_c} \to r = 2.96 \times 10^{18}\,\text{m} = 313\ \text{ly}$$

ASSESS The universe is really not a very dense place at all. It's only dense in small, very widely separated regions such as planets and solar systems and galaxies and so on.

47. **INTERPRET** In this problem we want to find the size and the ground-state energy of a muonic atom.

    DEVELOP We can use the results for the Bohr atom, with $m_\mu = 207m_e$ replacing m_e (see Equations 34.12a and 34.13). Thus,

    $$a_\mu = \frac{\hbar^2}{m_\mu ke^2} = \frac{\hbar^2}{207m_e ke^2} = \frac{1}{207}a_0$$

    and

    $$E_n^{(\mu)} = -\frac{ke^2}{2a_\mu}\left(\frac{1}{n^2}\right) = -\frac{ke^2}{2(a_0/207)}\left(\frac{1}{n^2}\right) = 207E_n$$

    EVALUATE (a) For the ground state ($n = 1$), $r_1 = a_0/207 = (0.0529 \text{ nm})/207 = 256 \text{ fm}$.
    (b) The ground-state energy is $E_1^{(\mu)} = 207E_1 = 207 \times (-13.6 \text{ eV}) = -2.81 \text{ keV}$.

    ASSESS Comparing with the ground-state hydrogen atom, the size of the muonic atom is smaller by a factor of 207, and its ground-state energy is also lower (more negative) by a factor of 207.

49. **INTERPRET** We use the Doppler shift of a galaxy's hydrogen-β line to find the recession speed of the galaxy and then the distance to the galaxy using Hubble's law.

    DEVELOP The Doppler shift, from chapter 14, is $f' = \frac{f}{1+\frac{v}{c}}$. The actual frequency of the galaxy source is $f = \frac{c}{\lambda}$ where $\lambda = 486.1 \text{ nm}$; and the measured frequency is $f' = \frac{c}{\lambda'}$ where $\lambda' = 495.4 \text{ nm}$. We will solve the Doppler shift for v, and use Hubble's law $v = H_0 d$ to find d.

    EVALUATE
    (a)

    $$\frac{c}{\lambda'} = \frac{\frac{c}{\lambda}}{1+\frac{v}{c}} \rightarrow 1 + \frac{v}{c} = \frac{\lambda'}{\lambda} \rightarrow v = c\left(\frac{\lambda'}{\lambda} - 1\right) = 0.0191c = 5.74 \text{ km/s}$$

    (b)

    $$d = \frac{v}{H_0} = 1.90 \times 10^8 \text{ ly} = 190 \text{ Mly}$$

    ASSESS The speed we obtain using the Doppler equation is a small fraction of the speed of light, so we are justified in using the nonrelativistic Doppler formula.

51. **INTERPRET** This problem is about using the width of the measured energy distribution to estimate the lifetime of a particle.

    DEVELOP In the energy-time uncertainty relation (Equation 34.16), $\Delta t = \tau$ can be taken to be the lifetime of the particle (i.e., the time available for the measurement) and $\Delta E = \Gamma$ to be its width (i.e., the spread in measured rest-energies), so $\tau = \hbar/\Gamma$. (This is, in fact, the definition of the width.)

    EVALUATE For the Z^0, a width $\Gamma = 2.5 \text{ GeV}$ implies a lifetime

    $$\tau = \frac{\hbar}{\Gamma} = \frac{6.582 \times 10^{-16} \text{ eV} \cdot \text{s}}{2.5 \text{ GeV}} = 2.6 \times 10^{-25} \text{ s}$$

    ASSESS These extremely short-lived particles are called "resonances." The lifetime of these particles is on the order of 10^{-23} s.

53. **INTERPRET** We compare the age of the universe obtained from two different values of the Hubble constant, using the procedure shown in Example 39.3.

    DEVELOP In Example 39.3 we see that the age of the universe is $t = \frac{1}{H_0}$. The two values of the Hubble constant given are $H_0' = 17 \text{ km/s/Mly}$ and $H_0 = 22 \text{ km/s/Mly}$.

    EVALUATE We use $t = \frac{1}{H_0}$: for $H_0 = 22 \text{ km/s/Mly}$, $t = 13.6 \text{ Gy}$ as shown in the example, and for $H_0' = 17 \text{ km/s/Mly}$, $t' = 17.6 \text{ Gy}$.

    ASSESS The larger our estimate of the Hubble constant, the lower the age of the universe.

55.   **INTERPRET**   What will be the value of the Hubble constant when the universe is 60 Gy old?

   DEVELOP   We use $t = \frac{1}{H_0}$ for the age of the universe, but we solve for H_0 using $t = 60$ Gy.

   EVALUATE $H_0 = \frac{1}{t} = 5.0$ km/s/Mly.

   ASSESS   The Hubble constant is not really a constant: it just seems like it is because it changes rather slowly at the current age of the universe.

55.   **INTERPRET**   What will be the value of the Hubble constant when the universe is 60 Gy old?

   DEVELOP   We use $t = \frac{1}{H_0}$ for the age of the universe, but we solve for H_0 using $t = 60$ Gy.